Smart Sensors at the IoT Frontier

Hiroto Yasuura • Chong-Min Kyung
Yongpan Liu • Youn-Long Lin
Editors

Smart Sensors at the IoT Frontier

Springer

Editors
Hiroto Yasuura
Kyushu University
Fukuoka, Japan

Yongpan Liu
Circuits and Systems Division
Tsinghua University, Beijing
Beijing, China

Chong-Min Kyung
Department of Electrical Engineering
Korea Advanced Institute of Science and Technology (KAIST)
Daejeon, South Korea

Youn-Long Lin
National Tsing Hua University
Hsinchu, Taiwan, Taiwan

ISBN 978-3-319-85634-6 ISBN 978-3-319-55345-0 (eBook)
DOI 10.1007/978-3-319-55345-0

Printed on acid-free paper

This Springer imprint is published by Springer Nature
The registered company is Springer International Publishing AG
The registered company address is: Gewerbestrasse 11, 6330 Cham, Switzerland

Contents

Introduction

Hiroto Yasuura

Internet of Things (IoT) has become a big trend in the ICT (information and communication technologies) field. In addition to smartphones, tablets, and personal computers, a wide range of items including daily necessities such as refrigerators, bathrooms, and air conditioners are directly connected to the Internet. Many of the new ICT-based services that create potentially large markets are expected to become available based on IoT.

One of the large and well-known examples of IoT activities is "Industrie 4.0" jointly developed by the German government/industry/academia. The goal is to connect all machines in the factory via the network to digitize the whole process in factory activities. It completely changes the style of the production process. In the normal manufacturing process, the structure of the process is carefully designed, but once it is built, it will be fixed for a certain period of time. By contrast, in Industrie 4.0, the process including the physical placement of the factory machine is changed dynamically referring to the data obtained by observing the activities of the process via the sensor network. Data includes not only the status of all the machinery in the factory but also the activities of workers in the factory, demand for products, and requests from customers. They are the fourth "industrial revolution," and production costs will be drastically reduced.

Similar activities are under way in several countries. The Industrial Internet Consortium (IIC) in the US, which was established by major US ICT companies, AT&T, Cisco, GE, IBM, and Intel, aims at digitalization of not only production processes but other social services such as medical services, energy services, etc. The Chinese government has also presented the plan "Made in China 2025 (MiC2025)," which is the road map of manufacturing industries in China. It aims to augment the Chinese industry in many aspects, and the key ideas include enhancement of innovation,

---

H. Yasuura (✉)
Kyushu University, Fukuoka, Japan
e-mail: yasuura.hiroto.117@m.kyushu-u.ac.jp

H. Yasuura et al. (eds.), *Smart Sensors at the IoT Frontier*,
DOI 10.1007/978-3-319-55345-0_1

quality/brand-power, environmental protection, etc. in the manufacturing. In the fifth Science and Technology Basic Plan, the Japanese government has proposed the concept of "Society 5.0," where advanced ICT improves every aspect of our life including industry, economics, health, transportation, education, etc. The plan emphasizes the fifth social paradigm change, which follows the "hunting and gathering society," "agrarian society," "industrial society," and "information society." Our society is becoming truly a "cyber-physical system," which is the mixture of the real world and the cyber world connected by IoT technology.

The background of this IoT trend is explained in several engineering contexts:

1. Huge and complicated networks, which cover thoroughly our world using a combination of huge bandwidth wired network and ubiquitous wireless network, have been realized with relatively low cost and with high throughput. This enables us to connect quite a large number of devices to the Internet.
2. Thanks to the recent progress of device technology, which realizes highly integrated, energy-efficient, and low-cost devices, a wide variety of sensors and apparatuses with network connection capability have been developed and available in the market.
3. A huge amount of data acquired by various sensors in smartphones and ICT devices, i.e., "big data," are being gathered and analyzed to extract valuable information using cloud services.
4. Since a huge number of smartphones and other display devices are popularly available, people require various types of information, especially personalized one, via the smartphones. A typical example is fine-grained prediction of sudden and heavy rain falls, which people want to avoid, and it can be produced by integrating various kinds of sensory information of climate condition and location. Others include transportation congestion, real-time availability of bus/taxi, etc.

The above points indicate that many progresses in ICT fields support recent development of the IoT. To understand what is IoT and what is coming in quite the near future, we should know what is currently going on in the field of IoT. This book has been designed to provide such current engineering aspects on IoT, especially from the viewpoint of smart sensing. In this book, we have covered wide areas of smart sensor technologies divided into three parts including smart devices, sensing methodology, and systems and applications. The topics described in each part are summarized as follows:

1 Part I Device Technology for IoT

In "Energy-Autonomous Supply-Sensing Biosensor Platform Using CMOS Electronics and Biofuel Cells (Niitsu)," the author presents a new method to build energy-autonomous semiconductor devices, which can solve the battery issue of electronic appliances and which are inevitable to fine-grained IoT systems. He has developed an energy-autonomous supply-sensing biosensor platform using CMOS electronics and biofuel cells, which is used in human health condition sensing for big data-based healthcare. The device enables low-voltage operation and a small footprint, even in a cost-competitive legacy CMOS technology. This work realizes converter-less energy-autonomous operation using a biofuel cell, which is ideal for disposable healthcare applications.

In "Smart Microfluidic Biochips: Cyber-Physical Sensor Integration for Dynamic Error Recovery (Yao et al.)," the authors describe the recent progress of digital microfluidic biochips, which are gaining increasing attention with promising applications for automating and miniaturizing laboratory procedures in biochemistry. Automated design of digital microfluidic biochips includes two major parts: fluidic-level synthesis and chip-level design. They describe how a digital microfluidic biochip is designed. Automatic control logic is also described, where cyber-physical sensors can be integrated for dynamic error recovery in real-life biochemical applications.

In "Reducing Timing Discrepancy for Energy-Efficient On-Chip Memory Architectures at Low-Voltage Mode (Chen)," the author describes a technique to reduce power consumption in processor systems, especially in SRAM cache memory, which is commonly used in modern processor systems. To reduce the power consumption, voltage scaling is an effective technique, but timing discrepancies between on-chip memory and CPU cores occur with the voltage scaling down, which significantly harms the system performance. These discrepancies are primarily caused by severe process variations of a few slow SRAM cells. This work addresses the issue of an 8Tr. SRAM cache and proposes some solutions to tolerate those slow cells to eliminate timing discrepancies.

In "Redesigning Software and Systems for Nonvolatile Processors on Self-Powered Devices (Xue)," the author presents how energy harvesting in circuits should be handled. The energy harvesting is quite an important aspect of wearable devices and other very small-scaled systems. The author develops a method to utilize nonvolatile processors (NVP), which can back up the volatile state before the battery energy is used up and which can resume the program execution when enough energy is supplied. The NVP is required in systems with energy harvesting, where the power supply tends to be unstable. Due to backup and resumption procedures resulted from power failures, the nonvolatile processor exhibits significantly different characteristics from traditional processors, necessitating a set of adaptive design and optimization strategies. The author provides an overview of the state-of-the-art NVP research including the software and system level.

2 Part II Sensing Technology for IoT

In "OEICs for High-Speed Data Links and Tympanic Membrane Transducer of Hearing Aid Device (Chen et al.)," the authors describe the design of photonics integrated with electronics (OEICs) for the applications in data-intensive optical links and tympanic membrane transducer of hearing aid devices. OEICs are expected to be one of the key enablers for emerging applications, covering from short distance sensing and data links to the backbones for the next-generation telecommunications network. In their chapter, very high-speed, fully integrated CMOS optical receivers incorporating on-chip photodetectors are presented first. Then, the authors present a novel architecture for signal and power transfer in a tympanic membrane transducer using OEIC, showing the feasibility to mechanically stimulate the tympanic membrane (TM) to improve sound quality.

In "Depth Estimation Using Single Camera with Dual Apertures (Park et al.)," the authors presented a new sensing, or imaging, method to acquire depth information or the distance to objects from the camera. The depth information is very useful to detect and to analyze events in the real world, and there are many depth sensors available, such as Microsoft Kinect. The uniqueness of the authors' method is its simplicity: only a one-shot image is captured with dual apertures. In their system, IR (infrared) light is captured through a small aperture, and only visible light is captured through a larger RGB-pass aperture. The difference of the aperture sizes causes the blur size difference between the sharp IR and blurry color components, which is the clue to estimate the depth.

In "Scintillator-Based Electronic Personal Dosimeter for Mobile Application (Cho et al.)," an electronic personal dosimeter (EPD) which measures the energy spectrum and the personal dose rate in radiation exposure environment is presented. This device is composed of a compact radiation sensor to detect gamma ray; an integrated circuit of preamplifier, peak holder, etc.; and a software to calculate the personal dose from the measured spectrum. The CsI(Tl)-coupled pin-diode is used as a compact spectroscopic radiation sensor to measure the energy spectrum for the radioisotope identification or the activity analysis. To optimally design the size of the compact radiation sensor to be used as an accessary of mobile personal devices, the authors have determined a guideline such that the sensor must satisfy the international criteria of angular response, as well as have the maximum value of a figure of merit which is a product of the geometric detection efficiency and the energy resolution.

In "LED Spectrophotometry and Its Performance Enhancement Based on Pseudo BJT (Choi et al.)," the authors present a LED-based spectrophotometry, which can be implemented in a small feature size with relatively small cost and can provide a suitable way to integrate the optical spectrometer into the smart and mobile sensor systems. In addition, recent advances in LED technology extend a wavelength selection window of LED from a deep ultraviolet region to an infrared region. In this work, a guide to set up the LED-PD system is provided for LED spectrophotometry covering a device selection, driving circuit composition and applications. As applications of LED spectrophotometry for the bio- and chemical sensor, some examples including the water pollution and glucose sensor are discussed.

3 Part III System and Application

In “An Air Quality and Event Detection System with Life Logging for Monitoring Household Environments (Cho),” the author presents a system of indoor air quality measurement and event detection to monitor the household environment. The system is for relaxing the problems of disease caused by indoor air pollution and of stress caused by indoor noise generated on upper floors. It uses multiple sensors and microphones to measure indoor air quality and indoor noise and simultaneously maintains the measured data in internal memory and on Internet server. It can act as an indoor life logger or indoor black box. The author presents a hardware design and software architecture for a new system that incorporates digital hardware, analogue circuits, and a network including communication protocols.

In “Mobile Crowdsensing to Collect Road Conditions and Events (Aihara et al.),” the authors present a mobile sensing framework for collecting personal-based road and traffic situation. In their framework, crowdsourcing, i.e., a mechanism to obtain required data/information from a lot of individuals through Internet services, is the key. They have developed a smartphone application with cloud service, with which the road and traffic situation, such as occurrences of frozen road, road construction, and traffic accident, is observed by a lot of people. An interesting feature is a driving recorder that collects not only sensor data but also videos recorded from the driver’s point of view, and the acquired data are used to extract roadside phenomena.

In “Sensing and Visualization in Agriculture with Reasonable Smart Devices (Okayasu et al.),” the authors explain how IoT improves the efficiency of agricultural works and the quality of agricultural products. There is a big trend of smart agriculture in the world, but their activities are unique in the sense that their technology is for small-scaled or medium-scaled farms. There are quite a lot of small-scaled farms, especially in Japan and several other countries, especially in Asia, who produce high-quality agricultural products spending time and effort. To make their farming process more efficient and to reduce the labor, ICT support is a promising approach, but, in such small-scaled farms, the cost of ICT becomes a problem. Therefore, they are developing their tools using affordable smart devices such as low-price microcomputers and sensors and open-source software to reduce the cost.

In “Analyzing e-Book-Based Educational Big Data in Kyushu University (Ogata et al.),” the authors explain several activities of “learning analytics,” which means acquisition, or “sensing,” of learners’ activities and analysis of acquired data to improve the efficiency of teaching and learning. Kyushu University has introduced the BYOD (bring your own personal device) policy for all students and provided campus-wide high-speed broadband wireless Internet access. This infrastructure enables students to browse e-book materials before, during, and after lectures. Analyzing the detailed access logs of the e-books, teachers can understand how the students comprehend the lectures and how their teaching processes are effective to the students, which becomes very important information to improve the course materials and the teaching method.

In "Security and Privacy in IoT Era (Arias et al.)," the authors present security and privacy issues in IoT, which are very important and urgent issues. Thanks to recent development of small, low-power devices with network connectivity and wearable devices, automated home and industrial systems are loaded with sensors, collect information from their surroundings, process it, and relay it to remote locations for further analysis. But the process raises security and privacy concerns. The authors evaluate the security of these devices from an industry point of view, concentrating on the design flow, and catalogue the types of vulnerabilities. They also present an in-depth evaluation of popular IoT devices, such as the Google Nest Thermostat and the Nike+ Fuelband SE Fitness Tracker, in daily setting.

Unfortunately, due to page limitation, we might miss other important and interesting topics. However, we think this editorial book helps the readers to understand the current situation of IoT and to inspire innovation in the IoT era, which improves the efficiency and the comfort of our coming sustainable society.

Part I
Device Technology for IoT

Energy-Autonomous Supply-Sensing Biosensor Platform Using CMOS Electronics and Biofuel Cells

Kiichi Niitsu

1 Introduction

Ensuring stable energy is one of the most important challenges in the current wearable and implantable healthcare devices associated with big-data-based analysis (Fig. 1). To address this issue, many attempts such as developments of battery, wireless power delivery, and energy harvesting have been reported. Although the technical improvement has been rapid, none of these methods fully satisfy the requirement. Battery is unsuitable for use near a human body due to its inherent danger. Wireless power delivery requires large-size power-receiving antenna. Energy harvesting is unstable for healthcare application. Additionally, the latter two approaches require area-consuming power management unit such as power receiver and AC-DC converter that increase cost.

To satisfy the requirement, biofuel cells are intensely developed such as for transdermal iontophoresis patch [1] and brain-machine interface [2]. Biofuel cells are safe, stable, and do not require an antenna or an AC-DC converter. Additionally, the value of the obtained energy from a human body can be used as biosensing data, and, thus, sensor electrodes and front ends become unnecessary. Among the biofuel cell types, the organic biofuel cell [1] is especially promising because it is cheap and environment friendly, which enables disposable healthcare. However, the output supply voltage of a biofuel cell is usually lower than 0.4 V, and conventional circuits cannot operate using a biofuel cell without power management circuits such as up-converter. Thus, new circuit technique must be developed for converterless operation.

K. Niitsu (✉)
Department of Electrical Engineering and Computer Science, Nagoya University, C3-1(631), Furo-Cho, Chikusa-Ku, Nagoya, 464-8603, Japan
e-mail: niitsu@nuee.nagoya-u.ac.jp

H. Yasuura et al. (eds.), *Smart Sensors at the IoT Frontier*,
DOI 10.1007/978-3-319-55345-0_2

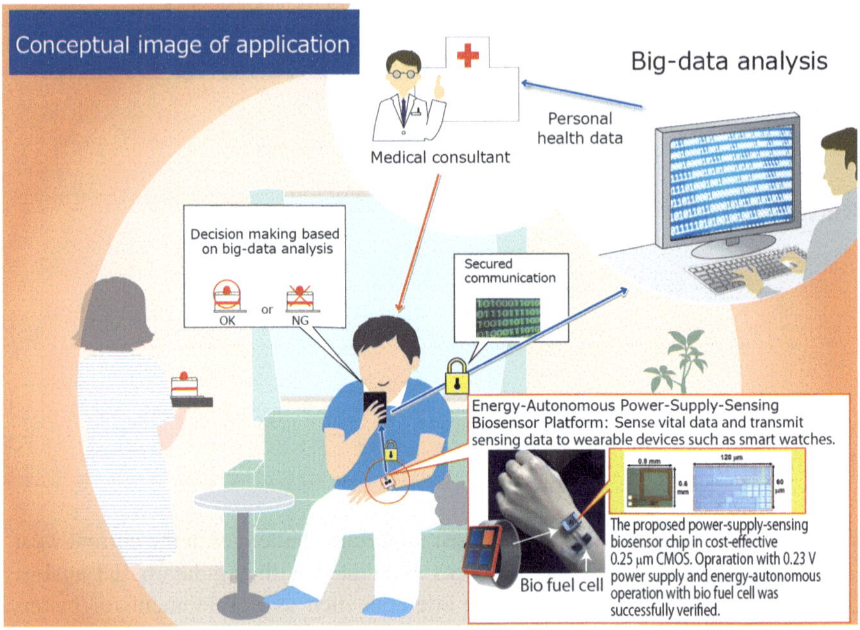

Fig. 1 Conceptual image of the application of the proposed work. The target application is big-data-based healthcare. The proposed energy-autonomous biosensor transmits vital data to the wearable device

The current chapter introduces a supply-sensing biosensor platform using a biofuel cell and a 0.23-V 0.25-μm zero-V_{th} all-digital CMOS supply-controlled ring oscillator (SCRO) with a current-driven pulse-interval-modulated inductive-coupling transmitter (Fig. 2). Compared with conventional architecture [3, 4], the occupied area and required power can be dramatically reduced. To realize transmit sensing data to the wearable device without power-hungry security protection circuits, a proximity inductive-coupling transmitter is employed.

To verify the effectiveness of the proposed approach, a test chip was fabricated using the cost-competitive legacy 0.25-μm CMOS technology. The measured results show successful operation with a 0.23-V power supply, which is the lowest supply voltage ever reported. In addition to the chip functional test, energy-autonomous operation using a biofuel cell was successfully demonstrated.

This chapter is organized as follows: the proposed energy-harvesting and biosensor platform is introduced in Sect. 2. The design of the prototype CMOS sensor and the measurement setup are summarized in Sect. 3. Sections 4 and 5 present the measurement results and demonstration of the energy-autonomous operation. Section 6 concludes this chapter.

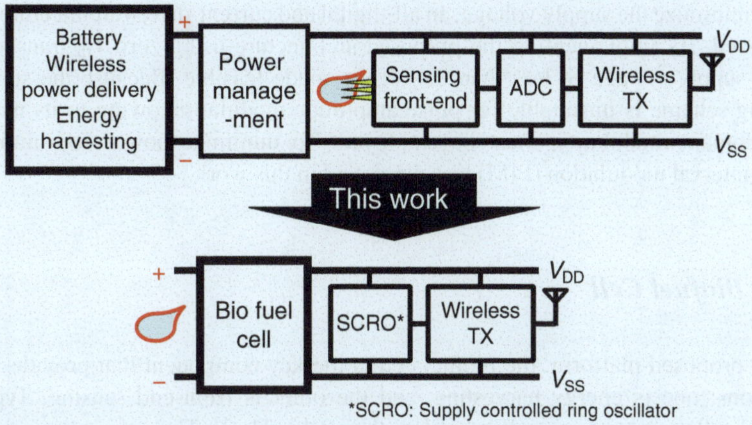

Fig. 2 Performance comparison with the state-of-the-art proximity communications

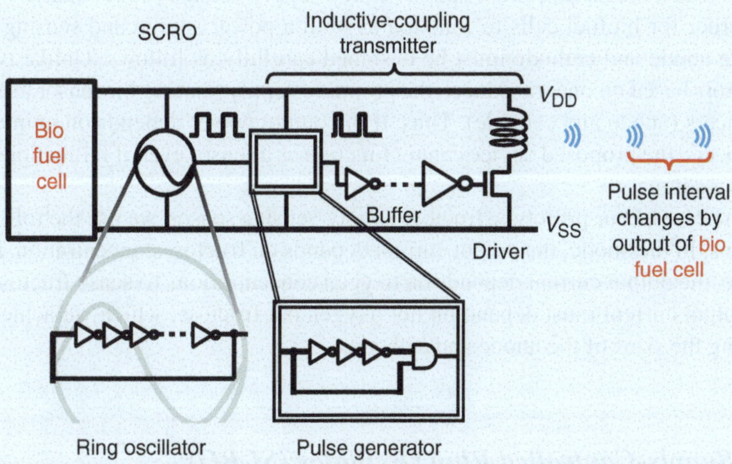

Fig. 3 Circuit diagram of the proposed supply-sensing biosensor

2 Supply-Sensing Biosensor Platform

2.1 Principle of Supply-Sensing Biosensor Platform

Figure 3 shows the circuit diagram of the proposed supply-sensing biosensor platform. The proposed sensor platform consists of three parts: biofuel cells, SCRO, and an inductive-coupling transmitter. By eliminating the area-hungry power management circuits, sensing front-end circuit, and power-hungry analog-to-digital converters (ADC), the occupied area and required power can be dramatically reduced.

To minimize the supply voltage, an all-digital and current-driven architecture was employed. By implementing the proposed architecture using zero-V_{th} transistors, a low-supply voltage of less than 0.4 V was made feasible. Because the supply-sensing scheme is unsuitable for pulse amplitude modulation owing to its nature, time-domain modulation must be employed. To minimize power consumption, pulse-interval modulation (PIM) was employed in this work.

2.2 Biofuel Cell

In the proposed platform, the biofuel cell is the key component that provides two functions: one is energy harvesting, and the other is front-end sensing. Typical biofuel cells can generate voltage of less than 0.4 V [1, 2]. Thus, to realize energy-autonomous operation without area-consuming power management circuits such as up-converters, the circuits must operate with a supply voltage of less than 0.4 V.

In order for biofuel cells to function as both a power source and sensing front end, the anode and cathode must be designed carefully as follows. Unlike typical biosensors based on one transducer, the proposed supply-sensing biosensor uses two transducers (anode and cathode). Thus, if the output power depends on unintended transducers, the proposed device cannot function as a sensor even if it functions well as power source.

In the case of our prototype fructose supply-sensing sensor, we use the following reactions. In the anode, the output current depends on fructose concentration. In the cathode, the output current depends on oxygen concentration. To sense fructose, the total output current must depend on not oxygen but fructose, which we achieve by adjusting the sizes of the anode and cathode.

2.3 Supply-Controlled Ring Oscillator (SCRO)

To realize PIM, the supply voltage must be modulated to pulse interval. To enable low-voltage operation, we implemented an SCRO. The SCRO consists of normal ring oscillator with PMOS and NMOS. The number of stages was determined by considering the trade-off between area overhead and power consumption. In this work, to minimize the occupied area, the number of stages of inverter gates was designed to be as small as possible while maintaining effective operation.

2.4 Inductive-Coupling Transmitter

For the wireless transmitter, we implemented a current-driven inductive-coupling transmitter. By considering the limited power budget, proximity communication without any security protection was adopted. Proximity communication can be

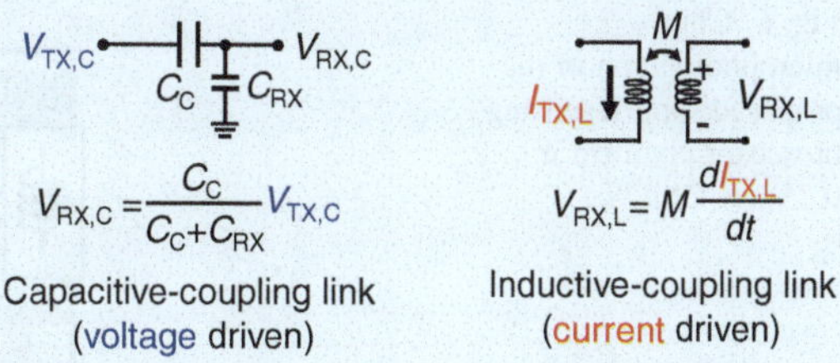

Fig. 4 Operating principle of the proximity communication

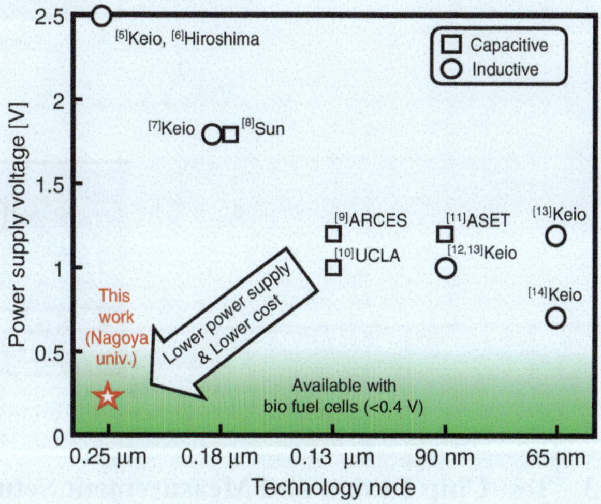

Fig. 5 Performance comparison with the state-of-the-art proximity communications

categorized into two approaches: one is the capacitive-coupling link, and another is the inductive-coupling link. Figure 4 shows their conceptual operating principle. The obtained voltage at the receiver side of capacitive link is determined by the ratio of the coupled capacitance to the total capacitance. Thus, a received voltage that is higher than the transmit voltage cannot be obtained.

In contrast, the received voltage in the inductive-coupling link is determined by a multiple of the slew rate of the transmit current and the mutual inductance. A high received voltage can be obtained even with a low supply-voltage transmitter. Biofuel cells can generate larger current at lower voltage owing to its characteristic; thus, the current-driven inductive-coupling link is preferred.

To minimize power supply voltage while enjoying the advantage of the inductive-coupling link, the proposed inductive-coupling transmitter was designed to be as simple as possible, as shown in Fig. 3, which consisted of a pulse generator, buffers, driver, and inductor. The pulse generator consisted of an inverter chain and an AND gate, which converts the clock signal to a low-duty pulse signal.

Figure 5 shows the performance comparison of the proposed method with the state-of-the-art proximity communications. The literature shows a trade-off between power supply voltage and technology node. The lowest supply voltage was 0.7 V [14] for the clock-based synchronous inductive-coupling link, and none of the conventional proximity communications could satisfy the requirement. This work achieved the lowest supply using the most cost-competitive technology node.

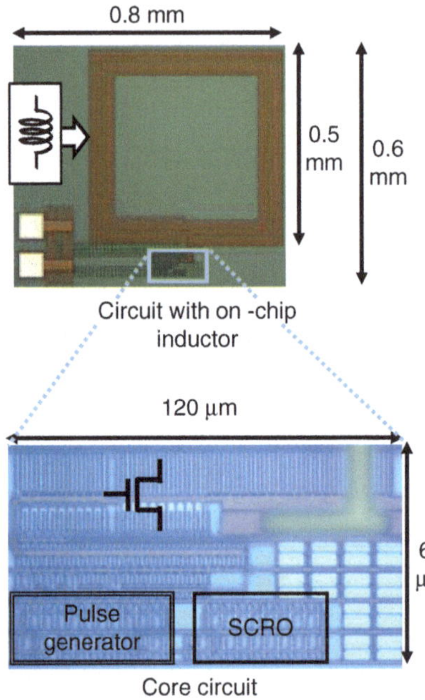

Fig. 6 Chip microphotograph of the proposed supply-sensing biosensor platform in 0.25-μm CMOS

3 Test Chip Design and Measurement Setup

To verify the effectiveness of the proposed approach, a test chip was fabricated using a cost-competitive 0.25-μm CMOS technology. Figure 6 shows the microphotograph of the test chip. The occupied footprints of the core circuit without an on-chip inductor and the entire circuit with an on-chip inductor were 60×120 μm and 0.6×0.8 mm, respectively. The test chip was assembled in a ceramic package. The diameter of the inductor is 0.5 mm.

The measurement setup is shown in Fig. 7. Only two electrical signals, namely, V_{DD} and V_{SS}, were supplied from the power supply (Keysight Technologies, E3632A). To verify the transmitter operation using magnetic detection, a magnetic field probe (Langer, H-Field probe MFA-K 0.1-12, 0.1–6 GHz) and a bias tee (Langer, Bias Tee) were employed. The waveform was obtained using a sampling oscilloscope (Keysight Technologies, DSO6102A).

4 Measurement Results

Figure 8 shows the measured current-consumption dependence on the supply voltage. The operation under a 0.23-V power supply was successfully verified. The 0.23-V supply is sufficiently low for energy-autonomous operation using a

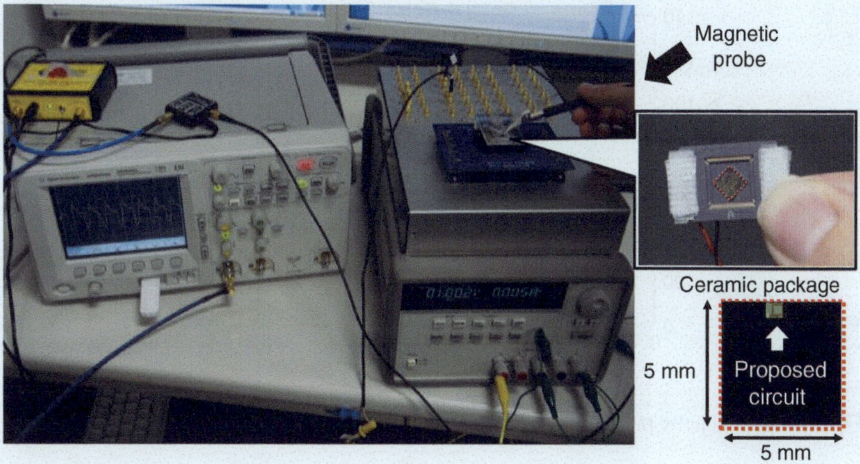

Fig. 7 Measurement setup

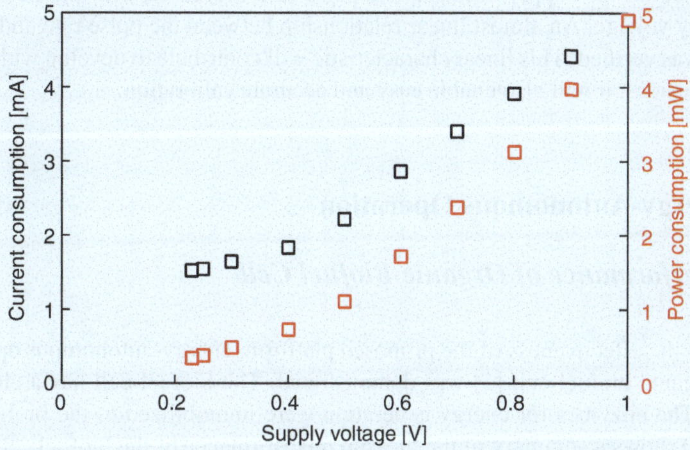

Fig. 8 Measured current-consumption dependence on the supply voltage. A 0.23-V operation was verified

biofuel cell. This is the lowest supply voltage ever reported among proximity communications. Because the drain current of a zero-V_{th} transistor is proportional to the square of the gate-source voltage, V_{GS}, from 0 to 0.4 V and is proportional to V_{GS} from 0.4 V, the current-consumption trend changes at 0.4 V.

The measured current consumption at 0.23-V power supply is 1.52 mA. The measured power is 0.35 mW. This power can be obtained by a 1 cm^2 of biofuel cell [1]. Owing to the conservative design, the current consumption in this technology node was not minimized. By optimizing the design parameters, further power reduction can be realized.

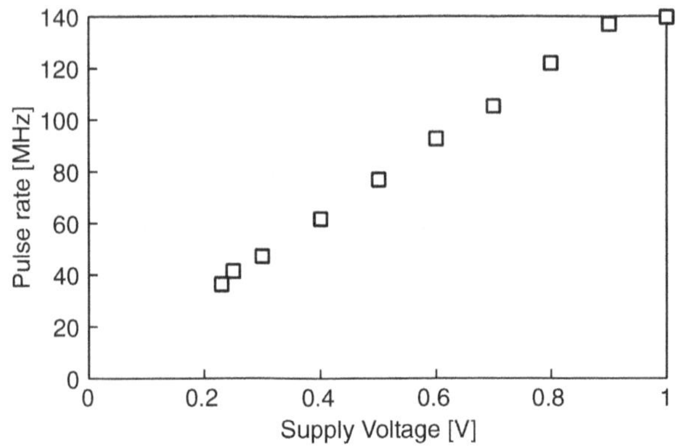

Fig. 9 Measured pulse rate dependence on the supply voltage

Figure 9 shows the frequency of the output pulse from the magnetic field dependence on the supply voltage. The pulse rate decreased with the decrease in the supply voltage. An almost linear relationship between the pulse rate and supply voltage was verified. This linear characteristic will contribute to develop wide-range sensor. Besides, it will also enable easy and accurate calibration.

5 Energy-Autonomous Operation

5.1 Performance of Organic Biofuel Cell

To verify the effectiveness of the proposed platform, energy-autonomous operation using organic biofuel cell [1] was demonstrated. The biofuel cell has a cloth-like feature. The enzymes for energy generation were immobilized to the biofuel cell. Figure 10 shows a summary of the measured performance of the biofuel cell.

The biofuel cell generates energy from fructose. The size of the biofuel cell is 1-cm square. The peak power can be obtained at 0.3–0.4 V, which is a typical characteristic of biofuel cells [1, 2]. The measurement was performed using a three-electrode system (BSA, 730C electrochemical analyzer).

5.2 Demonstration of Energy-Autonomous Biosensing

Figure 11 shows a successful energy-autonomous operation using biofuel cell. By dipping the biofuel cell into a fructose solution, the circuit transmitted a magnetic field, and its waveform appeared on the oscilloscope. This work is the first demonstration of an energy-autonomous proximity transmission using biofuel cells.

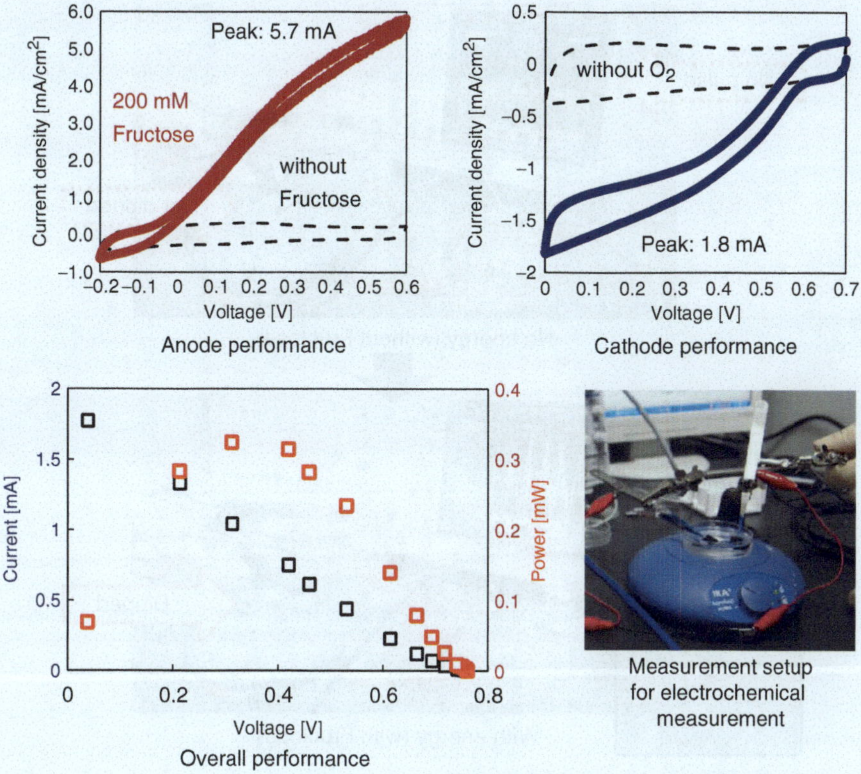

Fig. 10 Measured performance of the biofuel cell and its measurement setup

Figure 12 shows a summary of the energy-autonomous operation. Figure 12a shows the output voltage and current of the biofuel cell as a function of fructose concentration. The voltage and current increased as the fructose concentration increased. Figure 12b shows the measured pulse rate of the output magnetic field from the proposed biosensor platform. The pulse rate increased as the fructose concentration increased. These results agree well with the performance of the platform shown in Fig. 9. From these measured results, we have successfully confirmed the feasibility of the proposed energy-autonomous supply-sensing biosensor platform.

6 Discussion

This chapter demonstrates that the proposed supply-sensing sensor platform can be applied to fructose sensing. But, the proposed sensor platform can be applied to wide applications. Today's CMOS bioelectronics enables various kinds of biosensing [15–25]. Thus, by combining the proposed sensor platform and the present

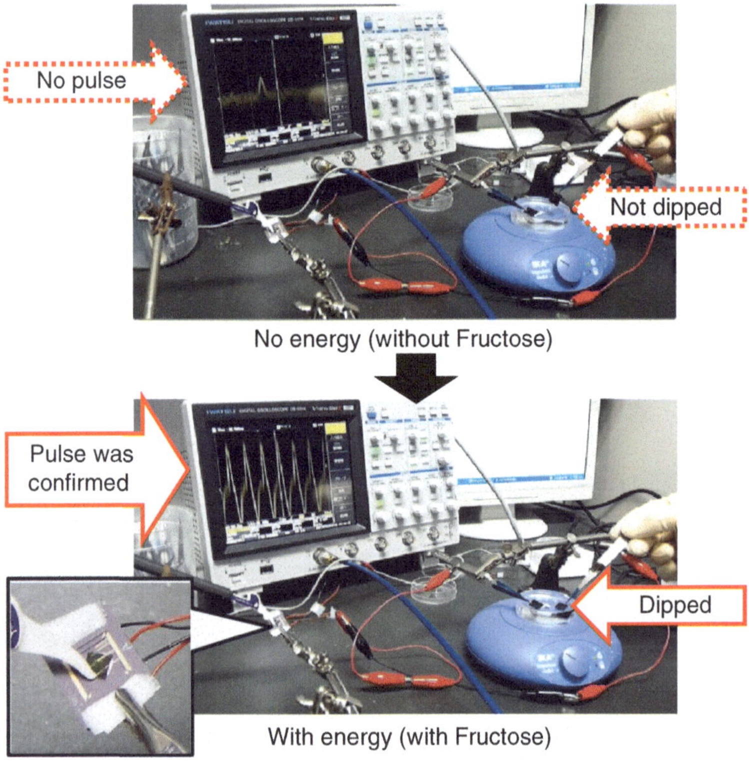

Fig. 11 Demonstration of energy-autonomous operation using organic biofuel cell. Only when dipping the biofuel cell into fructose solution, the proposed biosensor transmits magnetic pulse

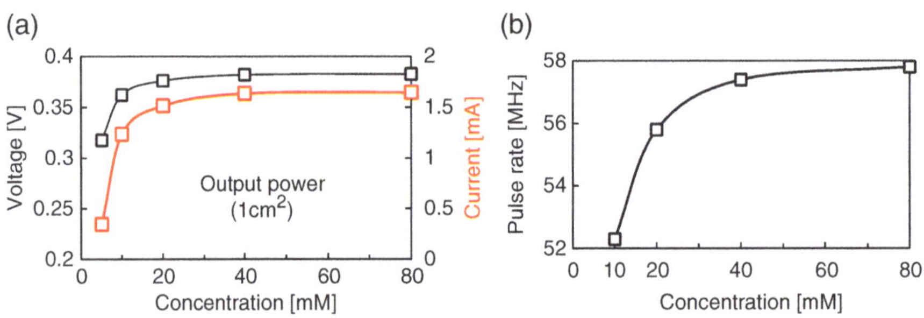

Fig. 12 Measured output voltage from biofuel cell (**a**) and measured pulse rate dependence on fructose concentration (**b**)

CMOS bioelectronics, energy-autonomous smart biosensor can be emerged. Furthermore, the proposed technique is familiar with CMOS scaling because it utilizes time-domain processing [26–30] and current-driven inductive-coupling inter-chip communication [31–37]. Future CMOS scaling will enhance the feasibility and effectiveness of the proposed sensor platform.

7 Conclusion

An energy-autonomous, disposable, supply-sensing biosensor platform has been demonstrated. The platform is based on a biofuel cell and a zero-V_{th} all-digital CMOS supply-controlled ring oscillator with a current-driven pulse-interval-modulated inductive-coupling transmitter that enables operation under low-power supply voltage using legacy CMOS technology. The measurement using a 0.25-μm CMOS prototype chip successfully demonstrated wireless transmission under a 0.23-V power supply, which is the lowest power supply proximity transmitter ever reported. Additionally, an energy-autonomous operation using an organic biofuel cell was successfully verified.

Acknowledgments This research was financially supported by JST, PRESTO, by a Grant-in-Aid for Scientific Research (S) (Nos. 20226009, 25220906, 26220801), Grants-in-Aid for Young Scientists (A) (No. 16H06088) from the Ministry of Education, Culture, Sports, Science and Technology of Japan, by the Strategic Information and Communications R&D Promotion Programme (Nos. 121806006, 152106004) of the Ministry of Internal Affairs and Communications, Japan, by TOYOTA RIKEN, and by The Nitto Foundation. The fabrication of CMOS chips was supported by Taiwan Semiconductor Manufacturing Co., Ltd. (TSMC, Taiwan), and the VLSI Design and Education Center (VDEC), University of Tokyo in collaboration with Synopsys, Inc. and Cadence Design Systems, Inc.

References

1. Ogawa, Y., Nishizawa, M., et al.: Organic transdermal iontophoresis patch with built-in biofuel cell. Adv. Healthc. Mater. **4**(4), 506–510 (2015)
2. Rapoport, B.I., et al.: A glucose fuel cell for implantable brain–machine interfaces. PLoS ONE. **7**(6), e38436 (2012)
3. Liao, Y.-T., et al.: A 3-μW CMOS glucose sensor for wireless contact-lens tear glucose monitoring. IEEE J. Solid-State Circ. **47**(1), 335–344 (2012)
4. Komori, H., Niitsu, K., Nakazato, K., et al.: An extended-gate CMOS sensor array with enzyme-immobilized microbeads for redox-potential glucose detection. In: IEEE Biomedical Circuits and Systems Conf, pp. 464–467 (2014)
5. Miura, N., et al.: A 195Gb/s 1.2W 3D-stacked inductive inter-chip wireless superconnect with transmit power control scheme. In: Proc. IEEE ISSCC, pp. 264–265 (2005)
6. Iwata, A., et al.: A 3D integration scheme utilizing wireless interconnections for implementing hyper brains. In: Proc. IEEE ISSCC, pp. 368–369 (2007)
7. Miura, N., et al.: A 1 Tb/s 3 W inductive-coupling transceiver for 3D-stacked inter-chip clock and data link. IEEE J. Solid State Circuits. **42**(1), 111–122 (2007)

8. Hopkins, D., et al.: Circuit techniques to enable 430Gb/s/mm^2 proximity communication. In: Proc. IEEE ISSCC, pp. 368–369 (2007)
9. Fazzi, A., et al.: 3D capacitive interconnections with mono- and bi-directional capabilities. In: Proc. IEEE ISSCC, pp. 356–357 (2007)
10. Gu, Q., et al.: Two 10Gb/s/pin low-power interconnect methods for 3D ICs. In: Proc. IEEE ISSCC, pp. 448–449 (2007)
11. Daito, M., et al.: Capacitively coupled non-contact probing circuits for membrane-based wafer-level simultaneous testing. In: Proc. IEEE ISSCC, pp. 144–145 (2010)
12. Niitsu, K., et al.: A 65fJ/b inter-chip inductive-coupling data transceivers using charge-recycling technique for low-power inter-chip communication in 3D system integration. IEEE Trans. Very Large Scale Integration (VLSI) Syst. pp. 1285–1294 (2012)
13. Niitsu, K., et al.: An inductive-coupling link for 3D integration of a 90nm CMOS processor and a 65nm CMOS SRAM. In: Proc. IEEE ISSCC, pp.480–481 (2009)
14. Miura, N., et al.: A 0.55 V 10 fJ/bit inductive-coupling data link and 0.7 V 135 fJ/cycle clock link with dual-coil transmission scheme. IEEE J. Solid State Circ., 965–973 (2011)
15. K. Niitsu, A. Kobayashi, Y. Ogawa, M. Nishizawa, K. Nakazato: An energy-autonomous, disposable, big-data-based supply-sensing biosensor using bio fuel cell and 0.23-V 0.25-μm zero-V_{th} all-digital CMOS supply-controlled ring oscillator with inductive transmitter. In: Proc. IEEE Biomed. Circ. Syst. Conf. pp. 595–598 (2015)
16. Niitsu, K., Ota, S., Gamo, K., Kondo, H., Hori, M., Nakazato, K.: Development of microelectrode arrays using electroless plating for CMOS-based direct counting of bacterial and HeLa cells. IEEE Trans. Biomed. Circ. Syst. **9**(5), 607–619 (2015)
17. Kuno, T., Niitsu, K., Nakazato, K.: Amperometric electrochemical sensor array for on-chip simultaneous imaging. Jpn. J. Appl. Phys. **53**, 04EL01 (7 pages) (2014)
18. Ishihara, H., Niitsu, K., Nakazato, K.: Analysis and experimental verification of DNA Single Base polymerization detection using CMOS FET-based redox potential sensor Array. Jpn. J. Appl. Phys. **54**(4S), 04DL05. (6 pages) (2015)
19. Niitsu, K., Yoshida, K., Nakazato, K.: Design and experimental demonstration of low-power CMOS magnetic cell manipulation platform using charge recycling technique. Jpn. J. Appl. Phys. **55**(3S2), 03DF13. (4 pages) (2016)
20. Tanaka, S., Niitsu, K., Nakazato, K.: A low-power inverter-based CMOS level-crossing a/D converter for low-frequency biosignal sensing. Jpn. J. Appl. Phys. **55**(3S2), 03DF10. (7 pages) (2016)
21. Yamaji, Y., Niitsu, K., Nakazato, K.: Design and experimental verification of low-voltage two-dimensional CMOS electrophoresis platform with 32×32 sample/hold cell Array. Jpn. J. Appl. Phys. **55**(3S2), 03DF07. (5 pages) (2016)
22. Niitsu, K., Kuno, T., Takihi, M., Nakazato, K.: Well-shaped microelectrode Array structure for high-density CMOS amperometric electrochemical sensor array. IEICE Trans. Electron. **E99-C**(6), 663–666 (2016)
23. K. Gamo, K. Niitsu, K. Nakazato: Noise-immune current-integration-based CMOS amperometric sensor platform with 1.2 μm x 2.05 μm electroless-plated microelectrode array for robust bacteria counting. In: Proc. IEEE Biomed. Circ. Syst. Conf. pp. 539–542 (2015)
24. K. Niitsu, A. Kobayashi, Y. Ogawa, M. Nishizawa, K. Nakazato. An energy-autonomous, disposable, big-data-based supply-sensing biosensor using Bio Fuel Cell and 0.23-V 0.25-μm Zero-Vth all-digital CMOS supply-controlled ring oscillator with inductive transmitter. In: Proc. IEEE Biomed. Circ. Syst. Conf. pp. 595–598 (2015)
25. S. Ota, K. Niitsu, H. Kondo, M. Hori, K. Nakazato: A CMOS sensor platform with 1.2 μm × 2.05 μm electroless-plated 1024 × 1024 microelectrode array for high-sensitivity rapid direct bacteria counting. In Proc. IEEE Biomedical Circuits and Systems Conf. pp. 460–463 (2014)
26. Niitsu, K., Sakurai, M., Harigai, N., Yamaguchi, T.J., Kobayashi, H.: CMOS circuits to measure timing jitter using a self-referenced clock and a cascaded time difference amplifier with duty-cycle compensation. IEEE J. Solid State Circuits. **47**(11), 2701–2710 (2012)
27. Niitsu, K., Harigai, N., Yamaguchi, T.J., Kobayashi, H.: A feed-forward time amplifier using phase detector and variable delay line. IEICE Trans. Electron. **E96-C**(6), 920–922 (2013)

28. Niitsu, K., Harigai, N., Kobayashi, H.: Design methodology for determining the number of stages in a cascaded time amplifier to minimize area consumption. IEICE Electron. Exp. **10**(11), 20130289 (2013)
29. Niitsu, K., Harigai, N., Yamaguchi, T.J., Kobayashi, H.: A low-offset cascaded time amplifier with reconfigurable inter-stage connection. IEICE Electron. Exp. **11**(10), 20140203 (2014)
30. Niitsu, K., Osawa, Y., Hirabayashi, D., Kobayashi, O., Yamaguchi, T.J., Kobayashi, H.: A CMOS PWM transceiver using self-referenced edge detection. IEEE Trans. Very Large Scale Integration (VLSI) Syst. **23**(6), 1145–1149 (2015)
31. Niitsu, K., Kang, S., Kulkarni, V.V., Ishikuro, H., Kuroda, T.: A 14 GHz AC-coupled clock distribution scheme with phase averaging technique using Sigle LC-VCO and distributed phase interpolators. IEEE Trans. Very Large Scale Integr (VLSI) Syst. (TVLSI). **19**(11), 2058–2066 (2011)
32. Niitsu, K., Sugimori, Y., Kohama, Y., Osada, K., Irie, N., Ishikuro, H., Kuroda, T.: Analysis and techniques for mitigating interference from power/signal lines and to SRAM circuits in CMOS inductive-coupling link for low-power 3D system integration. IEEE Trans. Very Large Scale Integr (VLSI) Syst. **19**(10), 1902–1907 (2011)
33. Niitsu, K., Kohama, Y., Sugimori, Y., Kasuga, K., Osada, K., Irie, N., Ishikuro, H., Kuroda, T.: Modeling and experimental verification of misalignment tolerance in inductive-coupling inter-Chip link for low-power 3D system integration. IEEE Trans. Very Large Scale Integr (VLSI) Syst. **18**(8), 1238–1243 (2010)
34. Saen, M., Osada, K., Okuma, Y., Niitsu, K., Shimazaki, Y., Sugimori, Y., Kohama, Y., Kasuga, K., Nonomura, I., Irie, N., Hattori, T., Hasegawa, A., Kuroda, T.: 3-D system integration of processor and multi-stacked SRAMs using inductive-coupling link. IEEE J. Solid-State Circ. **45**(4), 856–862 (2010)
35. Niitsu, K., Yuxiang, Y., Ishikuro, H., Kuroda, T.: A 33% improvement in efficiency of wireless inter-chip power delivery by thin film magnetic material for three-dimensional system integration. Jpn. J. Appl. Phys. **48**, 04C073. (5 pages) (2009)
36. Niitsu, K., Miura, N., Inoue, M., Nakagawa, Y.O., Tago, M., Mizuno, M., Sakurai, T., Kuroda, T.: Daisy chain transmitter for power reduction in inductive-coupling CMOS link. IEICE Trans. Electron. **E90-C**(4), 829–835 (2007)
37. Niitsu, K., Miura, N., Inoue, M., Nakagawa, Y., Tago, M., Mizuno, M., Ishikuro, H., Kuroda, T.: 60% power reduction in inductive-coupling inter-Chip link by current-sensing technique. Jpn. J. Appl. Phys. **46**(4B), 2215–2219 (2007)

Smart Microfluidic Biochips: Cyberphysical Sensor Integration for Dynamic Error Recovery

Hailong Yao, Qin Wang, and Tsung-Yi Ho

1 Background

Thanks to the electrowetting-on-dielectric (EWOD) technology, digital microfluidic biochip (DMFB) has emerged as a revolutionary platform for automating and miniaturizing laboratory process in biochemistry [1–5]. DMFB is also called as Lab-on-a-Chip (LoC), in the sense that the whole traditional biochemistry laboratory can functionally shrink onto a biochip, with all the biochemistry experiments conducted within the single chip. In such an LoC platform, droplets of microliter or even nanoliter sizes are manipulated on the surface of a 2-D array of electrodes using the electrowetting technology [5]. Compared with the traditional laboratory procedures, DMFB greatly reduces the analysis time and the sample and reagent consumption and thus has many promising applications in biochemical analysis including enzymatic assays, DNA sequencing, cell-based assays, and immunoassays [5–8]. Besides, DMFBs can both be dynamically reconfigured for different types of sequential experiments and be used for multiplexed assays at the same time. These merits significantly enhance the flexibility and throughput.

DMFBs are based on the electrowetting technology [5], which controls the wetting behavior of a polarizable or conductive liquid droplet by an electric field, so as to control the movement of the droplet in the four directions on a plane (i.e., north, south, west, and east). Figure 1 shows an exemplary schematic of a DMFB. In Fig. 1a, the cross-sectional view is given. By applying a sequence of actuation voltages to the control electrodes, the droplets between the top and bottom plates

H. Yao (✉) • Q. Wang
Tsinghua University, Beijing 100084, P. R. China
e-mail: hailongyao@tsinghua.edu.cn; woodythu@163.com

T.-Y. Ho
National Tsing Hua University, Hsinchu, Taiwan
e-mail: tyho@cs.nthu.edu.tw

H. Yasuura et al. (eds.), *Smart Sensors at the IoT Frontier*,
DOI 10.1007/978-3-319-55345-0_3

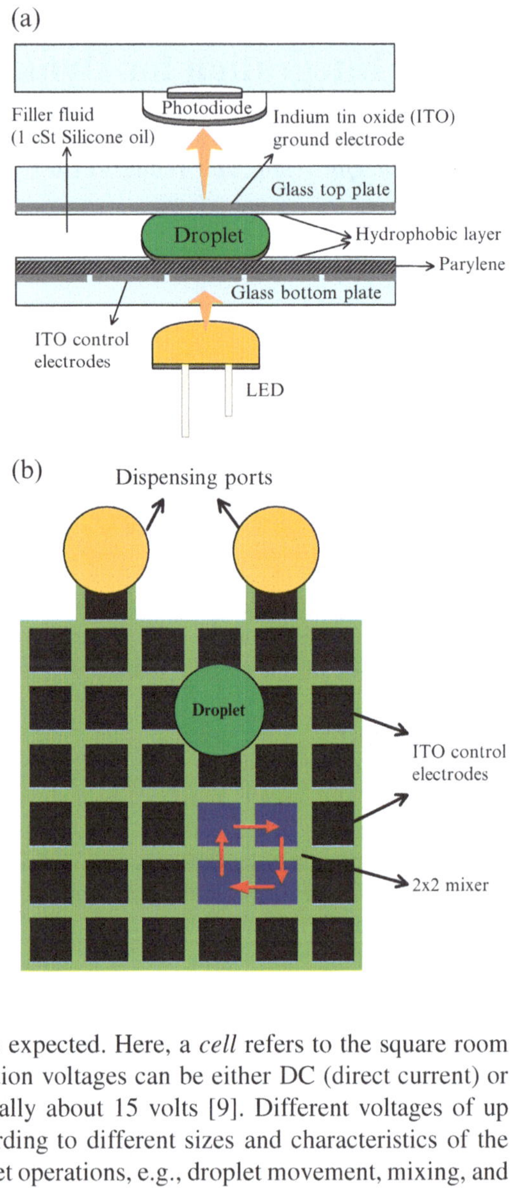

Fig. 1 Schematic of a digital microfluidic biochip [2, 5]: (**a**) cross-sectional view, and (**b**) top view

will move along adjacent cells as expected. Here, a *cell* refers to the square room of a control electrode. The actuation voltages can be either DC (direct current) or AC (alternating current) of typically about 15 volts [9]. Different voltages of up to 70 volts may be applied according to different sizes and characteristics of the droplets as well as different droplet operations, e.g., droplet movement, mixing, and splitting [10].

Utilizing the electrowetting technology, automatic biochemical experiments can be performed in a programmed way using controllers such as Arduino [11], Raspberry PI [12], and FPGA (field programmable gate array) boards. Different sample and reagent droplets can be transported to the same cell for mixing and then transported to another cell for detection. A typical method for detection is to use LED and photodiode detector as shown in Fig. 1a. Figure 1b shows the top view of the DMFB's 2-D electrode array. The dispensing ports are used to input/output the droplets. As shown in the figure, a droplet is so large that droplets sitting on horizontally/vertically/diagonally adjacent electrodes will automatically mix together. Therefore, droplets are not allowed to be adjacent unless they are planned to mix together. A mixer can be formed by adjacent cells to completely mix two droplets. Figure 1b shows a 2 × 2 mixer, i.e., a 2 × 2 electrode array circularly actuated by switching voltages. There are also other modules for the biochemical experiments, such as the mixers of other sizes, the storage cell, the detector cell, etc. For different types of sequential experiments, these modules can be dynamically reconfigured to different places by rescheduling the actuation voltages on the electrodes. As the size of DMFB becomes larger and the bioassay becomes more complicated, manual design is no longer competent for valid chip design solution with fast turnaround time. Therefore, computer-aided design (CAD) methods are becoming necessary for automated design of droplet paths, droplet scheduling along the computed paths, wire interconnection for the underlying electrodes, as well as their automatic control logics.

2 Automated Design Flow for Digital Microfluidic Biochips

Figure 2 shows the typical CAD flow for DMFBs, which consists of two main stages: (1) fluidic-level synthesis and (2) chip-level design [13]. Given the input experiment specification represented by a directed acyclic graph (DAG), which is also called as a *sequencing graph*, the fluidic-level synthesis stage computes the droplet routing paths and the droplet scheduling results along their paths. This stage typically includes the following steps:

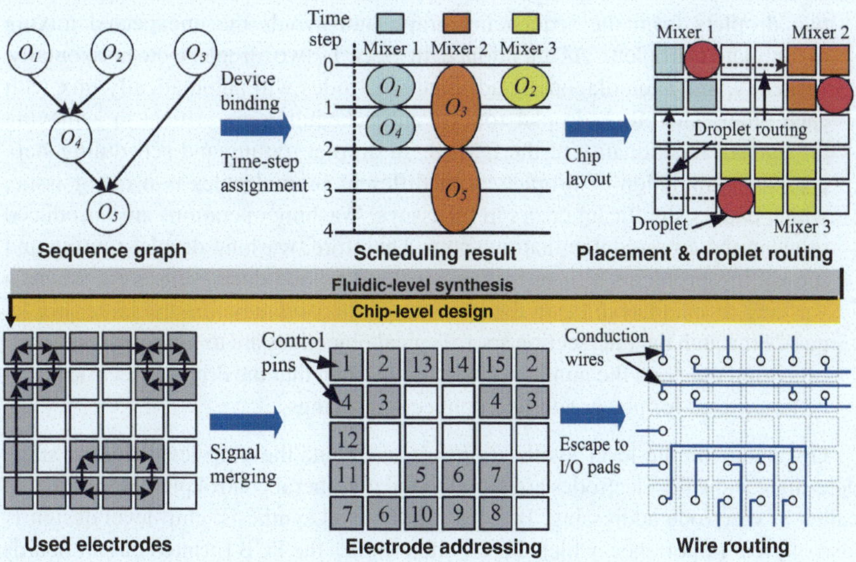

Fig. 2 Regular CAD flow of DMFBs [13]

1. Device binding and operation scheduling step: This step binds/maps each biochemical operation O_i to a functional module (e.g., Mixer j) and schedules the order of the operations when there are limited numbers of functional modules. As shown in Fig. 2b, because there is only a single Mixer 1, operations O_1 and O_4 have to be scheduled sequentially. Similarly, O_3 and O_5 are scheduled sequentially because of the single mixer Mixer 2. Different functional modules may take different time for performing the target operation. Thus, the operations are scheduled in the manner of clock cycles, where the clock period may be determined by the minimum operation time of all the functional modules.
2. Module placement: When the operations are binded to functional modules, we need to plan the positions of the modules according to their interconnection denoted in the sequencing graph. As many functional modules, such as mixers, are formed by electrode array with regular switching voltages on the electrodes, the underling electrodes of different modules can be dynamically reconfigured for different uses. In other words, the functional modules only exist at specific positions on the DMFB for a certain period of time. Therefore, the module placement problem in DMFB is a 3-D placement problem where the X- and Y-axis are for the position of the module and the Z-axis is for the period of time. Typical objective of the module placement is to minimize the weighted sum of the volume of the placed cube of all the modules and the length of the interconnection between modules. The interconnection denotes the paths where droplets move from one module to another. After module placement, the temporal positions of the modules are determined.
3. Droplet routing and scheduling: When the positions of the modules are determined, the droplet routing paths are computed according to the interconnection information, and the droplets are scheduled along their paths. The scheduling of the droplets is necessary, which guarantees the expected mixing between two droplets from the sequencing graph and avoids the unexpected mixing of unrelated droplets. As mentioned in Sect. 1, two droplets on horizontally, vertically, and even diagonally adjacent electrodes will automatically mix with each other. Therefore, droplet routing and scheduling are critical in achieving the correct functionality of the DMFB. In droplet routing and scheduling step, cross-contamination of droplets with different biomolecules is a major issue, which causes significant errors in bioassays. Washing operations are introduced to clean the cross-contamination spots. Therefore, washing droplet routing and scheduling problems also need to be well addressed during this step, where a washing droplet has to clean the prior droplet's residue before the latter droplet passes through the intersection spot. Typical objectives are to minimize the assay execution time and the number of used cells, such that the driving electrodes can be minimized for power and interconnection savings.

Given the fluidic-level synthesis result as input, the chip-level design stage determines how the electrodes are wired to the peripheral control pins, which is also called as electrode addressing. Besides fluidic-level synthesis, chip-level design is also of great importance, which directly determines the PCB (printed circuit board) fabrication cost and reliability. If the wires for electrode addressing fail to be routed, additional PCB routing layers are needed, which will unavoidably increase the fabrication cost. Besides, chip-level design significantly affects DMFB's reliability, which is a critical issue in future portable point-of-care devices. Therefore, the routability and reliability challenges in the chip-level design stage need to be addressed. This stage typically includes following steps:

1. Mark used electrodes: After the fluidic-level synthesis result, droplet routing paths are determined. To control the movement of the droplets in a programmable way, the underlying electrodes along the paths need to be connected to the peripheral electrical pads via *control pins*, where the time-varying voltages are injected by the controller. This is called *electrode addressing*. Only those electrodes with droplets passing by need to be driven by the actuation voltages. These electrodes are called *used electrodes*. Those electrodes without droplets passing by will be removed during fabrication. During this step, used electrodes are marked for electrode addressing.
2. Electrode addressing: The mapping between the electrodes and the control pins is called *electrode addressing*. There are two types of electrode addressing schemes: (1) *direct addressing* and (2) *broadcast addressing*. DMFBs in early stages use direct addressing, where each electrode is driven by an independent peripheral control pin. However, the large chip size nowadays makes direct addressing infeasible due to large number of electrodes and limited number of control pins. The DMFBs with constrained number of control pins are called *pin-constrained DMFBs* (*PDMFBs*). Broadcast addressing scheme is required for PDMFBs, where each control pin may drive multiple electrodes as long as the assay executes correctly.
3. Wire routing: When the electrode addressing solution is determined, the wire routing process is performed to compute the conduction wires between each used electrode and the designated control pin. In direct addressing scheme, the wire routing problem is the same as the PCB escape routing problem. Whereas in broadcast addressing scheme, multiple electrodes will be routed to a single control pin using either minimum spanning tree (MST) or rectilinear Steiner tree (RST). We use a *net* to represent the set of electrodes and control pin to be interconnected. In state-of-the-art algorithms, electrode addressing step and wire routing are typically performed interactively and iteratively for enhanced solution quality.

3 Fluidic-Level Synthesis

3.1 Droplet Routing and Cross-Contamination

In the past decade, noticeable advances have been made in fluidic-level synthesis methods for DMFBs, including device binding, operation scheduling, module placement, and droplet routing [14–19]. Among the different steps in the automated design flow, droplet routing is a most important stage, which determines the final routing paths for droplets between reservoirs/dispensing ports, optical detectors, etc., and thus determines the correctness and performance (execution time) in implementing the assays. Previous droplet routing methods mostly focus on two basic routing constraints [14, 16–19]: (1) fluidic constraint to avoid unexpected mixing of two droplets during their transportation and (2) timing constraint to satisfy the maximum allowed transportation time of a droplet. A typical objective is to minimize the number of cells used for droplet routing, such that the number of driving electrodes can be minimized for power and interconnection savings.

The above-mentioned basic constraints do not consider the cross-contamination issue. Cross-contamination occurs between sequential droplet routes on their intersection spots. As functional (i.e., sample or reagent) droplets leave residues on cells (electrodes) along their paths, cross-contamination occurs when the routing paths have intersections. Although there is filler fluid (e.g., silicone oil) between the top and bottom plates, it is still unavoidable for functional droplets to leave residues along their paths, which causes significant contamination issue. This is especially true for many types of proteins and heterogeneous immunoassays, because proteins tend to adsorb the hydrophobic surface. As a result, the particles and liquid residues will probably lead to cross-contamination. Such cross-contamination will cause significant errors in assay outcome.

Therefore, routing paths of different *nets*[1] should ideally be disjoint from each other to avoid the number of cross-contamination spots. When disjoint routing paths are not available, which is very common due to the single routing layer, the so-called washing droplets are introduced for cleaning the prior droplets' residue before the latter droplet passes through the intersection spot [20]. Several droplet-routing methods have been proposed to consider the cross-contamination issue [21–25]. However, the above works have oversimplified assumptions that the washing droplets have unlimited washing capacity. In fact, the washing capacity of a washing droplet will decrease when residues are washed away from the electrodes. Thus, the capacity constraint for a washing droplet needs to be considered [26].

In [27], the integrated functional and washing droplet routing flow considering the realistic washing capacity constraint is proposed. Functional routing and washing routing are simultaneously considered to resolve the routing conflicts. When the washing droplet is heading toward a specific cross-contamination spot, it should avoid the cells with residues of other functional droplets as much as possible. When the residues are unavoidable, the washing capacity will be consumed accordingly. In congested DMFB designs, certain cross-contamination spot may be surrounded by so many functional paths that the washing capacity of a small washing droplet may be exhausted before it reaches the spot for residue cleaning. Thus, larger washing droplets with larger capacity are also adopted to wash those congested spots.

[1]In the droplet routing context, a *net* refers to a set of electrodes to be connected, among which there may be more than one source electrodes and a single target electrode.

Table 1 Comparison between the proposed method and existing works

Methods	❶	❷	❸	❹	❺
[21]	Yes	No	No	No	No
[22]	Yes	No	No	No	No
[23]	Yes	No	Yes	No	No
[24]	Yes	No	N/A	No	No
[25]	Yes	No	No	No	No
[26]	Yes	Yes	No	Yes	No
[27]	Yes	Yes	Yes	Yes	Yes

❶ Use washing droplets

❷ With washing capacity limit

❸ Transport functional and washing droplets concurrently

❹ Consider washing capacity consumption of cross-contamination spot

❺ Consider washing capacity consumption of all residues

N/A: Not clearly stated in the paper

Table 1 shows the difference between the existing works in cross-contamination avoidance for digital microfluidic biochips. Next we will describe the functional and washing droplet routing methods in more detail.

3.2 *Problem Formulation of Functional and Washing Droplet Routing*

Figure 3 illustrates an example showing the practical issues in the droplet routing and scheduling step, where washing droplets have realistic capacity constraint. In the figure, a washing droplet w is dispensed from the wash reservoir on the top left corner. In the experiments, we adopt the same configuration as [23] that there are four wash reservoirs at the four corners. In Fig. 3, the washing droplet w will clean the cross-contamination spot caused by functional (i.e., sample/reagent) droplets D_3 and D_4. However, the washing path intersects with the functional paths of droplets D_1 and D_2, respectively. Thus, it is possible for the residue of D_1 and D_2 to consume w's washing capacity, even if the routing paths are carefully synchronized. Here, we call the above issue as *routing conflicts* between functional and washing droplets. If we want to keep the washing droplet clean on its way, then we have to make D_1 and D_2 wait until w passes the washing-capacity-consumption spots. That may result in timing constraint violations on functional droplets D_1 or D_2. The same issue happens to D_3 and D_4. As the washing capacity of w is limited, we need to avoid the possible washing-capacity-consumption spots as many as possible, in order to wash more cross-contamination spots. Another important issue is that w needs to reach the cross-contamination spot after the first functional droplet passes the spot, as well as before the second functional droplet reaches the spot. Only in this way can the

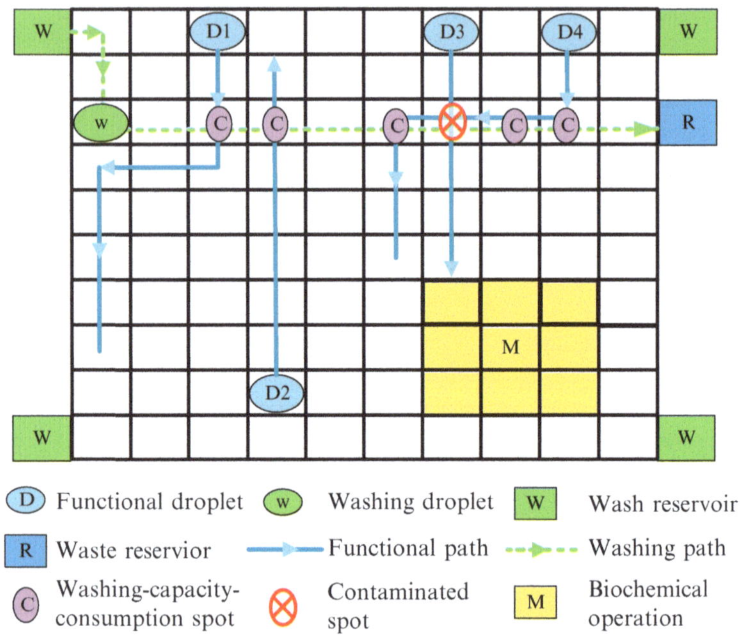

Fig. 3 Practical issues in washing operation with realistic capacity constraint

washing operation be meaningful. Both the above important washing issues need to be effectively addressed. Now we state the problem formulation of the droplet routing problem considering realistic washing operations.

There are four constraints in contamination-aware functional and washing droplet routing: (1) the fluidic constraint, (2) the timing constraint, (3) the contamination constraint, and (4) the washing capacity constraint. We assume (x_i^t, y_i^t) represents where droplet D_i is located at time t. The fluidic constraint is used to prevent unexpected mixing between two droplets of different nets during droplet transportation. Then the *static* and *dynamic* fluidic constraints between different droplets D_i and D_j can be stated as follows:

$$|x_i^t - x_j^t| > 1 \quad \text{or} \quad |y_i^t - y_j^t| > 1 \tag{1}$$

$$\begin{aligned} &|x_i^{t+1} - x_j^t| > 1 \quad \text{or} \quad |y_i^{t+1} - y_j^t| > 1 \\ or \quad &|x_i^t - x_j^{t+1}| > 1 \quad \text{or} \quad |y_i^t - y_j^{t+1}| > 1 \end{aligned} \tag{2}$$

We introduce the timing constraint to denote the maximum allowed transportation time of a droplet from its source to target, which is mainly used to ensure the bioassay's overall execution time. Typically, the shorter the routing paths of the droplets and the less waiting time for the droplets due to scheduling, the faster the bioassay execution time. Due to the high complexity of the simultaneous functional and washing droplet routing process, tight timing constraints may need to be relaxed for finding a feasible solution. To reduce the overall computation complexity, the whole functional and washing droplet routing problem is partitioned into a series of subproblems. Assume the maximum allowed transportation time is τ for all the functional and washing droplets in each subproblem. For any type of droplet D_i with source spot (x_i^S, y_i^S) and destination spot (x_i^D, y_i^D), the timing constraint is formulated as

$$\begin{aligned} (x_i^t, y_i^t) &= (x_i^S, y_i^S) \quad \text{for } t = 0 \\ (x_i^t, y_i^t) &= (x_i^D, y_i^D) \quad \text{for } t = \tau \end{aligned} \tag{3}$$

We introduce the contamination constraint to prevent cross-contamination between functional droplets. The liquid residue left by the first droplet should be washed away before the second droplet passes through the intersection spot. Therefore, the contamination constraint enforces the relative arriving times of the two functional droplets and the washing droplet at their intersection spot. Assume functional droplets D_1 and D_2 pass cross-contamination spot S_i at t_1 and t_2, respectively. Without loss of generality, assume $t_1 < t_2$, and assume a washing droplet passes S_i at t_w to wash the residue for avoiding cross-contamination. Then besides the fluidic constraints Eqs. (1) and (2), the functional and washing droplets should also satisfy the contamination constraint on S_i defined as

$$t_w > t_1 \quad \text{and} \quad t_2 > t_w \tag{4}$$

In real applications, the washing droplet gets dirty after several washing operations. Therefore, realistic washing capacity constraint needs to be considered, where the threshold is set for the washing droplets denoting the maximum allowed number of contaminated spots that a droplet could wash. Let ζ represent the washing capacity limit of a typical washing droplet. Assume a washing droplet washes N_o ordinary spots with residues and N_c cross-contamination spots before getting dirty. Then the realistic washing capacity constraint for the washing droplet is

$$N_o + N_c \leq \zeta \tag{5}$$

The contamination-aware functional and washing droplet routing problem of a DMFB can be formulated as follows:

Input: A list of nets to be connected, a set of washing droplets, a set of routing blockages, a set of reservoirs, the timing constraint, and the washing capacity constraint.

Objective: Compute the feasible routing and scheduling solution for all nets without violating the constraints, while minimizing the weighted sum of execution time, the number of cross-contamination spots, and the number of used cells for routing.[2]

Constraint: Fluidic constraint (Eqs. (1) and (2)), timing constraint (Eq. (3)), contamination constraint (Eq. (4)), and the capacity constraint of the washing droplet (Eq. (5)).

3.3 Algorithm Overview

Figure 4 shows the overall flow of [27], which consists of five major steps: (1) functional routing, (2) functional droplets routing compaction, (3) cross-contamination spots analysis, (4) washing routing, and (5) functional and washing droplets routing

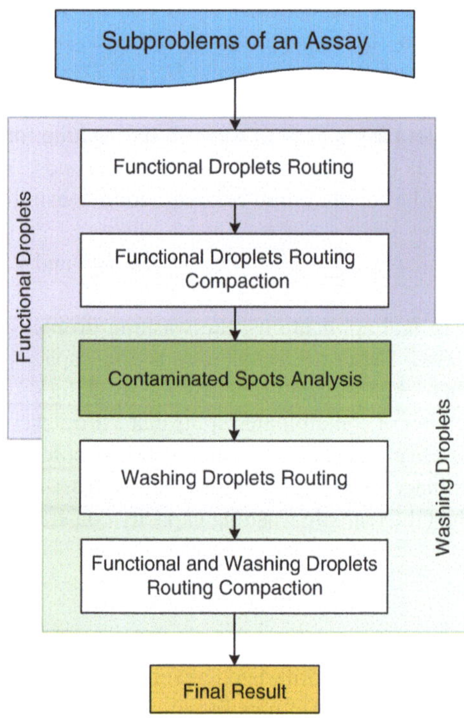

Fig. 4 Functional and washing droplet routing flow in [27]

[2]The number of used cells should be minimized for better reliability, because each used cell needs to be driven by the corresponding electrode. The less the number of working electrodes, the less probability for functional errors and thus the better reliability. Here, functional errors refer to the wrong control logic either due to the errors in control pins or errors in the wires connecting electrodes to the control pins.

compaction. In functional routing stage, the routing paths for the nets from their source cells to their target cells are computed, while minimizing the path length and the number of path intersections. During functional droplet routing compaction, an effective compaction algorithm is proposed to simultaneously schedule all the routing paths step by step, optimizing the overall execution time. The contaminated spots analysis step obtains the coordinates and the desired washing time-interval of each cross-contamination spot. Then, in the process of washing routing, the information of the cross-contamination spots is used to determine the washing order and compute the routing paths of the washing droplets. Then, a washing duration relaxation method is applied to expand the lifetime of the cross-contamination spots without violating the specified timing constraint. After that, the washing order decision technique is proposed to construct the routing paths for washing droplets, while considering the realistic washing capacity constraint. Finally, a routing compaction procedure is proposed to schedule all the functional and washing paths simultaneously for the final solution. The notations used in the following subsections are given in Table 2.

3.4 Functional Routing and Compaction

During functional routing procedure, the routing paths for the set of nets are computed separately for each subproblem. Then the routing compaction procedure simultaneously schedules the routing paths. During functional routing, the number of path intersections needs to be minimized, because each intersection spot needs a washing droplet for the cleaning task. The less the number of path intersection spots, the less washing tasks will be required. Therefore, the objective of functional routing is to find the routing paths with minimized lengths and number of intersections.

Table 2 Notations used in the proposed algorithms

Notations	Meaning
$\mathcal{D}$	List of functional droplets
D_i	The *i*th functional droplet
$\mathcal{W}$	List of washing droplets
w_i	The *i*th washing droplet
$\mathcal{S}$	List of cross-contamination spots
S_i	The *i*th cross-contamination spot
t	The current clock cycle
$\mathcal{P}$	List of functional paths
P_i	The *i*th functional path
T_c	Timing constraint for a subproblem

3.4.1 Functional Path Routing

In the proposed flow, the routing paths of functional droplets are first computed. The functional routing method is based on the classic A* searching algorithm (i.e., the Lee-style maze routing with the A* cost function). An A* search algorithm was proposed in [28], which allows for simultaneous motion of multiple droplets and thus is able to obtain globally optimal solution. However, the runtime may not be endurable for large designs due to the exponentially increasing solution space. As mentioned in Sect. 3.2, timing constraint, fluidic constraint, and contamination constraint need to be observed. Although the functional droplets will be scheduled later to satisfy those constraints, good functional routing solutions will facilitate the scheduling process and help avoid constraint violations.

For fluidic constraint, droplets cannot be horizontally, vertically, or diagonally adjacent to each other at any time during transportation, except for those that they are expected to be mixed together. Rescheduling of the droplets (i.e., stalling one droplet to make way to the other droplets) may not always resolve the fluidic-constraint violations. We present to compute *nonadjacent* routing paths for different droplets to guarantee the fluidic constraint. Figure 5 shows an example, where different droplet routing paths for droplet D_2 have different effects on droplet D_1. In Fig. 5a, the two routing paths are adjacent to each other, which makes the fluidic-constraint violation between D_1 and D_2 unavoidable even with droplet scheduling. In Fig. 5b, a different solution of D_2 obtains nonadjacent routing paths, which easily avoids the fluidic-constraint violation even without the need for droplet scheduling. To obtain nonadjacent routing paths, in the proposed routing method, the surrounding cells of routed paths are set as *used*. In this way, the A* searching algorithm will be encouraged to choose unused cells, which preferably computes nonadjacent droplet routing paths.

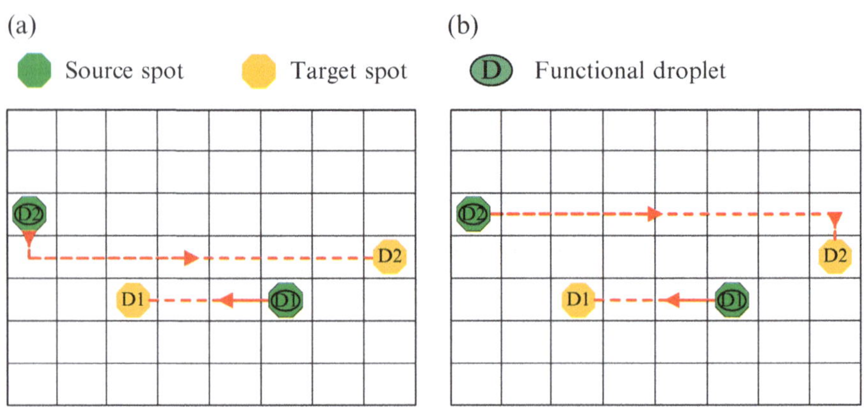

Fig. 5 Adjacent vs. nonadjacent routing paths: (**a**) due to adjacent routing paths, fluidic-constraint violation between droplets D_1 and D_2 cannot be resolved by droplet scheduling, and (**b**) using nonadjacent routing paths, there are no fluidic-constraint violations and no need for droplet scheduling

The timing constraint gives an upper-bound threshold on droplets' transportation time along their paths. This constraint is used to ensure the total execution time of an assay. During A* searching algorithm, those paths that violate the timing constraint are pruned away to avoid timing constraint violation. As a result, the proposed routing method will choose paths with used cells rather than long paths violating the timing constraint.

For avoiding violations to the cross-contamination constraint, it would be helpful to reduce the number of path intersections for saving the washing efforts. The proposed routing algorithm is modified to avoid path intersections as many as possible. For each already routed functional path, all the cells along the path are set as *used*. Then higher routing cost can be set to the used cells to avoid the path intersections. In A* searching algorithm, the routing cost of the current searching cell c_i is computed as follows:

$$\begin{aligned} F(c_i) &= G'(c_i) + H(c_i) \\ G'(c_i) &= G(c_i) + C_u \times U(c_i) \end{aligned} \tag{6}$$

where $G(c_i)$ denotes the path length from the source cell to c_i, $H(c_i)$ denotes the estimated path length from c_i to the target cell, $U(c_i)$ is a binary (0/1) variable denoting whether c_i is set as used, and C_u is the user-defined parameter for the cost of selecting a used cell. Typically, C_u is set to be 4 for choosing a used cell, i.e., when the routing path has to detour more than 4 cells, it will prefer to choose a used cell instead.

3.4.2 Path Ordering

Cross-contamination occurs when different functional droplets pass the same cell. To successfully clean the cell at the cross-contamination spot, a washing droplet should arrive at the spot within the time interval between two sequentially arriving functional droplets. This time interval is called as *washing duration* for each cross-contamination spot, which represents the feasible washing interval for the washing operation.

Figure 6 shows an example of a potential deadlock between the functional paths, where a feasible washing solution does not exist. In Fig. 6a, there are three functional paths crossing each other at cross-contamination spots S_1, S_2, and S_3, with corresponding washing durations (T_1, T_2), (T_3, T_1'), and (T_2', T_3'). The washing durations are computed according to the actual path lengths. For example, for cross-contamination spot S_2, functional droplet D_3 reaches the spot earlier than D_1, which results in the washing duration (T_3, T_1'), i.e., a washing droplet is needed to wash S_2 after D_3 passes through the spot and before D_1 reaches the spot. When the washing droplet cannot reach S_2 on time, we need to fall back and stall the latter droplet D_1. In congested designs, there may not be a good place for D_1 to stall halfway without violating the fluidic constraint. Therefore, the safe position to stall D_1 is at its source

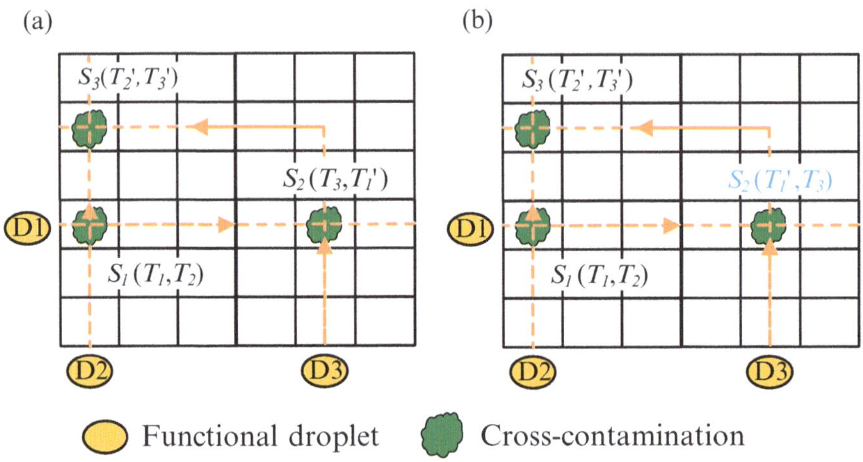

Fig. 6 Potential washing deadlock and path ordering for resolving the deadlock: (**a**) washing deadlock without considering path ordering, i.e., any droplet may arrive earlier at the cross-contamination spot, and (**b**) washing deadlock is resolved by path ordering $D_1 < D_2 < D_3$, i.e., at the cross-contamination spots, D_1 is required to arrive earlier than D_2 and D_2 earlier than D_3

position. However, when D_1 is stalled at its source position, cross-contamination spot S_1 will be affected. At S_1, a washing droplet is needed to wash the spot after D_1 passes the spot and before D_2 reaches the spot. As a result, D_2 also needs to fall back and stall at its source position, which in turn affects cross-contamination spot S_3. Then, to successfully wash S_3, D_3 needs to fall back at its source position, which will postpone the washing of S_2. In summary, the washing of S_2 affects S_1, S_1 affects S_3, and S_3 affects S_2, i.e., a deadlock is formed that cannot be resolved.

A *path ordering* method is proposed to resolve the potential washing deadlocks. As shown in Fig. 6b, the potential washing deadlock can be resolved by path ordering $D_1 < D_2 < D_3$, i.e., at any cross-contamination spot, D_1 is required to arrive earlier than D_2 and D_2 is required to arrive earlier than D_3. In this case, the washing duration of S_2 is changed to (T_1', T_3) because $T_1' < T_3$. Then stalling any droplet at its source position does not introduce deadlocks. In an extreme case, we may stall the droplets such that D_2 (D_3) waits at its source position until D_1 (D_2) reaches its target. Therefore, a valid washing solution is always guaranteed. Any path ordering solution along with the updated washing durations can be used to resolve such deadlocks.

Another issue that affects the droplet scheduling is the fluidic constraint on the source and target positions of the functional droplets. In each subproblem, source and target positions of functional droplets are typically located inside the 2-D biochip array. Therefore, fluidic constraint also needs to be satisfied for functional droplets at their source/target positions. Figure 7 shows an example, where D_1 is located at its target position and D_2 is at its source position. The shaded cells denote the blockages caused by D_1 and D_2 according to the fluidic constraint. To avoid

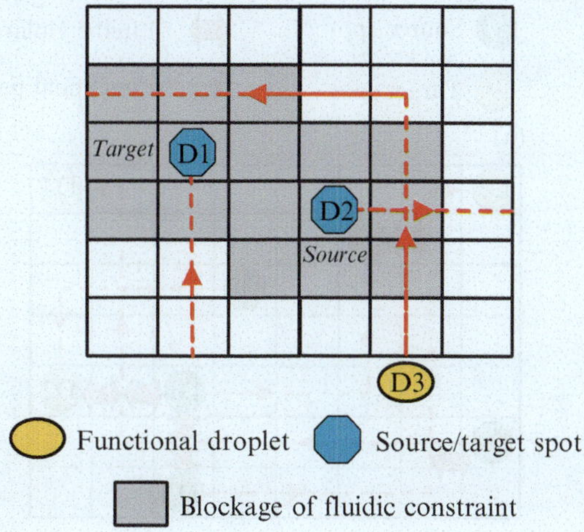

Fig. 7 Fluidic constraint for functional droplets at source/target positions

---

Algorithm 1: Functional Path Order Computation Algorithm (called in Algorithm 2).

Input: List of functional paths $\mathcal{P}$.
Output: The sorted functional paths.
Construct a directed acyclic graph *DAG* for paths $\mathcal{P}_1$ satisfying the path ordering rule;
Perform *topological sorting* on *DAG* to obtain an ordering of $\mathcal{P}_1$;
Sort the remaining paths in $\mathcal{P}_2 = \mathcal{P} - \mathcal{P}_1$ in non-ascending order of their lengths;
Perform *mergesort* on $\mathcal{P}_1$ and $\mathcal{P}_2$ according to their lengths.

---

unexpected droplet mixing, D_3 cannot pass the shaded cells of D_2 unless D_2 leaves its source position first. Besides, D_3 cannot pass the shaded cells of D_1 unless D_1 stalls somewhere without reaching its target to let D_3 pass first. Therefore, we have the following path ordering rules: **Droplet A needs to be scheduled earlier than droplet B if any of the following conditions are satisfied (1) A's source position blocks B's routing path and (2) B's target position blocks A's routing path.**

When the functional paths are successfully computed, Algorithm 1 is proposed to sort all functional paths. First, the path ordering rule is examined for all the source/target positions of the functional droplets. Then, a directed acyclic graph *DAG* is constructed on the related paths as follows: when functional droplet D_1 needs to be scheduled earlier than functional droplet D_2, two nodes V_1 and V_1 will be added into *DAG* corresponding to the paths of D_1 and D_2, and a directed edge will be added from V_1 to V_2. Please note that it is possible to have cycles in the constructed graph. The following methods can be used to remove the cycles: (1) rip-up and rerouting based on the negotiation strategy [29, 30], (2) routing concession method [14], and (3) placement refinement based on virtual topology for deadlock-free routing solutions [31]. In the experiments, the rip-up and rerouting method successfully resolves all the cycles. Figure 8 shows an example, where the constructed directed graph (Fig. 8b) for the original functional paths (Fig. 8a) contains cycles. To remove the cycles, we iteratively rip-up and reroute each

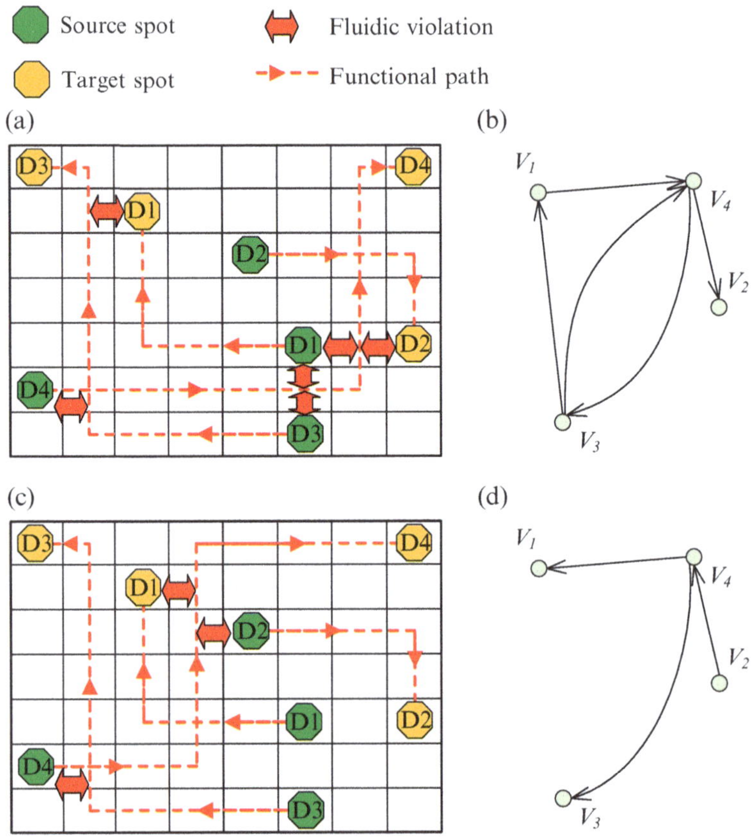

Fig. 8 Rip-up and rerouting for cycle removal in *DAG*: (**a**) original functional paths, (**b**) directed graph corresponding to (**a**), (**c**) functional paths after rip-up and rerouting path of D_4, and (**d**) *DAG* corresponding to (**c**) without cycles

functional path belonging to the cycles until they could be eliminated without introducing new cycles. To avoid obtaining the same routing path as the original one, the router sets the conflicting cells along the original path with larger routing cost. Figure 8c shows a solution by ripping up and rerouting the path of D_4. During rip-up and rerouting, higher routing cost is set to cells along the original path having fluidic violations with D_1's source spot, D_2's target spot, and D_3's source spot. When the new path is computed as shown in Fig. 8c, the new corresponding *DAG* without any cycle is shown in Fig. 8d.

Next, topological sorting will be performed on *DAG* to obtain an ordering of paths $\mathcal{P}_1$ [32]. The remaining paths $\mathcal{P}_2$ are sorted according to their path lengths. The longer the path length is, the smaller is the order for the corresponding droplet. Finally, the two sorted list of functional paths are merged together according to their lengths by *mergesort*. The topological sorting algorithm on $DAG(V, E)$ runs in

Algorithm 2: Functional path ordering and washing duration computation algorithm

Input: Lists of functional paths $\mathcal{P}$ and cross-contamination spots $\mathcal{S}$.

Output: The scheduled paths with feasible washing durations at the cross-contamination spots.

```
Sort the paths in 𝒫 by Algorithm 1;
Set each functional path P_i ∈ 𝒫 a sorted order o_i;
while true do
    Set ordered ← true;
    for j = 1 to |𝒮| do
        Find first and second functional paths P_1 and P_2 related to cross-contamination spot S_j;
        Compute the arrival times T_1 and T_2 at S_j corresponding to P_1 and P_2;
        Obtain the order values of P_1 and P_2 as o_1 and o_2, respectively;
        if o_1 < o_2 and T_1 + 1 < T_2 then
            Set the duration of S_j as (T_1, T_2);
        else if o_1 > o_2 and T_1 > T_2 + 1 then
            Set the duration of S_j as (T_2, T_1);
            Switch between the first and second functional paths for S_j;
            Set ordered ← false;
        else
            if o_1 < o_2 then
                Stall P_2 at its source position by T_1 − T_2 + 3;
                Set the duration of S_j as (T_1, T_1 + 3);
            else
                Stall P_1 at its source position by T_2 − T_1 + 3;
                Set the duration of S_j as (T_2, T_2 + 3);
                Switch between the first and second functional paths for S_j;
            Set ordered ← false;
    if ordered = true then
        break;
```

time $O(|V| + |E|)$. $DAG(V, E)$ is typically a sparse graph in the experiments, i.e., $|E| \approx |V|$. In Line 3, functional paths $\mathcal{P}_2$ are sorted in $O(|\mathcal{P}_2| \cdot log|\mathcal{P}_2|)$ time. Then in Line 4, the one-pass mergesort procedure on $\mathcal{P}_1$ and $\mathcal{P}_2$ runs in $O(|\mathcal{P}_1| + |\mathcal{P}_2|)$ time. Therefore, the overall time complexity of Algorithm 1 is $O(|\mathcal{P}_2| \cdot log|\mathcal{P}_2|)$.

When the functional paths and their related droplets are sorted in order, for the cross-contamination spots, we iteratively stall the droplet with larger order value to relax the washing durations. For any cross-contamination spot, we only allow the droplet with smaller order to pass the spot earlier. In this way, the above-mentioned deadlocks and fluidic-constraint violations can be successfully avoided. In an extreme case, we can sequentially schedule functional droplets one by one according to their orders, and wash away all the cross-contamination spots of prior functional droplets before the latter functional droplet starts out. Therefore, the proposed path ordering method always guarantees a feasible washing solution.

Algorithm 2 shows the proposed functional path ordering and washing duration computation algorithm to compute the washing durations with the potential washing deadlocks avoided. The proposed algorithm first sorts the functional paths to obtain the orders and then iteratively checks the washing duration of each cross-contamination spot. For each cross-contamination spot, the first and second droplets passing through the spot are checked according to their assigned orders. The corresponding washing duration is also examined. If there are any violations in the assigned order and/or washing duration, the function path with higher-order value will be stalled. The iteration continues until all the cross-contamination spots are valid. As mentioned above, the sorting step by Algorithm 1 takes $O(|\mathcal{P}_2| \cdot log|\mathcal{P}_2|)$ time. And in the worst case, the iterative checking on the cross-contamination spots takes $O(|\mathcal{S}|^2)$. Therefore, Algorithm 2 runs in $O(|\mathcal{P}_2| \cdot log|\mathcal{P}_2| + |\mathcal{S}|^2)$ time.

3.4.3 Functional Path Compaction

When the droplets are sorted and scheduled, there may still be unexpected mixing between functional droplets. Therefore, the compaction process is proposed to obtain the further scheduled solution for the movement of each droplet. At each time step, the droplet can either move forward one cell along the routing path or stall at the current cell. During the movements of the droplets, unexpected droplet mixing must be avoided. Furthermore, the overall execution time needs to be minimized to finish the bioassay as soon as possible. To achieve the above objectives, an effective compaction algorithm is proposed to schedule all the routing paths simultaneously. Compared with the previous compaction approach [23], a new feature of our method is that the conflicts between droplets are resolved in a global manner.

Algorithm 3 shows our routing compaction algorithm. Our simultaneous approach checks the conflicts between droplets for each step of droplet movement. If fluidic-constraint violation occurs between two functional droplets, the one with larger droplet order value will be chosen to fall back and wait (Lines 4–7 in Algorithm 4). In Algorithm 4, a preferred stall position is computed such that it has no violations with any other functional paths, i.e., the stall of the droplet will not block in the way of other droplets. Therefore, the violations between droplets can be iteratively resolved.

Now we analyze the time complexity of the proposed algorithm. The outer loop counts t from 0 to T_c. The inner loops are used to check the fluidic constraint for each pair of droplets. Therefore, the time complexity of inner loops is $O(n^2)$, where n denotes the number of paths in the subproblem. The path scheduling method in Algorithm 4 takes $O(K)$ time, where K denotes the path length in the worst case. Furthermore, when we solve one conflict of two paths, the algorithm will be restarted. Assume that the number of restarts is n_r. Then the overall time complexity of the algorithm is $O(n_r \cdot T_c \cdot n^2 \cdot K)$.

Algorithm 3: Routing compaction algorithm

Input: List of functional paths $\mathcal{P} = \{P_1, P_2, \cdots, P_n\}$ and timing constraint T_c.
Output: The scheduled paths $\mathcal{P}$ without violations.

```
Set loopFlag ← true;
Set n to be the number of functional paths in P (|P|);
while loopFlag do
    Set restart ← false;
    for t = 0 to T_c do
        Move each droplet forward by one cell along its path;
        if all the droplets reach their targets then
            Set loopFlag ← false;
            break;
        for i = 0 to n do
            for j = i + 1 to n do
                if P_i and P_j violate fluidic constraint at t then
                    Set restart ← true;
                    Call Algorithm 4 for P_i and P_j;
            if restart = true then
                break;
        if restart = true then
            break;
    if loopFlag = false then
        break;
    Reset the functional paths in P;
```

3.5 *Washing Droplet Routing*

After the above-mentioned functional routing stage, a simultaneous washing droplet routing and compaction method is proposed to clean the cross-contamination spots. Unlike previous works, our washing droplet routing method considers the realistic washing capacity constraint and the routing conflicts with functional droplets. It is necessary to perform washing routing and functional routing simultaneously, because otherwise a separate washing routing stage after each functional droplet's routing will greatly increase the overall assay execution time. Moreover, disjoint paths are not always available and often with large detouring path length. Therefore, simultaneous washing and functional routing is of great importance for avoiding cross-contamination and enhancing the performance of assay execution.

In the following subsections, the washing duration relaxation method is first proposed, which enlarges the feasible washing durations on the cross-contamination spots, thus facilitating the washing operations. Then the washing droplet routing method is proposed, which determines the washing order on the cross-contamination spots and computes the corresponding washing paths, while satisfying the timing constraint and minimizing the total path length. After comput-

Algorithm 4: Path scheduling algorithm (called in Algorithm 3)

Input: List of functional paths $\mathcal{P}$, conflict paths P_i and P_j, and current clock t.
Output: The scheduled paths P_i' and P_j'.

1 Set $id \leftarrow -1$;
2 Set $pos \leftarrow -1$;
3 Compute the droplet order o_i and o_j for P_i and P_j, respectively;
4 **if** $o_i < o_j$ *and* $P_j[t]$ *is not at source spot* **then**
5   Set $id \leftarrow j$;
6 **else**
7   Set $id \leftarrow i$;
8 **if** $id = i$ **then**
9   Set $s_i \leftarrow 0$;
10   **for** $k = t - 1$ *to* 1 **do**
11   **if** $P_i[k]$ *has no conflict with any other path* **then**
12   Set $s_i \leftarrow k$;
13   **break**;
14   Set $pos \leftarrow s_i$;
15 **else**
16   Compute s_j for P_j similar to Lines 9–14;
17 Append 3 stalls to P_{id} at pos.

ing all the washing paths, the washing paths and the functional paths are compacted/scheduled, where the arrival order of droplets at the cross-contamination spots is adjusted to successfully finish the washing tasks without contamination violations.

3.5.1 Washing Duration Relaxation

The initial washing duration for a cross-contamination spot after functional routing and compaction can be represented as follows:

$$T_{\text{washing}} = (T_{\text{early}}, T_{\text{late}}) \tag{7}$$

where T_{early} represents the arrival time of the first functional droplet (e.g., D_1 in Fig. 9) and T_{late} represents the arrival time of the second functional droplet (e.g., D_2 in Fig. 9). However, the washing droplets may not be able to finish the cleaning task within the designated washing duration. One possible reason is that the cross-contamination spot is too far away from the washing reservoir. To solve this problem, the algorithm in [22] seeks to relax the washing duration by delaying the arrival time of the latter functional droplet. However, the delayed functional droplet may violate the timing constraint. A washing duration relaxation method is proposed to guarantee the timing constraint. Let T_{used} be the time used for transporting the second functional droplet from its source cell to the target cell. Let T_c be the timing constraint. Then the maximum allowed relaxation time of the cross-contamination

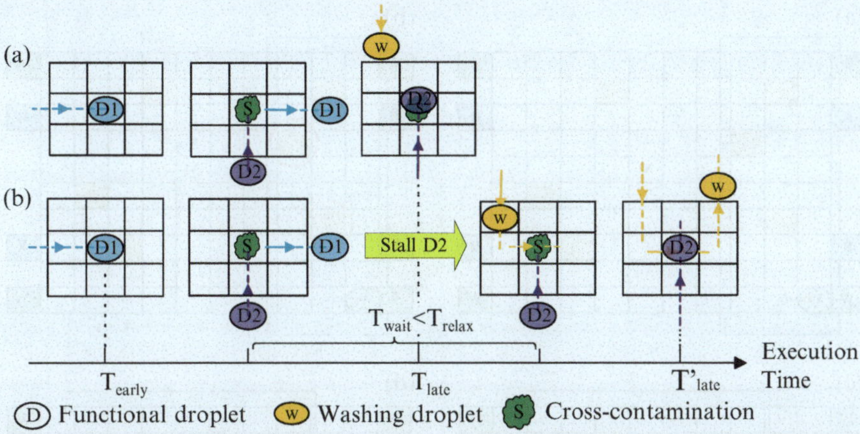

Fig. 9 Two functional droplets cross the same cell, forming cross-contamination spot S: (**a**) washing droplet fails to clean the cross-contamination spot on time, and (**b**) by stalling droplet D_2, the residue is successfully washed by w without droplet routing conflicts

spot is $T_{\text{relax}} = T_c - T_{\text{used}}$. So the relaxed washing duration for a cross-contamination spot is computed as follows:

$$T_{\text{washing}}^{\text{relax}} = (T_{\text{early}}, T_{\text{late}} + T_{\text{relax}}) \tag{8}$$

Figure 9 illustrates the washing duration relaxation process. In Fig. 9a, functional droplets D_1 and D_2 arrive at cross-contamination spot S on time T_{early} and T_{late}, respectively. But the washing droplet w fails to reach S in the washing duration T_{washing} to finish its cleaning task. So we need to stall D_2 before it arrives at S and let w wash away the residue first. In Fig. 9b, T_{wait} is the time for stalling D_2, which should not exceed T_{relax}. After the adjustment, the new arrival time of D_2 is $T'_{\text{late}} > T_{\text{late}}$. Moreover, T'_{late} does not exceed $T_{\text{late}} + T_{\text{relax}}$ because $T_{\text{wait}} \leq T_{\text{relax}}$, which ensures the timing constraint. Thus, the relaxed washing duration for each cross-contamination spot facilitates the scheduling of the functional and washing droplets and avoids the timing constraint violation.

3.5.2 Washing Order Decision and Washing Path Computation

In real applications, after cleaning a certain number of cross-contamination spots, the washing droplet will become saturated and cannot wash anymore. Moreover, the washing droplet has to clean the cross-contamination spots within the required washing duration. Therefore, the washing order of the cross-contamination spots is critical, which determines the number of spots a washing droplet could successfully clean without violating the timing constraint and capacity constraint. A method is proposed to compute the washing order and the washing paths concurrently.

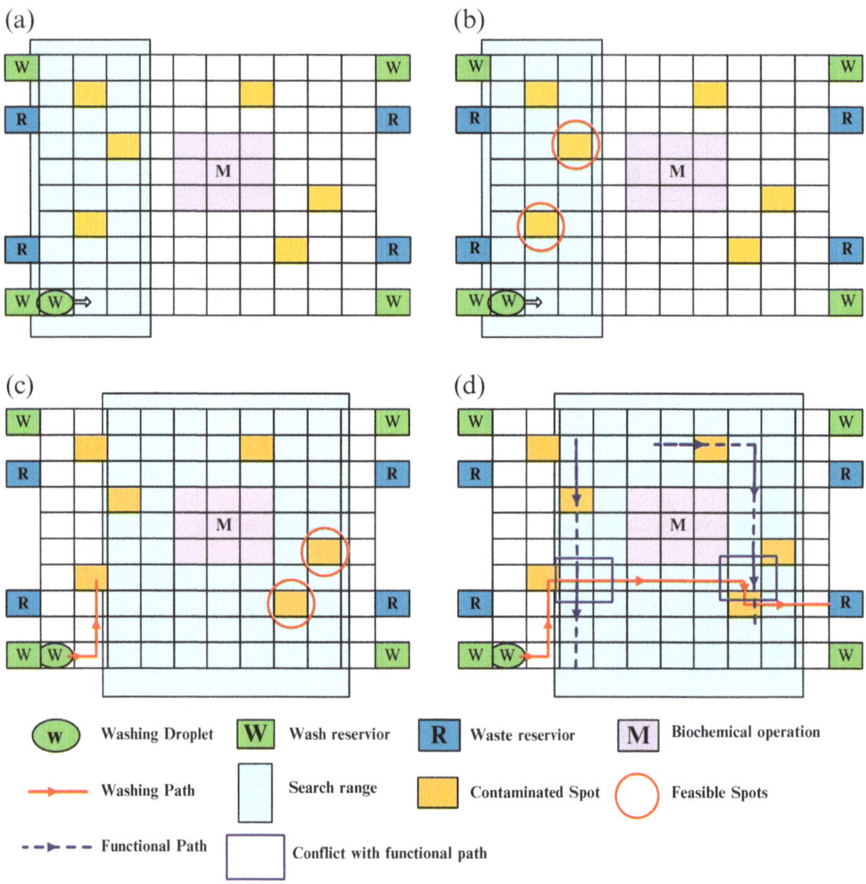

Fig. 10 Washing order decision method for washing droplet routing: (**a**) washing droplet starts from the source with the searching range initialized, (**b**) two feasible cross-contamination spots are found satisfying the washing duration, (**c**) washing droplet moves to the best cross-contamination spot chosen from the candidates and a new searching operation starts, and (**d**) finish the washing path construction when the washing capacity constraint is met until reaching the biochip boundary

Figure 10 shows the washing droplet routing process. One washing droplet is dispensed from the wash reservoir. Then the feasible cross-contamination spots are searched in several neighboring columns (e.g., 3) of the biochip array (Fig. 10a). Here, feasible cross-contamination spots refer to the spots with feasible relaxed washing durations that the washing droplet can reach in time. In Fig. 10b, two feasible cross-contamination spots are obtained as candidates. Then one of these spot candidates is chosen as the washing target according to the following equation:

$$\mathrm{Cost}_S = \alpha \cdot \frac{L}{L_c} + \beta \cdot \frac{T_{\mathrm{early}}}{T_c} \tag{9}$$

where L represents the length of the routing path from the washing droplet's current position to the cross-contamination spot,[3] T_{early} means the arriving time of the first functional droplet as defined above, T_c means the timing constraint as defined above, and L_c means the designated length constraint. We assume the droplets move one cell at each clock cycle, and set L_c to be equal to T_c. The cross-contamination spot with the minimum cost Cost_S is chosen as the intermediate routing target (see Fig. 10c). The intrinsic idea of Eq. (9) is to choose the cross-contamination spot both close to the current washing droplet's position and with small T_{early}. It is easy to understand that a close spot helps reduce the path length such that the spot does not need to wait long for washing. Moreover, the smaller the T_{early} is, the earlier the contamination happens, and thus the washing droplet does not need to wait long to wash away the generated residue. In this case, after washing the spot, more time is left for the washing droplet to clean other cross-contamination spots. α and β are user-defined parameters, which are set to be 2 and 0.5 in the experiments, respectively.

As shown in Fig. 10c, after one cross-contamination spot is determined along with the routing path, a new searching area (denoted as the shaded rectangle) is constructed to find the next set of feasible cross-contamination spots. This time the searching area could be modified to be larger according to the number of feasible candidates in the area. Besides, the crossings between the washing path and the existing functional paths are recorded. Such crossings may result in washing capacity consumption for the washing droplet. Thus, we need to subtract the consumption from the washing capacity. The searching and recording process is repeated until the biochip boundary is reached or the washing capacity is exhausted. Figure 10d shows an example of a complete washing path from wash reservoir to waste reservoir. It has two routing conflicts with existing functional paths, where each conflict possibly consumes one washing capacity. Then a new washing droplet is dispensed from another wash reservoir in clockwise order to clean the remaining cross-contamination spots. The process is repeated until all the cross-contamination spots are finished.

The washing droplet routing algorithm is summarized in Algorithm 5. The algorithm iteratively dispenses washing droplets from the reservoirs for the cleaning task until no cross-contamination spots are left. First, we initialize the washing droplet and prepare to record its routing path (Lines 3–6). Then, in Line 7, a for-loop is entered to iteratively check the searching areas to wash as many feasible cross-contamination spots as possible. In Line 8, the set of cross-contamination spots in current searching area are computed. In Line 9, Algorithm 6 is called to compute the feasible cross-contamination spots from the testing spots.

In Algorithm 6, the testing spots are iteratively checked. For each testing spot, the compatibility in the related path orders is first checked. The idea is to iteratively

[3] A* routing algorithm (i.e., Lee-style maze routing with the A* cost function) is used to compute the routing paths of the washing droplet from its current position to the candidate cross-contamination spots.

Algorithm 5: Washing droplet routing algorithm

Input: List of cross-contamination spots $\mathcal{S}$ with relaxed washing duration and list of functional paths $\mathcal{P}_{\mathcal{F}}$.

Output: List of routing paths $\mathcal{P}_{\mathcal{W}}$ for the washing droplets.

1 **while** *$\mathcal{S}$ is not empty* **do**
2   Dispense a new washing droplet w_k from one of the reservoirs;
3   Initialize washing path P_k for w_k;
4   Set current spot S_c to w_k's current position;
5   Set next spot $S_n \leftarrow NULL$;
6   Set $o_1 \leftarrow -\infty$, $o_2 \leftarrow +\infty$;
7   **for** *Searching area R_i in searching order* **do**
8   Compute the set of testing cross-contamination spots $\mathcal{S}_{\mathcal{T}} = \{S_i | S_i \in \mathcal{R}_i\}$;
9   Call Algorithm 6 to compute feasible spots $\mathcal{S}_{\mathcal{F}}$ from $\mathcal{S}_{\mathcal{T}}$ with the washing paths;
10   **if** $|\mathcal{S}_{\mathcal{F}}| = \emptyset$ **then**
11   **continue**;
12   Compute the best spot from $\mathcal{S}_{\mathcal{F}}$ according to Eq. (9) and assign it to S_n;
13   Obtain the washing path $P_{c,n}$ from S_c to S_n;
14   Accumulate w_k's washing capability consumption along path $P_{c,n}$;
15   **if** *w_k's washing capacity is violated* **then**
16   **break**;
17   Find the first and second functional paths P_1 and P_2 related to S_n;
18   Set $o_1 \leftarrow max\{o_1, order(P_1)\}$, $o_2 \leftarrow min\{o_2, order(P_2)\}$;
19   Append path $P_{c,n}$ to the end of P_k;
20   Set $S_c \leftarrow S_n$;
21   Remove S_n from $\mathcal{S}$;
22   **if** *P_k is not empty* **then**
23   Route from S_c to one of the waste reservoir;
24   Append the routed path to the end of P_k;
25   Append P_k to $\mathcal{P}_{\mathcal{W}}$;

squeeze the order values of the two related functional paths. In this way, an order can be assigned to the washing droplet without introducing deadlocks between washing and functional paths (see Sect. 3.6). After checking the path orders, the washing path P_i is computed for spot S_i. Then in Lines 8–15, fluidic constraints are checked between washing path P_i and the source/target positions of all functional paths. As stated in Sect. 3.4.2, the orders of the paths need to be sorted carefully to observe the fluidic constraint. The washing paths should follow the same path ordering rule. If P_i passes the checking process, it will be scheduled for the washing duration required at S_i. The scheduling method is similar to Lines 8–17 in Algorithm 4. Finally, a valid cross-contamination spot along with the washing path is found and stored.

Then in Line 12 of Algorithm 5, the best destination is chosen from the feasible spots based on Eq. (9), and the corresponding washing path is obtained. Next, the washing capacity consumption is computed and checked. If the washing path is valid, we will update the path order values, append the washing path to the end of washing path list, and delete the finished cross-contamination spot. Finally, the washing path to the waste reservoir is computed for discarding the washing droplet.

Algorithm 6: Feasible cross-contamination spot computation algorithm (called in Algorithm 5)

Input: Lists of testing cross-contamination spots $\mathcal{S}_\mathcal{T}$ and functional paths $\mathcal{P}_\mathcal{F}$, current spot S_c, and path orders o_1 and o_2.
Output: Lists of feasible cross-contamination spots $\mathcal{S}_\mathcal{F}$ and washing paths $\mathcal{P}'_\mathcal{W}$.

for $i = 1$ *to* $|\mathcal{S}_\mathcal{T}|$ **do**
  Set cross-contamination spot $S_i \leftarrow \mathcal{S}_\mathcal{T}[i]$;
  Find the first and second functional paths P_1 and P_2 related to S_i;
  Set $o'_1 \leftarrow max\{o_1, order(P_1)\}$, $o'_2 \leftarrow min\{o_2, order(P_2)\}$;
  if $o'_1 \geq o'_2$ **then**
    continue;
  Compute the washing path P_i from S_c to S_i;
  Set *flag* $\leftarrow$ *true*;
  for $j = 1$ *to* $|\mathcal{P}_\mathcal{F}|$ **do**
    if *P_j's source position violates fluidic constraint with P_i and $order(P_j) > o'_1$* **then**
      Set *flag* $\leftarrow$ *false*;
      break;
    if *P_j's target position violates fluidic constraint with P_i and $order(P_j) < o'_2$* **then**
      Set *flag* $\leftarrow$ *false*;
      break;
  if *flag* $=$ *true* **then**
    Schedule P_i according to the washing duration of S_i;
    Append S_i to the end of $\mathcal{S}_\mathcal{F}$;
    Append P_i to the end of $\mathcal{P}'_\mathcal{W}$;

Please note that when there are more than one washing droplet dispensed from the same reservoir, the latter washing droplet is delayed by 2 clock cycles to avoid unexpected droplet mixing.

Now we analyze the time complexity of Algorithm 5. The cross-contamination spots are first sorted according to their column indices. Therefore, to find the feasible cross-contamination spots, we only need to scan the columns sequentially in the designated searching area. Let w and h denote the width and height of the biochip array, respectively. And let $|\mathcal{S}|$ represent the number of cross-contamination spots. The time complexity of sorting and searching for feasible cross-contamination spots is $O(|\mathcal{S}|)$ using bucket sort. The routing paths of the washing droplet are computed using A* routing (Lee-style maze routing with the A* cost function), where the worst-case time complexity is $O(k \cdot w \cdot h)$. Here k represents the average number of routing paths for each cross-contamination spot. In Algorithm 6, the checking process for fluidic-constraint violations takes $O(|\mathcal{P}_\mathcal{F}|)$ time, and the path scheduling process for the washing paths takes $O(K)$ time, where K denotes the worst-case path length. In the worst case, each washing droplet can only clean one cross-contamination spot in its washing path, i.e., the algorithm will be finished in $|\mathcal{S}|$ rounds. Therefore, the overall time complexity for one subproblem is $O(|\mathcal{S}| \cdot k \cdot w \cdot h \cdot (K + |\mathcal{P}_\mathcal{F}|))$.

3.6 Simultaneous Functional and Washing Path Compaction

When the washing paths are computed, there may still be fluidic-constraint violations between washing and functional routing paths. Therefore, a final compaction step is performed on all the functional and washing paths. To avoid the deadlock problem mentioned in Sect. 3.4.2, we insert each washing path into the sorted functional paths with an order value between o_1 and o_2 computed in Algorithm 5. Then, all the functional and washing paths are sorted and each path has a new order. Finally, Algorithm 3 is called to compact all the paths simultaneously. When there are routing violations, the conflicting path (either functional or washing path) with the higher-order value will be stalled. Besides, to guarantee the washing duration constraint for the cross-contamination spots, the washing feasibility is validated when the droplets reach those spots during clock forwarding from 0 to T_c. When the first functional droplet or washing droplet is stalled to make washing impossible, the latter droplet(s) will be stalled accordingly. The merits of having an order for each droplet is that, whenever a conflict occurs, we only need to select the path with higher order to stall without worrying about the deadlock issue.

Theorem 1 *Using the proposed path ordering method, Algorithm 3 will always converge with a feasible functional and washing routing solution.*

Proof The proposed path ordering method first assigns each functional path an order. Then according to the washing relation, each washing path is also given an order value as follows. Assume the washing path P_k washes two cross-contamination spots S_1 and S_2. And assume the first functional path passing through S_1 is $P_{1,1}$ and the second one $P_{1,2}$. Similarly, assume the corresponding functional paths for S_2 are $P_{2,1}$ and $P_{2,2}$, respectively. Let the corresponding orders of the paths be o_k, $o_{1,1}$, $o_{1,2}$, $o_{2,1}$, and $o_{2,2}$, respectively. From Algorithms 5 and 6, we have $o_1 = max\{o_{1,1}, o_{1,2}\}$ and $o_2 = min\{o_{2,1}, o_{2,2}\}$. Therefore, the order of P_k is set to be o_k satisfying $o_1 < o_k < o_2$, i.e., P_k is inserted in between the functional paths without affecting the original sequential order. There are three cases when we stall a path: (1) if the path is the first one to pass some cross-contamination spots, all the related washing paths and second functional paths must have higher-order values and need to be stalled; (2) if the path is a washing path, all the related second functional paths must have higher-order values and need to be stalled; (3) if the path is the second one to pass some cross-contamination spots, no related paths need to be stalled because it has the highest-order value. Therefore, the deadlock shown in Fig. 6 will never occur. In an extreme case, the functional and washing droplets can be walked to their targets one by one without concurrency, which guarantees a feasible functional and washing routing solution. Therefore, by stalling the paths according to their order values, Algorithm 3 will always converge with a feasible solution.

Figure 11 shows an illustrative example of the functional and washing droplet routing process for the example in Fig. 8c. Assume the order of functional droplets is $(D_2 < D_4 < D_1 < D_3)$. And assume washing droplets w_1 and w_2 are computed to

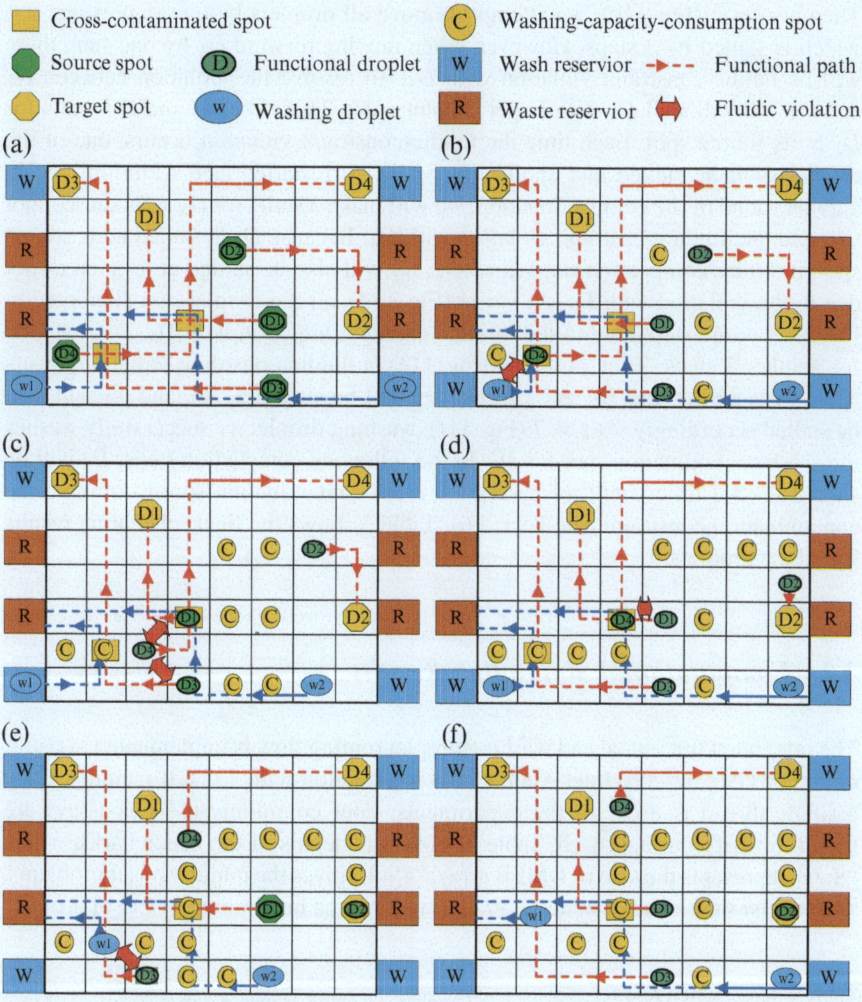

Fig. 11 An illustrative example: (**a**) initial status with computed washing paths at time $t = 0$, (**b**) compaction at time $t = 1$, (**c**) compaction at time $t = 2$, (**d**) compaction at time $t = 4$, (**e**) compaction at time $t = 6$, and (**f**) compaction at time $t = 7$

wash the cross-contamination spots of (D_4, D_3) and (D_4, D_1). According to Sect. 3.6, a feasible order for all the droplets is $(D_2 < D_4 < w_1 < w_2 < D_1 < D_3)$. Figure 11a shows the initial status with computed washing paths at time $t = 0$. At $t = 1$ (Fig. 11b), all the droplets are attempted to forward by one step. However, when moving forward w_1 by one step, there will be fluidic-constraint violations between w_1 and D_4. To resolve the violation between w_1 and D_4, we will stall w_1 with larger droplet order as shown in Algorithm 4. Therefore, we make three stalls for w_1 at the wash reservoir. All the remaining droplets are successfully transported by one step.

Then at $t = 2$ (Fig. 11c), we attempt to move all droplets by one step except w_1, which is stalled by 3 steps. However, when moving forward D_1 by one step, there will be fluidic-constraint violation with D_4. To resolve the violation between D_1 and D_4, we will stall D_1 with larger droplet order. Therefore, we make 3 stalls for D_1 at its source spot. Each time the fluidic-constraint violation occurs, one of the droplets will be stalled and another loop will be restarted (see Algorithm 3). In another round of the compaction loop, we will make 3 stalls for D_3 at its source spot to avoid the fluidic-constraint violation with D_4. Because D_3 is stalled at its source spot, in future compaction steps at $t = 2$, w_2 will also be stalled at its source spot due to the violation with D_3. At $t = 4$ (Fig. 11d), all the droplets are attempted to forward by one step. Due to the fluidic-constraint violation with D_4, D_1 is stalled for another 3 steps. Then at $t = 6$ (Fig. 11e), a fluidic-constraint violation occurs again between w_1 and D_3. As a result, D_3 will be stalled again, and w_2 will also be stalled accordingly. At $t = 7$ (Fig. 11f), washing droplet w_1 successfully washes the cross-contamination spot for D_3. In the following compaction steps, D_1 will be stalled several times until w_2 passes the cross-contamination spot to observe the contamination constraint (see Sect. 3.6). Table 3 shows the final scheduling results for all the droplets.

3.7 *Computational Simulation Results*

The integrated functional and washing droplet routing flow is implemented in C++ on a 2.60 GHz 32-core Intel Xeon Linux workstation with 132 GB memory. Only a single thread is used for the experiments. Four commonly used bioassays are tested to verify our approach. Table 4 shows the details of the benchmarks, where "Size" represents the size of DMFB array, "#Sub" gives the number of subproblems, "#Net" gives the number of nets, "#D_{max}" records the maximum number of droplets

Table 3 Compaction results of the example in Fig. 11. "1" represents moving the droplet forward by one step, and "0" represents stall the droplet at current position. Total 22 steps are given, i.e., from $t = 1$ to $t = 22$

Droplets	Control sequence
D_1	0 0 0 0 0 0 0 0 0 0 0 0 0 0 0 1 1 1 1 1 1 0
D_2	1 1 1 1 1 0 0 0 0 0 0 0 0 0 0 0 0 0 0 0 0 0
D_3	0 0 0 0 0 0 1 1 1 1 1 1 1 1 1 1 1 0 0 0 0 0
D_4	1 1 1 1 1 1 1 1 1 1 1 1 0 0 0 0 0 0 0 0 0 0
w_1	0 0 0 1 1 1 1 1 1 0 0 0 0 0 0 0 0 0 0 0 0 0
w_2	0 0 0 0 0 0 1 1 1 1 1 1 1 1 1 1 1 0 0 0 0 0

Table 4 Statistics of the routing benchmarks

Circuit	Size	#Sub	#Net	#D_{max}	#Reservoir
In-vitro_1	16 × 16	11	28	5	4
In-vitro_2	14 × 14	15	34	6	4
Protein_1	21 × 21	64	181	6	4
Protein_2	13 × 13	68	137	6	4

within one subproblem, and "#Reservoir" denotes the number of wash reservoirs. In the experiments, the washing capacity constraint for each washing droplet is set to be 4. Besides, to fully test the performance of the proposed washing flow, the functional paths are allowed to have intersections between each other.

In the first experiment, we compute the number of washing droplets violating the capacity constraint. This experiment verifies the importance of washing capacity constraint and the effectiveness of our method. Table 5 shows the comparison results of our routing flow with vs. without the washing capacity constraint. "#Cont." gives the number of cross-contamination spots, "#$\mathcal{W}_{\text{vio}}$" gives the number of washing droplets that conducted the washing task with violated capacity constraint, "#$\mathcal{W}$" gives the total number of used washing droplets, "Error" gives the error rate calculated by "#$\mathcal{W}_{\text{vio}}$/#$\mathcal{W}$," "$\mathcal{S}_{\text{fail}}$" gives the number of cross-contamination spots that fail to be washed, "#UC" gives the number of used cells for routing, "T_r" gives the execution time for bioassays (i.e., the number of clock cycles), and "CPU" gives the CPU time in seconds (s). The results show that our work is effective with significant improvement, which reduces all the error rates to 0. Without the capacity constraint, overall there are 67% invalid washing droplets violating the capacity constraint. Using our algorithm, all the washing operations are valid within the capacity limit. From the results, there are also some cross-contamination spots that fail to be washed. In those cases, there are so many functional paths in the way blocking the washing droplet and consuming its capacity that the washing capacity is exhausted before reaching the cross-contamination spot. In such cases, a larger washing droplet could be adopted to wash the congested cross-contamination spots (see Fig. 12 for details).

In the second experiment (Table 6), we compare our approach (the capacity limit is removed) with state-of-the-art contamination-aware droplet routing method in [23], which does not consider the washing capacity constraints. The method in [23] seems to perform better than our proposed washing droplet routing method. That is because the minimum cost circulation problem formulation is used to schedule optimal and correct wash operations. However, the problem we are addressing in this chapter is much more difficult than the one in [23]. In our problem, (1) washing droplets have realistic washing capacity constraints, and (2) functional and washing droplets are transported simultaneously, while the realistic washing capacity consumptions are considered for all residues along the path (i.e., not only residues at the intersection sites as in [23]). The problem is so difficult that there is not an easy way to modify the method in [23] and formulate our problem as a minimum cost circulation problem. Based on the above considerations, the overhead (i.e., number of used cells and the execution time) is reasonable. Besides, the runtime of our method is much faster with 28× speedup in average.

In the third experiment, we compare two approaches of constructing the washing paths. The first method finds the washing paths by diagonal searching. That is, the next destination spot is found for washing droplets in the diagonal direction in the 2-D biochip array. The second method (our proposed method) finds the washing paths by horizontal searching. That is, the next destination spot is found in the horizontal direction. The results in Table 7 show that the horizontal searching has

Table 5 Computational simulation results w/ vs. w/o washing capacity limit

Bioassay	#Cont.	Without capacity limit							With capacity limit						
		$\#\mathcal{W}_{vio}$	$\#\mathcal{W}$	Error	$\mathcal{S}_{fail}$	#UC	T_r	CPU	$\#\mathcal{W}_{vio}$	$\#\mathcal{W}$	Error	$\mathcal{S}_{fail}$	#UC	T_r	CPU
In-vitro_1	31	7	8	88%	0	548	293	0.02	0	31	0%	14	571	444	0.03
In-vitro_2	34	7	10	70%	0	617	351	0.03	0	34	0%	7	582	432	0.04
Protein_1	69	19	23	83%	0	3042	1591	0.08	0	69	0%	9	3438	1724	0.13
Protein_2	75	11	25	44%	0	2040	1336	0.05	0	75	0%	7	2046	1318	0.08
Total	209	44	66	67%	0	6247	3571	0.18	0	209	0%	37	6637	3918	0.29

Fig. 12 Computational simulation results on different sizes of the washing droplets

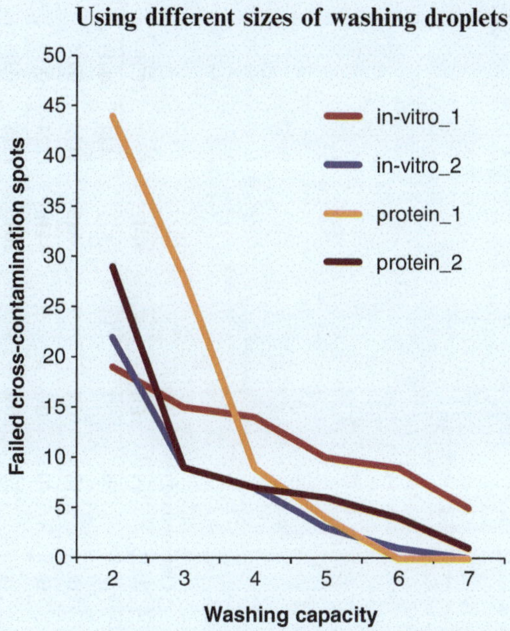

Table 6 Comparison result between [23] and our method

	[23]				Our method without capacity limit			
Bioassay	#Cont.	#UC	T_r	CPU	#Cont.	#UC	T_r	CPU
In-vitro_1	21	351	193	0.58	31	548	293	0.02
In-vitro_2	5	281	191	0.39	34	617	351	0.03
Protein_1	82	2213	1394	2.58	69	3042	1591	0.08
Protein_2	61	1362	1108	1.49	75	2040	1336	0.05
Total	169	4207	2886	5.04	209	6247	3571	0.18

a better performance in CPU time than diagonal direction. This is because the horizontal searching has a smaller searching range in each step and thus is more efficient than the diagonal one. Moreover, the horizontal searching method results in fewer failed cross-contamination spots. We attribute this to the fact that horizontal searching method has more flexibility in the Y-axis (i.e., searching both up and down) than the monotone diagonal searching. Since congested cross-contamination spots are generated during functional routing, the merits of the additional searching flexibility in horizontal searching method become notable.

Figure 12 shows the computational simulation results using different sizes of washing droplets. From the figure, using a larger washing droplet, i.e., with larger washing capacity, the cross-contamination spots are more likely to be successfully washed away. However, with small washing droplets, some cross-contamination spots fail to be successfully washed. This is because of the fact that it is usual for some functional paths to surround a specific cross-contamination spot and

Table 7 Diagonal searching vs. horizontal searching

Bioassay	#Cont.	Our method (diagonal)							Our method (horizontal)						
		$\#\mathcal{W}_{\text{vio}}$	$\#\mathcal{W}$	Error	S_{fail}	#UC	T_r	CPU	$\#\mathcal{W}_{\text{vio}}$	$\#\mathcal{W}$	Error	S_{fail}	#UC	T_r	CPU
In-vitro_1	31	0	6	0%	21	481	443	0.08	0	9	0%	14	571	444	0.03
In-vitro_2	34	0	2	0%	32	420	460	0.09	0	12	0%	7	582	432	0.04
Protein_1	69	0	6	0%	59	2362	2330	0.40	0	40	0%	9	3438	1724	0.13
Protein_2	75	0	21	0%	32	1846	1666	0.16	0	32	0%	7	2046	1318	0.08
Total	209	0	35	0%	144	5109	4899	0.73	0	93	0%	37	6637	3918	0.29

consume a certain number of washing capacity before the washing droplet could reach the spot. In such cases, it is possible to use multiple small washing droplets to wash a single cross-contamination spot. But that would result in delayed execution time. Therefore, this chapter proposes to use large washing droplets to perform the washing tasks for congested cross-contamination spots. From the figure, all the cross-contamination spots in benchmarks protein_1 and in vitro_2 can be successfully washed with a larger washing droplet of washing capacity 7. For a washing droplet with washing capacity greater than 7, the washing droplet would be so large that it will take multiple electrodes in space. That is left for future works.

4 Chip-Level Design

4.1 Electrode Addressing and Wire Routing

Besides fluidic-level synthesis, chip-level design is also of great importance, which directly determines the PCB (printed circuit board) fabrication cost and reliability. If the wires for electrode addressing fail to be routed, additional PCB routing layers are needed, which will unavoidably increase the fabrication cost. Besides, chip-level design significantly affects DMFB's reliability, which is a critical issue in future portable point-of-care devices. Therefore, this subsection mainly addresses the routability and reliability challenges in the chip-level design stage, where the major focus is on electrode addressing and wire routing.

As mentioned in Sect. 2, there are two types of electrode addressing scheme, i.e., direct addressing and broadcast addressing. Broadcast addressing is superior to direct addressing in terms of the number of required control pins. The controller generates the *actuation sequences* to the control pins for driving the electrodes, which are essentially sequences of voltage values: (1) value "1" for logic high value, (2) value "0" for logic low value, and (3) "X" denotes a don't-care value which can either be "1" or "0" without affecting the designated droplet movements. For correctly controlling the movement of the droplets, each working electrode is assigned an actuation sequence. In [33], Xu et al. presented a compatible graph to model the compatibility in actuation sequences between electrodes, where compatible electrodes can share the same control pin. Figure 13a shows an example of broadcast addressing. Assume that the actuation sequences of the electrodes (s for short) are as follows: (1) $s(e_1)$ = "01X01X110X," (2) $s(e_2)$ = "0X00111X01," and (3) $s(e_3)$ = "01X0X111X1." Then the three electrodes are compatible with each other, and a single control pin with actuation sequence "0100111101" can correctly drive all the three electrodes simultaneously. Therefore, control pin CP_1 is introduced to drive the three electrodes (e_1, e_2, and e_3). Manhattan wires are routed for connecting the control pin and the electrodes on the *escape routing layer*, which actually form a Steiner tree. Please note that there is typically a single escape routing layer, and hence wires cannot cross each other. When there are routing failures, an

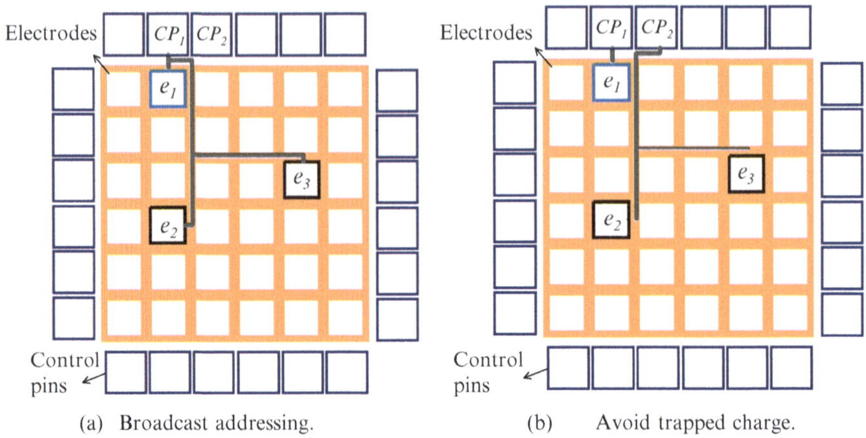

Fig. 13 Broadcast addressing and the trapped charge problem in a digital microfluidic biochip: (**a**) Broadcast addressing without considering the trapped charge problem. (**b**) Enhanced electrode addressing considering trapped charge for improved reliability

additional routing layer will be required with increased fabrication cost. Therefore, the electrode addressing and routing is critical in reducing the total manufacturing cost.

Another critical issue with broadcast addressing is the trapped charge problem [34–36]. Different electrodes require different driving voltages for different types of droplet operations, e.g., droplet dispensing from input reservoir may require 60–80 volts, while droplet transportation may require at least 10–20 volts [37]. If a control pin drives two electrodes, one for droplet dispensing and one for transportation, then the minimum driving voltage needs to be 60–80 volts for effectively driving both the two electrodes. In that case, charge is trapped in the dielectric insulating layer around the electrode for droplet transportation, due to excessive applied voltage. The trapped charge reduces the electrowetting force and thus causes wrong assay results and even permanent dielectric breakdown. For applications such as patient health monitoring, clinical diagnosis, etc., reliability is of great importance [38]. The reliability issue is even more critical in future portable point-of-care devices. Therefore, the trapped charge issue should be avoided in broadcast addressing, i.e., electrodes with different preferred driving voltages should avoid sharing the control pin as much as possible. Figure 13b shows an example of electrode addressing to avoid the trapped charge problem, where electrode e_1 is assumed to require much higher voltage than e_2 and e_3. So the three electrodes must not share a single control pin. As a result, another control pin CP_2 is used to drive e_2 and e_3, and e_1 is driven independently by CP_1. For minimizing the number of control pins, e_1 may also share the control signal with other electrodes requiring high voltages.

For chip-level design, the works in [39] and [40] presented to improve routability by simultaneous electrode addressing and wire routing. And the work in [41] presented to use decluster and reroute approach rather than rip-up and reroute to improve the routability. However, the above works do not consider the reliability issue and thus may not be practical for real applications. Regarding the reliability issue, Huang et al. presented a method to optimize the maximum actuation time on the electrodes for better reliability [42]. However, with appropriate actuation voltage, high actuation time may not be critical in causing the reliability issue. Yeh et al. presented the first work to address the trapped charge issue with the minimum cost maximum flow formulation [36], which is an extension of [39]. The presented network flow algorithm reduces the number of control pins without appropriate consideration of the routing requirement. As a result, routability may be an issue in the presented method. In [43], the first routability- and reliability-driven chip-level design method based on the SVM (support vector machine) classifier is presented. The SVM-based classifiers effectively improve routability in two aspects: (1) routability between the electrodes in each cluster and (2) routability between the clusters and the control pins. Experimental results show that the presented method obtains 100% routing completion rate for all the benchmarks. Moreover, the reliability issue induced by the trapped charge problem is also effectively addressed. The presented method will be discussed in more detail in the following subsections.

4.2 Problem Formulation of Electrode Addressing and Wire Routing

Two major problems in chip-level design need to be considered early in the electrode addressing stage.

1. Routability: Routing is not a trivial task because there is typically a single routing layer in chip-level design. And routing failures will cause additional routing layers, which may dramatically increase the fabrication cost.
2. Reliability caused by the trapped charge issue: When an electrode is driven by excessive actuation voltage due to inappropriate control signal sharing, chip malfunction or even dielectric breakdown may occur. Thus, the trapped charge problem must be addressed during electrode addressing.

The routability- and reliability-driven chip-level design problem can be stated as follows:

Given: (1) A set of electrodes $E = \{e_1, e_2, \ldots, e_n\}$; (2) the actuation sequences $S = \{s_1, s_2; \ldots, s_n\}$ corresponding to the electrodes in E; (3) the preferred voltage values $V = \{v_1, v_2, \ldots, v_n\}$ corresponding to the electrodes in E; (4) a threshold voltage value V_{th}, above which the driving voltage tends to cause the trapped charge problem; (5) the maximum number of allowed control pins $C_{\max}$ for external controller; and (6) the design rules of wire routing.

Find: A feasible routing solution from all the electrodes in E to the control pins with minimized total routing cost.

Subject to: (1) Control pin constraint: the number of used control pins must be less or equal to $C_{\max}$. (2) Routing constraint: each electrode is successfully routed to a control pin without any design rules violations. (3) Broadcast-addressing constraint: the actuation sequences of the electrodes within the same cluster must be compatible with each other. (4) Voltage constraint: for each cluster of electrodes, the driving voltage at the corresponding control pin should not be less than the preferred voltage of any member electrode.

For the trapped charge problem, we use the same measurement model as [36]. In the model, a variable TC_i is introduced to represent the trapped charge on electrode e_i due to excessive driving voltage. TC_i is defined as

$$TC_i = \begin{cases} v_i^* - max(V_{th}, v_i), & v_i^* \geq V_{th} \\ 0, & v_i^* < V_{th} \end{cases} \tag{10}$$

where v_i^* and v_i represent the actual driving voltage and the preferred voltage for electrode e_i, respectively. TC_i represents the trapped charge on e_i due to excessive driving voltage.

Based on Eq. (10), the overall cost of the trapped charge problem, denoted as TC, is computed as

$$TC = max\{TC_i | e_i \in E\} \tag{11}$$

Then the total routing cost considering the trapped charge problem is computed as

$$C = \alpha \cdot |CP| + \beta \cdot WL + \gamma \cdot TC \tag{12}$$

where $|CP|$ represents the total number of used control pins, WL represents the total wire length, and TC is for trapped charge as defined above. Here, α, β, and γ are user-defined parameters.

4.3 Algorithm Overview

Figure 14 presents the overall flow of the chip-level design method. There are five major steps, i.e., compatible graph construction, electrode addressing, cluster routing, escape routing, and rip-up and rerouting. First of all, a compatible graph is constructed according to the actuation sequences of electrodes. In the following stages, the electrodes within each cluster are interconnected first and then are routed to the control signals by escape routing. When necessary, rip-up and rerouting along with declustering are performed to improve the routing completion rate.

Here, the SVM-based strategy is proposed in electrode addressing module, which randomly generates a set of candidate clustering solutions first. Then a ranking model based on SVM is used to obtain a set of clustering solutions with higher

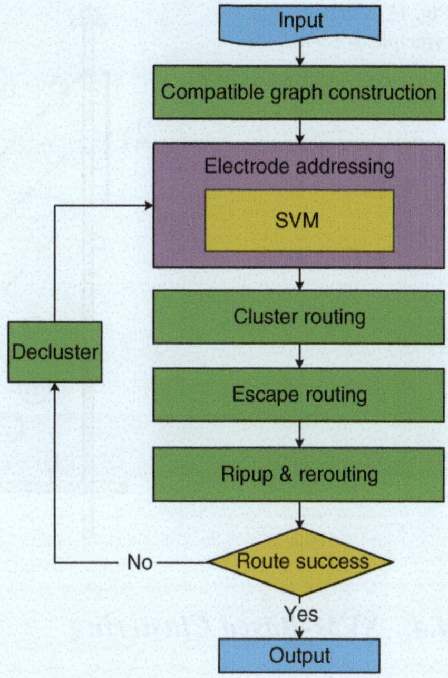

Fig. 14 Design flow of our approach

Table 8 Notations used in our approach

Notations	Meaning
C_N	Number of clusters in a clustering solution
C_{N_i}	Number of clusters belong to quadrant i
$\|E\|$	Number of electrodes
C_S	Total area of a chip
P_C	Number of clusters with only a single electrode
T_B	Bounding box area of the chip
T_{B_i}	Bounding box area of quadrant i
T_O	Total area of bounding box overlap for the whole chip
T_{O_i}	Area of bounding box overlap for quadrant i
T_{P_i}	Number of electrodes in cluster i
B_{P_i}	Number of electrodes on the edge of the chip in cluster i
O_{L_i}	Overlap area for bounding box of cluster i
B_{B_i}	Area of bounding box of cluster i
$v^*_{C_i}$	Actual driving voltage for cluster i
v_i	Preferred voltage for electrode e_i
V_{th}	Threshold voltage

ranking score. It is claimed that any searching algorithm for better clustering solutions can be adopted with the SVM ranking model. Table 8 presents the variables used in the following subsections along with their meanings.

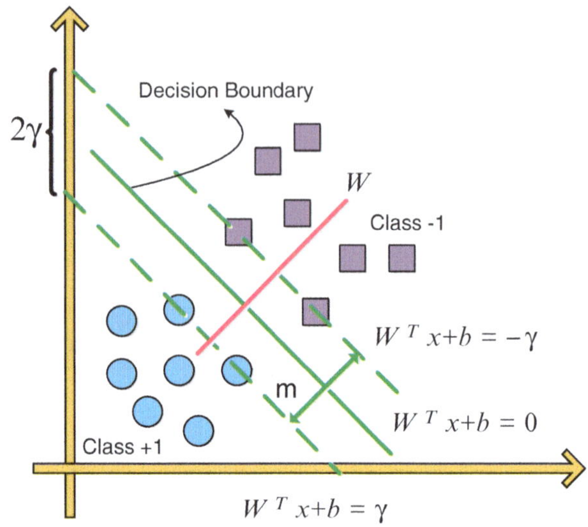

Fig. 15 Fundamental principle of SVM [44]

4.4 SVM-Based Clustering

There are two key steps in chip-level design flow, i.e., electrode addressing and wire routing. A big design gap exists between the two steps, which results in degraded routing solution. And inferior electrode addressing solution may not correspond to a successful routing solution. In order to minimize this gap, a routing prediction model is proposed to assess the electrode addressing solution for enhanced routability and reliability. The intrinsic idea of the prediction model is based on SVM (Support Vector Machine). Figure 15 shows the fundamental principle of SVM [44]. To discriminate the two classes, a decision boundary is required, which should be far away from the data points of both classes. Consequently, the margin m should be maximized, which is computed as

$$m = \frac{2\gamma}{||W||} \tag{13}$$

where W is the normal vector of decision boundary and γ is a parameter related to the intercept of the boundary line.

SVM classifies sample data vectors by generating a boundary with maximum margin of different classes. The vectors forming boundaries are called as *support vectors*. By transforming the original problem into binary classification, multi-class classification and ranking problems can also be solved by SVM. Here, the SVM learner in [45] is adopted.

Figure 16 presents the training flow in the SVM-based electrode addressing method. In the flow, the clustering module first computes the compatible graph and then randomly generates a set of clustering solutions according to the compatible

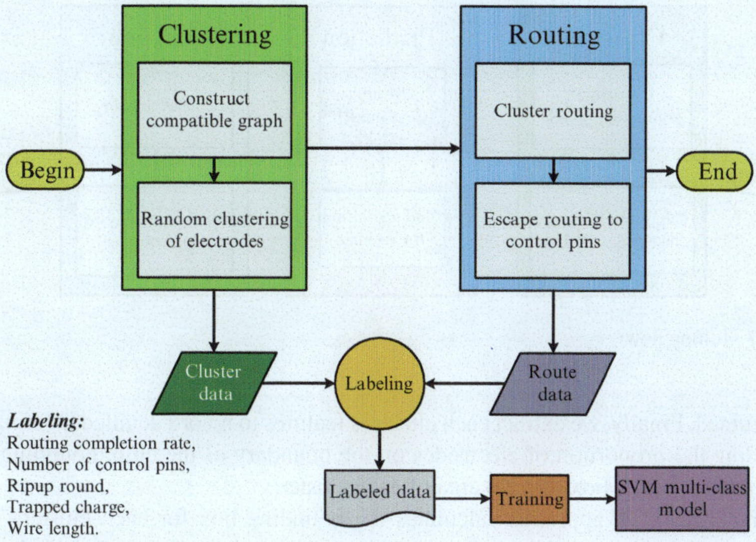

Fig. 16 Training flow

graph. Then the routing module computes the routing solutions for each clustering solution, which includes two major steps: (1) wire routing for each cluster and (2) escape routing from each cluster to control pins. In the clustering module, SVM features for each clustering solution are extracted as *cluster data*. And the *route data* are obtained from the routing module. The *cluster data* are labeled by the *route data*. The labeled data include wire length, routing congestion rate, number of used control pins, trapped charge, etc. The quality of a clustering solution is evaluated by Eq. (27). And the quality of electrode clustering solutions is classified into several levels according to *Score* value. When the training set including cluster data and route data are obtained, the SVM multi-class classifier is trained using the SVM learner in [45].

Figure 17 shows the SVM testing flow. After the training stage, the SVM-based multi-class classifier is obtained, which is used as the prediction module. In clustering module, candidate clustering solutions are randomly generated. Then the SVM-based prediction module is applied to obtain a certain number of clustering solutions with top ranking scores from the candidate solutions. In the experiments, around 5% of the candidate solutions are chosen. Finally, the routing solution is obtained from the routing module.

Feature extraction is an important step in SVM-based machine learning approaches. In the proposed approach, the features are obtained empirically with experimental calibration. The features could be divided into three parts: (1) general features, (2) context features, and (3) cluster features. The general features describe a clustering solution in the global view. The context features are used to represent the routing resource and congestion information when the clustering solution is

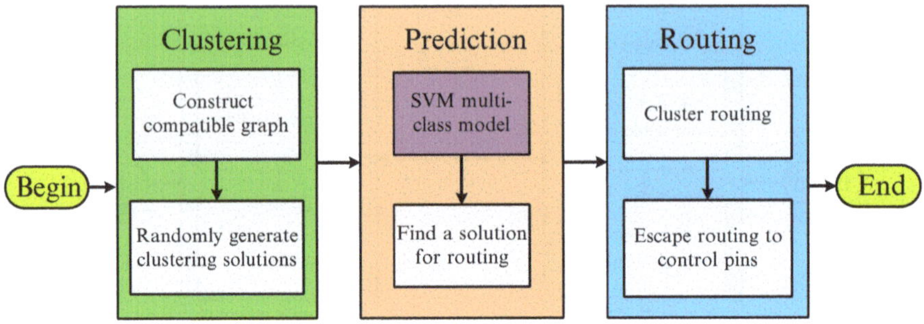

Fig. 17 Testing flow

determined. Finally, we extract each cluster's features to record detailed information including the proportion of electrodes on the boundary of the chip, bounding box area, and bounding box overlap area of each cluster.

First of all, our approach calculates the bounding box for each cluster. Then we obtain some basic information of a clustering solution: (1) number of clusters, (2) total area of bounding boxes, (3) number of clusters with a single electrode, and (4) total area of bounding box overlap. We use vector $G = (g_1, g_2, g_3, g_4)$ to represent the above general features. And the area of the chip and number of electrodes are used for normalization. In this way, the prediction model can be applied to different benchmarks. The definitions of the above features are as follows:

$$g_1 = \frac{C_N}{|E|}, \quad g_2 = \frac{C_S}{T_B}, \quad g_3 = \frac{P_C}{C_N}, \quad g_4 = \frac{C_S}{T_O} \tag{14}$$

Figure 18 presents an example of context feature extraction. To obtain the context features, we first compute the bounding box of each cluster and then divide the whole chip into four quadrants. If the center point of a bounding box is in quadrant i ($i \in \{1, 2, 3, 4\}$), we define that this cluster belongs to quadrant i. Each quadrant collects the information of clusters belonging to itself. As shown in Fig. 18, electrodes of the same color belong to the same cluster. The bounding box area and bounding box overlap area are computed separately for each quadrant. In the example, bounding boxes BB_1 and BB_2 belong to quadrant 1. BB_3, BB_4, and bounding box overlap O_{L_3} belong to quadrant 2. BB_5, BB_6, and bounding box overlap O_{L_6} belong to quadrant 3. Quadrant 4 has two clusters with a single electrode. Finally, the quadrants form a context feature vector denoted as vector $C = (P, R, N)$, which contains three parts defined as follows:

$$P = (p_1, p_2, p_3, p_4), \quad p_i = \frac{C_{N_i}}{C_N} \tag{15}$$

$$R = (r_1, r_2, r_3, r_4), \quad r_i = \frac{T_{B_i}}{T_B} \tag{16}$$

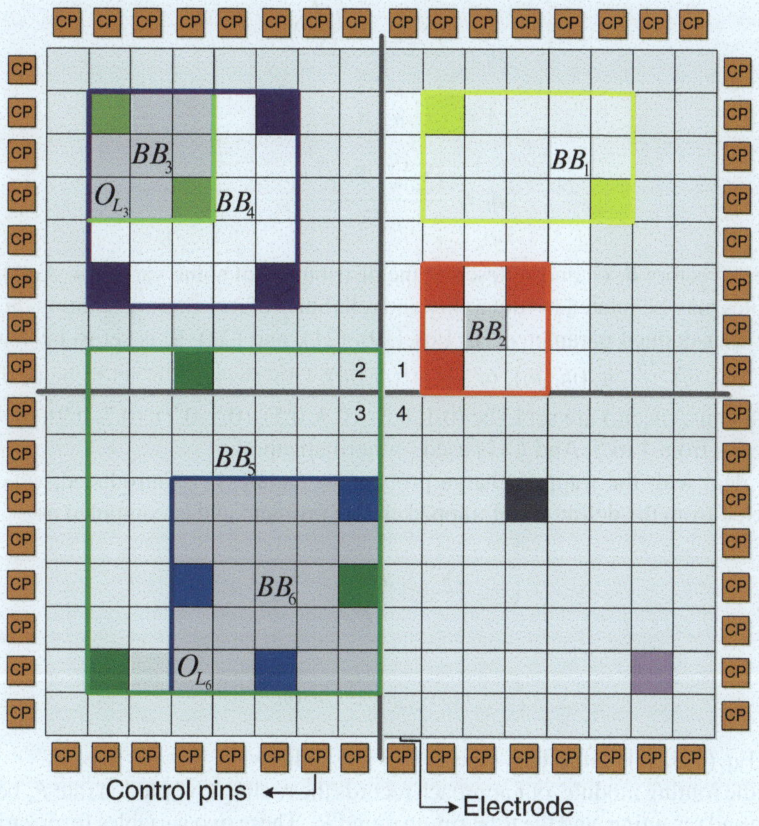

Fig. 18 Context feature extraction

$$N = (n_1, n_2, n_3, n_4), \quad n_i = \frac{T_{O_i}}{T_O} \tag{17}$$

where p_i denotes the proportion of clusters belonging to quadrant i, r_i denotes the proportion of bounding box area in quadrant i, and n_i represents the proportion of overlap area in quadrant i.

Cluster features describe a clustering solution regarding its routability especially for escaping routing from cluster to control pins. Vector $D = (B, O, A)$ represents the cluster features, where B, O, and A are defined as follows:

$$B = (b_1, b_2, b_3, b_4, b_5) \tag{18}$$

$$b_i = \frac{(\sum_{j=1}^{C_N} \mathcal{P}(\frac{B_{P_j}}{T_{P_j}}))}{C_N} \tag{19}$$

$$O = (o_1, o_2, o_3, o_4, o_5) \tag{20}$$

$$o_i = \frac{(\sum_{j=1}^{C_N} \mathcal{P}(\frac{O_{L_j}}{C_S}))}{C_N} \tag{21}$$

$$A = (a_1, a_2, a_3, a_4, a_5) \tag{22}$$

$$a_i = \frac{(\sum_{j=1}^{C_N} \mathcal{P}(\frac{B_{B_j}}{C_S}))}{C_N} \tag{23}$$

Here, vectors B, O, and A describe the distribution of some variables. And these variables may be related to routability and reliability of a clustering solution. m_i and n_i are user-defined parameters. In Eqs. (19), (21), and (23), $\mathcal{P}$ is set to be 1 when $\frac{B_{P_j}}{T_{P_j}} \in (m_i, n_i)$, $\frac{O_{L_j}}{C_S} \in (m_i, n_i)$, or $\frac{B_{B_j}}{C_S} \in (m_i, n_i)$. Otherwise, $\mathcal{P}$ is set to be 0. In the experiments, (m_i, n_i) are set to be (0.1, 0.3), (0.3, 0.5), (0.5, 0.7), (0.7, 0.9), (0.9, 1), where i is from 1 to 5. And C_S is used for normalization.

To deal with the trapped charge problem, a feature V is introduced, which is extracted from the definition of trapped charge problem and is computed as

$$V = \frac{(\sum_{i=1}^{C_N} \mathcal{P}(v_{C_i}^* > V_{th}))}{C_N} \tag{24}$$

$$v_{C_i}^* = max\{v_j | e_j \in cluster \quad i\} \tag{25}$$

In Eq. (24), $\mathcal{P}$ is set to be 1 when $v_{C_i}^* > V_{th}$. Otherwise, $\mathcal{P}$ is set to be 0.

In the routing module, our approach records the routing completion rate F_s before rip-up and rerouting, and the total rip-up round R_t. These two variables form variable R, which evaluates the routability of electrode addressing solution (see Eq. (26)). After the routing stage, we define a function *Score* to evaluate the quality of a clustering solution as follows:

$$R = \frac{\omega \cdot F_s}{\theta \cdot R_t} \quad (\omega + \theta = 1) \tag{26}$$

$$Score = \frac{R}{\alpha \cdot |CP| + \beta \cdot WL + \gamma \cdot TC} \cdot C_S \cdot E_C \tag{27}$$

where C_S and E_C are also used for normalization. ω and θ are user-defined parameters ($\omega + \theta = 1$) for balancing the importance of the two factors. Our approach classifies the clustering solutions into n classes according to the value of *Score*. In the experiments, ω is 0.7, and θ is 0.3. α, β, and γ are all set to be 1. Here, the parameters guarantee that the final routing completion rate enjoys higher superiority than rip-up rounds. And the total wire length, number of used control pins, and trapped charge are equally important.

Two different feature vectors $feature_1$ and $feature_2$ are designed and applied to train different SVM models, i.e., SVM_1 and SVM_2. In Sect. 4.6, we compare the experimental results of the two models. The two feature vectors can be represented as follows:

$$feature_1 = (G, C, V), \quad feature_2 = (G, C, V, D) \tag{28}$$

where D records the cluster data, i.e., proportion of electrodes along the boundary of the chip, bounding box area of a cluster, and bounding box overlap area of a cluster, which affect the overall routability. Experimental results show that, with feature D, SVM_2 has better performance than SVM_1 on routability and runtime.

4.5 Escape Routing to Control Pins

When the clusters are generated, the routing process consists of two major stages: (1) routing between the electrodes within each cluster and (2) escape routing from the clusters to the peripheral control pins. When all the clusters are successfully routed, the number of used control pins should be equal to the number of clusters. The objective of the routing process is to compute the routing tree connecting clusters of electrodes with properly selected control pins for minimized total wire length with enhanced routing completion rate.

For routing within a cluster of multiple electrodes, the minimum spanning tree (MST) is first constructed to determine the connection topology between the electrodes. When the MST edges are computed, the edges are sequentially routed one by one using the A* search algorithm [46]. Using randomly determined order for MST edges, there are three different cases: (1) routing between two electrodes, (2) routing between an electrode and a partially routed tree, and (3) routing between two partially routed trees. For the three different cases, we adopt different routing methods, i.e., point-to-point, point-to-path, and path-to-path routing algorithms. The modified multisource multi-target A* search algorithm enhances routability with reduced total wire length. For escape routing from clusters to the control pins, a similar multisource multi-target A* search algorithm is used, which simultaneously searches from all the routing grids along the routed tree of the cluster to all the available control pins.

After escape routing, the whole routing process will be finished if all the electrodes are successfully routed. However, routing failures may occur in congested designs. As a result, the declustering along with rerouting process is needed for improving the routing completion rate. In this stage, the blocking paths are identified and ripped up, which possibly declusters the original cluster into smaller ones. These smaller clusters are then routed to the control pins independently. The declustering and rerouting process is iterated, until all the electrodes are successfully routed or a predefined threshold value on number of routing iterations is reached.

Table 9 Statistics of benchmarks

Benchmark	Width	Height	Area	#E	Voltage(v)
Amino-acid-1	6	8	1008	20	50
Amino-acid-2	6	8	1008	24	50
Protein-1	13	13	3136	34	50
Protein-2	13	13	3136	51	50
Dilution	15	15	4096	54	50
Multiplex	15	15	4096	59	50
Random-1	10	10	1936	20	50
Random-2	15	15	4096	30	50
Random-3	20	20	7056	60	50
Random-4	30	30	15,376	90	50
Random-5	50	50	41,616	100	50
Random-6	50	50	41,616	100	50
Random-7	60	60	59,536	150	50

4.6 *Experimental Results*

We have implemented our routability- and reliability-driven chip-level design flow in C++ and tested it on a 2.40 GHz 16-core Intel Xeon Linux workstation with 40 GB memory. Only a single thread is used for the experiments.

Table 9 shows the details of the benchmarks, where "Width" and "Height" represent the size of a chip and "Area" denotes the actual routing area considering the routing grids between adjacent electrodes. There are 3 routing grids between the adjacent electrodes. "#E" gives the number of electrodes, and "Voltage" records the threshold voltage for trapped charge issue.

Table 10 shows the experimental results comparing the two prediction models SVM_1 and SVM_2, where "First" gives the routing completion rate immediately after the first round of routing without rip-up and rerouting. The final routing completion rates are all 100% after rip-up and rerouting with the iteration threshold set to be 50. "#Rip-up" represents the number of rip-up and rerouting iterations. The above factors are used to evaluate the routability of the electrode clustering solutions. "$|CP|$" denotes the number of used control pins, "WL" gives the total wire length, and "RT" records the total runtime. "$|CP|$," "WL," and "TC" are used to evaluate the reliability and manufacturing cost. From the results, SVM_2 obtains better solutions on routability than SVM_1. This is because SVM_2 includes more features than SVM_1, and these features are effective for routability prediction. In addition, SVM_2 is faster than SVM_1 because SVM_2 can obtain clustering solutions with better routability, and this effectively reduces the runtime consumption in rip-up and rerouting.

Table 10 Comparison between SVM_1 and SVM_2

	First		#Rip-up		$\|CP\|$		WL		TC (v)		RT (s)	
Benchmark	SVM_1	SVM_2	SVM_1	SVM_2	SVM_1	SVM_2	SVM_1	SVM_2	SVM_1	SVM_2	SVM_1	SVM_2
Amino-acid-1	88.31	83.27	1	2	12	12	289	279	11	14	0.66	0.37
Amino-acid-2	79.85	90.78	1	1	17	16	324	338	12	15	0.52	0.34
Protein-1	64.80	75.32	7	7	33	37	973	731	16	12	0.63	3.48
Protein-2	51.48	49.66	9	9	42	44	1190	1118	19	17	1.34	1.32
Dilution	34.54	50.58	9	9	41	42	1496	1373	18	18	3.12	3.52
Multiplex	88.48	84.64	6	6	48	47	1372	1440	15	14	2.05	1.34
Random-1	84.60	86.20	1	3	11	11	429	454	18	13	0.25	0.29
Random-2	71.48	69.85	5	3	25	20	979	889	19	11	1.86	1.52
Random-3	39.89	48.16	12	9	47	45	2459	2072	18	12	38.66	8.74
Random-4	27.98	47.25	9	17	54	77	3433	4829	15	16	191.26	239.64
Random-5	37.95	30.37	13	12	73	69	6154	6583	18	17	1201.54	982.47
Random-6	38.81	39.77	22	12	80	75	7455	8054	18	12	429.61	314.13
Random-7	25.13	39.21	26	27	118	115	12,331	11,398	17	18	1064.34	463.60
Avg.	56.41	61.16	9	9	46	47	2991	3043	16	15	225.53	155.44

5 Cyberphysical Sensor Integration for Dynamic Error Recovery

In digital microfluidic biochips, cyberphysical sensors can be integrated for monitoring the biochemical experiment process in real time. The monitored information are then analyzed to discover whether errors occur during the execution of the biological protocols on the biochip. When there are execution errors, feedbacks need to be sent to the controller to change the experimental plan for bypassing the errors. For example, during the droplet mixing and splitting process, it is possible that one droplet is much larger than the other one after splitting. In such cases, the data obtained from the sensor are needed to analyze the sizes of the droplets and check whether the unbalanced droplets are tolerable. If the errors are not tolerable, the alternative experimental plan specially designed for this error will be conducted to continue the experiment process. For example, in the alternative experimental plan, a new droplet may be generated for substituting the erroneous droplet. In this way, the biological execution error could be dynamically recovered. Based on the electrowetting technology, digital microfluidic biochips are often confronted with various sources of errors, such as dielectric breakdown, trapped charge issue, etc. The integration of the cyberphysical sensors makes the microfluidic biochips so *smart* that they could roll-forward when execution errors occur, thus improving reliability and error tolerance. Currently, different types of sensors used in biochips are as follows:

1. Optical sensor: Optical sensing system is sensitive and robust for most laboratory experiments. The micro-optical components can also be integrated onto the LoC platform, including the light source, lenses, wave guides, and detectors [47]. Besides, fluorescence sensing techniques are also popular for biochemical experiments where fluorophore can be attached to the droplets [48, 49]. With the light-emitting diode (LED), different droplets tagged with fluorophore emit light of different wavelengths. The photodiode is used to detect the changes in wavelengths. Figure 13a gives an example of the optical sensor with the photodetector.
2. Capacitive sensor: Capacitive sensing circuits are used to detect where a droplet is located on the DMFB. And even the volume of the droplet could be estimated from the change in capacitance [50]. The fundamental principle is the same as the capacitive touch sensing in touchpads. The ring-oscillator-based capacitive sensor can be designed to be very sensitive and accurate in detecting changes in the volume of the droplets.
3. CCD camera-based detector: A CCD camera could be placed on top of the DMFB or on top of a microscope over the DMFB when the droplets are too small to be well detected. In [51], the template matching method is presented to detect a droplet on the DMFB.

References

1. Balagadde, F.K., You, L., Hansen, C.L., Arnold, F.H., Quake, S.R.: Long-term monitoring of bacteria undergoing programmed population control in a microchemostat. Science **309**(5731), 137–140 (2005)
2. Chakrabarty, K., Su, F.: Digital Microfluidic Biochips. CRC Press, Hoboken (2006)
3. Whitesides, G.M.: The origins and the future of microfluidics. Nature **442**(7101), 368–373 (2006)
4. Yager, P., Edwards, T., Fu, E., Helton, K., Nelson, K., Tam, M.R., Weigl, B.H.: Microfluidic diagnostic technologies for global public health. Nature **442**(7101), 412–418 (2006)
5. Fair, R.B., Khlystov, A., Tailor, T.D., Ivanov, V., Evans, R.D., Griffin, P.B., Srinivasan, V., Pamula, V.K., Pollack, M.G., Zhou, J.: Chemical and biological applications of digital-microfluidic devices. IEEE Des. Test Comput. **24**(1), 10–24 (2007)
6. Srinivasan, V., Pamula, V.K., Fair, R.B.: An integrated digital microfluidic lab-on-a-chip for clinical diagnostics on human physiological fluids. Lab Chip **4**, 310–315 (2004)
7. Barbulovic-Nad, I., Yang, H., Park, P.S., Wheeler, A.R.: Digital microfluidics for cell-based assays. Lab Chip **8**, 519–526 (2008)
8. Srinivasan, V., Pamula, V.K., Paik, P., Fair, R.B.: Protein stamping for MALDI mass spectrometry using an electrowetting-based microfluidic platform. Opt. East **5591**, 26–32 (2004)
9. Dong, C., Chen, T., Gao, J., Jia, Y., Mak, P.-I., Vai, M.-I., Martins, R.P.: On the droplet velocity and electrode lifetime of digital microfluidics: voltage actuation techniques and comparison. Microfluid. Nanofluid. **18**(4), 673–683 (2015)
10. Bhattacharjee, B., Study of droplet splitting in an electrowetting based digital microfluidic system. Ph.D. Thesis, The University of British Columbia, Sept 2012
11. Arduino, Online available: https://www.arduino.cc/
12. Raspberry PI, Online available: https://www.raspberrypi.org/
13. Ho, T.-Y., Chakrabarty, K., Pop, P.: Digital microfluidic biochips: recent research and emerging challenges. In: Proceedings of International Conference on Hardware/Software Codesign and System Synthesis (CODES+ISSS), pp. 335–343 (2011)
14. Cho, M., Pan, D.Z.: A high-performance droplet routing algorithm for digital microfluidic biochips. IEEE Trans. Comput.-Aided Des. Integr. Circuits Syst. **27**(10), 1714–1724 (2008)
15. Su, F., Chakrabarty, K.: Unified high-level synthesis and module placement for defect-tolerant Microfluidic biochips. In: Proceedings of Design Automation Conference, pp. 825–830 (2005)
16. Su, F., Hwang, W., Chakrabarty, K.: Droplet routing in the synthesis of digital microfluidic biochips. In: Proceedings of Design, Automation and Test in Europe (DATE), pp. 1–6 (2006)
17. Xu, T., Chakrabarty, K.: Integrated droplet routing in the synthesis of microfluidic biochips. In: Proceedings of Design Automation Conference, pp. 948–953 (2007)
18. Yuh, P.-H., Yang, C.-L., Chang, Y.-W.: BioRoute: a network-flow-based routing algorithm for the synthesis of digital microfluidic biochips. IEEE Trans. Comput.-Aided Des. Integr. Circuits Syst. **27**(11), 1928–1941 (2008)
19. Yuh, P.-H., Sapatnekar, S.S., Yang, C.-L., et al.: A progressive-ILP-based routing algorithm for the synthesis of cross-referencing biochips. IEEE Trans. Comput.-Aided Des. Integr. Circuits Syst. **28**(9), 1295–1306 (2009)
20. Campàs, M., Katakis, I.: DNA biochip arraying, detection and amplification strategies. TrAC Trends Anal. Chem. **23**(1), 49–62 (2004)
21. Zhao, Y., Chakrabarty, K.: Cross-contamination avoidance for droplet routing in digital microfluidic biochips. In: Proceedings of Design, Automation and Test in Europe (DATE), pp. 1290–1295 (2009)
22. Zhao, Y., Chakrabarty, K.: Synchronization of washing operations with droplet routing for cross-contamination avoidance in digital microfluidic biochips. In: Proceedings of Design Automation Conference, pp. 635–640 (2010)

23. Huang, T.-W., Lin, C.-H., Ho, T.-Y.: A contamination aware droplet routing algorithm for the synthesis of digital microfluidic biochips. IEEE Trans. Comput.-Aided Des. Integr. Circuits Syst. **29**(11), 1682–1695 (2010)
24. Lin, C.C.Y., Chang, Y.-W.: Cross-contamination aware design methodology for pin-constrained digital microfluidic biochips. IEEE Trans. Comput.-Aided Des. Integr. Circuits Syst. **30**(6), 817–828 (2011)
25. Zhao, Y., Chakrabarty, K.: Cross-contamination avoidance for droplet routing in digital microfluidic biochips. IEEE Trans. Comput.-Aided Des. Integr. Circuits Syst. **31**(6), 817–830 (2012)
26. Mitra, D., Ghoshal, S., Rahaman, H., Chakrabarty, K., Bhattacharya, B.B.: On Residue Removal in Digital Microfluidic Biochips. In: Proceedings of the Great Lakes Symposium on VLSI, pp. 1–4 (2011)
27. Yao, H., Wang, Q., Shen, Y., Ho, T.-Y., Cai, Y.: Integrated functional and washing routing optimization for cross-contamination removal in digital microfluidic biochips. IEEE Trans. Comput.-Aided Des. Integr. Circuits Syst. **35**(8), 1283–1296 (2016)
28. Böhringer, K.F.: Modeling and controlling parallel tasks in droplet-based microfluidic systems. IEEE Trans. Comput.-Aided Des. Integr. Circuits Syst. **25**(2), 334–344 (2006)
29. McMurchie, L., Ebeling, C.: PathFinder: a negotiation-based performance-driven router for FPGAs. In: Proceedings of ACM Symposium on Field-Programmable Gate Arrays, pp. 111–117 (1995)
30. Yao, H., Ho, T.-Y., Cai, Y.: PACOR: practical control-layer routing flow with length-matching constraint for flow-based microfluidic biochips. In: Proceedings of IEEE/ACM Design Automation Conference (DAC), pp. 142–147 (2015)
31. Grissom, D., Brisk, P.: Fast online synthesis of digital microfluidic biochips. IEEE Trans. Comput.-Aided Des. Integr. Circuits Syst. **33**(3), 356–369 (2014)
32. Boost C++ Libraries. http://www.boost.org/
33. Xu, T., Chakrabarty, K.: Broadcast electrode-addressing for pin-constrained multi-functional digital microfluidic biochips. In: Proceedings of IEEE/ACM Design Automation Conference, pp. 173–178 (2008)
34. Verheijen, H.J.J., Prins, M.W.J.: Reversible electrowetting and trapping of charge: model and experiments. Langmuir **15**(20), 6616–6620 (1999)
35. Drygiannakis, A.I., Papathanasiou, A.G., Boudouvis, A.G.: On the connection between dielectric breakdown strength, trapping of charge, and contact angle saturation in electrowetting. Langmuir **25**(1), 147–152 (2009)
36. Yeh, S.-H., Chang, J.-W., Huang, T.-W., Ho, T.-Y.: Voltage-aware chip-level design for reliability-driven pin-constrained EWOD chips. In: Proceedings of IEEE/ACM International Conference on Computer-Aided Design, pp. 353–360 (2012)
37. Fair, R.: Digital Microfluidics: is a true lab-on-a-chip possible? Microfluid. Nanofluid. **3**(3), 245–281 (2007)
38. Chakrabarty, K.: Towards fault-tolerant digital microfluidic lab-on-chip: defects, fault modeling, testing, and reconfiguration. In: Transactions of the IRE Professional Group on Audio, pp. 329–332 (2008)
39. Huang, T.-W., Yeh, S.-Y., Ho, T.-Y.: A network-flow based pin-count aware routing algorithm for broadcast-addressing EWOD chips. IEEE Trans. Comput.-Aided Des. Integr. Circuits Syst. **30**(12), 1786–1799 (2011)
40. Chang, J.-W., Huang, T.-W., Ho, T.-Y.: An ILP-Based Obstacle-Avoiding Routing Algorithm for Pin-Constrained EWOD Chips. In: Proceedings of Asia and South Pacific design automation conference (ASP-DAC), pp. 67–72 (2012)
41. Liu, S.S.-Y., Chang, C.-H., Chen, H.-M., Ho, T.-Y.: ACER: an agglomerative clustering based electrode addressing and routing algorithm for pin-constrained EWOD chips. IEEE Trans. Comput.-Aided Des. Integr. Circuits Syst. **33**(9), 1316–1327 (2014)
42. Huang, T.-W., Ho, T.-Y., Chakrabarty, K.: Reliability-oriented broadcast electrode-addressing for pin-constrained digital microfluidic Biochips. In: Proceedings of IEEE/ACM International Conference on Computer-Aided Design, pp. 448–455 (2011)

43. Wang, Q., He, W., Yao, H., Ho, T.-Y., Cai, Y.: SVM-based routability-driven chip-level design for voltage-aware pin-constrained EWOD chips. In: Proceedings of International Symposium on Physical Design, pp. 49–56 (2015)
44. Cristianini, N., Shawe-Taylor, J.: An Introduction to Support Vector Machines. Cambridge University Press, Cambridge (2000)
45. Joachims, T.: Making large-scale SVM learning practical. In: Scholkopf, B., Burges, C., Smola, A. (eds.) Advances in Kernel Methods – Support Vector Learning. MIT Press, Cambridge (1999)
46. Hart, P.E., Nilsson, N.J., Raphael, B.: A formal basis for the heuristic determination of minimum cost paths. IEEE Trans. Syst. Sci. Cybern. **4**(2), 100–107 (1968)
47. Kuswandi, B., Nuriman, Huskens, J., Verboom, W.: Optical sensing systems for microfluidic devices: a review. Anal. Chim. Acta **601**(2), 141–155 (2007)
48. Srinivasan, V., Pamula, V.K., Pollack, M.G., Fair, R.B.: Clinical diagnostics on human whole blood, plasma, serum, urine, saliva, sweat, and tears on a digital microfluidic platform. In: Proceedings of International Conference on Miniaturized Chemical and Biochemical Analysis Systems, pp. 1287–1290 (2003)
49. Jokerst, N.M., Luan, L., Palit, S., Royal, M., Dhar, S., Brooke, M., Tyler, T.: Progress in chip-scale photonic sensing. IEEE Trans. Biomed. Circuits Syst. **3**(4), 202–211 (2009)
50. Hu, K., Hsu, B.N., Madison, A., Chakrabarty, K., Fair, R.: Fault detection, real-time error recovery, and experimental demonstration for digital microfluidic biochips. In: Proceedings of the Conference on Design, Automation and Test in Europe, pp. 559–564 (2013)
51. Luo, Y., Chakrabarty, K., Ho, T.-Y.: Error recovery in cyberphysical digital microfluidic biochips. IEEE Trans. Comput.-Aided Des. Integr. Circuits Syst. **32**(1), 59–72 (2013)

Reducing Timing Discrepancy for Energy-Efficient On-Chip Memory Architectures at Low-Voltage Mode

Po-Hao Wang and Tien-Fu Chen

1 Introduction

Voltage scaling is an effective method for saving energy in modern processor systems. In the past, caches limited the minimum operating voltage of systems because of the poor reliability and long latency of static random-access memory (SRAM) in low-voltage operations. To increase the cache reliability, numerous fault-tolerance caches, such as disabling [1], redundancy [2, 3], error correction code (ECC) designs [4, 5], and robust SRAM cell designs [6, 7], have been proposed. Unfortunately, most fault-tolerance designs necessarily sacrifice cache latency to increase reliability. Therefore, these designs are not suitable for latency-sensitive level 1 (L1) caches. To provide reliable access and dual-port access (better performance), robust 8T SRAM is widely used in on-chip memory such as modern L1 caches [8] and local memory without any fault-tolerance mechanism. The reliability issue of on-chip memory has been solved by using 8T SRAM; however, low-voltage environments cause on-chip memory to require long latency for access. This overly long access latency causes a timing discrepancy between a core and a latency-sensitive memory (L1 caches and local memory) that restricts the performance of the entire system, particularly in sub-threshold voltage operations.

Aggressive voltage scaling worsens timing discrepancy problems. Assuming the access cycle of an L1 cache is two cycles at normal voltage, for 0.5 V, the worst case of cache latency can be up to four cycles [9]. The gray and black lines in Fig. 1 represent the increasing latency of the core and cache, respectively, as the voltage is scaled down. When the voltage is decreased to a certain level, the cache is not able to be accessed correctly within the access cycle of normal-voltage operation

P.-H. Wang • T.-F. Chen (✉)
Department of Computer Science, National Chiao-Tung University, Hsinchu City, Taiwan
e-mail: tfuchen@gmail.com

H. Yasuura et al. (eds.), *Smart Sensors at the IoT Frontier*,
DOI 10.1007/978-3-319-55345-0_4

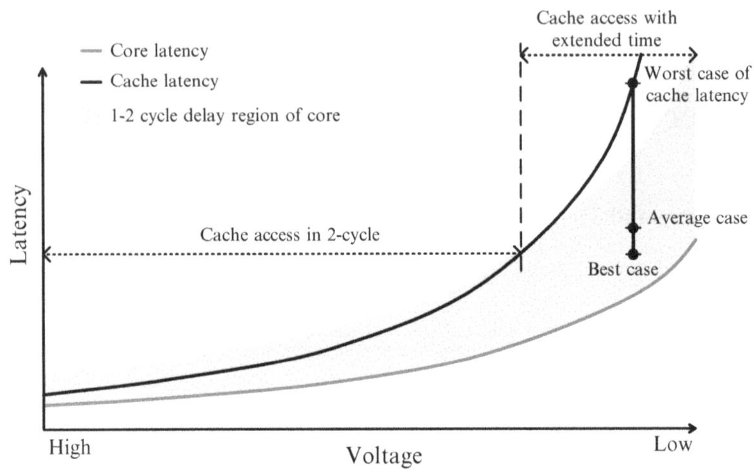

Fig. 1 Timing discrepancy between a core and a cache

(two cycles). Thus, the core needs to decrease its operating frequency or extend the access cycles of the cache. However, both of these methods impact the performance of the entire system.

The severe increase in timing discrepancy between a core and a cache is primarily caused by the severe process variations of slow SRAM cells. These slow cells increase the overall SRAM access latency. The three dots in the upper right part of Fig. 1 represent the best-, average-, and worst-case latencies of an SRAM cell. In the average case, the cache can be accessed correctly within the access cycle, which can catch up with the core's speed. Thus, only a few cells with long latency compromise the performance of the entire system. Figure 2 shows the delay distribution of SRAM cells at normal voltage and low voltage. Only a small fraction of the SRAM cells are slow. Nevertheless, the number of slow cells is increased by aggressive voltage decreases and technology node advancement. Therefore, tolerating access-time failure that occurred by slow cells to reduce the timing discrepancy will become a critical issue.

We observe that the value stored in 8T SRAM significantly influences the read latency of the cache. Based on this observation, we propose two different designs for on-chip local memory: zero-counting error detection code (ZC-EDC) and dynamic timing calibration SRAM (DTC-SRAM). Moreover, we propose three cache management strategies for better cache efficiency and tolerant ability of access-time failure: timing-aware LRU policy, bit-level failure-mask management strategy, and data allying management with a special wordline alliable SRAM.

In the remainder of this chapter, Sect. 2 discusses the impact of 8T SRAM in low voltage and details our observations. Section 3 shows the proposed designs for local memory in detail. Section 4 explains our cache management strategies based on the

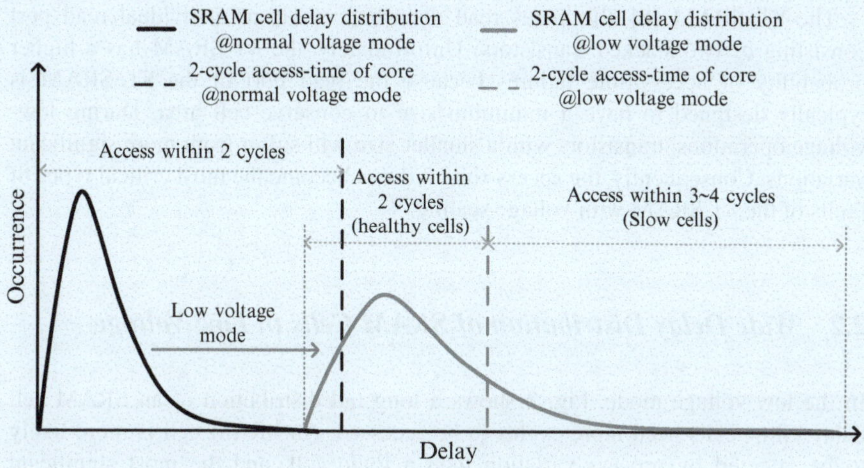

Fig. 2 Voltage scaling impact of SRAM cell delay distribution

memory designs we proposed. Section 5 introduces the experiment, evaluates our design, and estimates the overhead with different designs. Section 6 reviews related work, and Sect. 7 concludes the chapter.

2 Low-Voltage Influence on an 8T Cell

In the L1 cache and local memory of a modern processor system, the 8T cell has gradually replaced the 6T cell for low-voltage applications and dual-port access [8]. In this section, we present some observations on characteristics of 8T SRAM cells and discuss SRAM failure in low-voltage situations.

2.1 SRAM Faults on an 8T Cell

A fault model has been proposed to analyze the probabilities of various types of faults in the 6T SRAM [10] using voltage scaling. An analysis of this model revealed that there are four types of SRAM faults: read fault, write fault, access-time failure, and hold fault. Generally, the read fault is the primary problem encountered by the 6T SRAM and typically occurs when the stored value is affected by the bitline during the read operation. This issue incurs the degradation of the static noise margin and is referred to as the read disturbance. However, the fault probability of the access-time failures increases significantly when the SRAM is affected by the voltage drop or temperature.

The 8T SRAM [6] eliminates read disturbances via an individual read port consisting of two stacked transistors. Unfortunately, the 8T SRAM has a higher probability of access-time failures because the read port of the 8T SRAM is typically designed to have a minimum size to conserve cell area. During low-voltage operations, transistors with a smaller size will suffer from more significant variations. Consequently, the access-time failures become the most critical types of faults of the 8T SRAM with voltage scaling.

2.2 *Wide Delay Distribution of SRAM Cells in Low Voltage*

In the low-voltage mode, Fig. 2 shows a long tail distribution of an SRAM cell delay. Slow cells need more cycles to be accessed. An SRAM cell is more likely to be affected by process variation than a logic cell, and the most significant problem is access-time failure, which occurs when slow cells cannot complete their discharge in time due to variations. The logic part is not as vulnerable to slow cell problems, and the delay distribution is more balanced than with SRAM cells [11] because it is usually a series connected by logic gates and works one after one. Therefore, the total access time will be balanced by the gates on the path. Although a SRAM cell is stored or loaded independently, it is more vulnerable to access-time failure. To successfully access these slow cells with increased frequency, they require extending access cycles to complete their discharge and to allow the sense amplifier to determine the correct value. If these slow cells can be tolerated and accessed with total cycles close to normal cells, their performance can be improved.

2.3 *Effect of the Stored Value on the Latency*

Figure 3 shows the cell structure of an 8T SRAM. To perform a read operation, the read wordline (RWL) is activated, and the read bitline (RBL) is pre-charged. When reading “0,” the RBL is pulled down through the transistors M7 and M8. An access-time failure occurs when reading “0” if the RBL voltage drops too slowly for the sense amplifier to sense it in time. Contrarily, the datum “1” can be read via the RBL directly after pre-charging. Access-time failures will not occur because bitlines do not require any discharge time.

Figure 4 shows the read operation waveforms of slow cells and healthy cells with different stored values on an 8T SRAM. There is no critical issue with either healthy cells or slow cells when reading the value “1.” Because the bitline does not need to be discharged and the bitline voltage is always greater than the sense amplifier sensitivity, the sense amplifier will always sense the correct value “1.” However, when the value “0” is read, the value sensed by the sense amplifier at a shortened fetch point (SFP) is different for healthy cells and slow cells. For a healthy cell, the read bitline can discharge within sufficient time, and the bitline voltage is less than the sense amplifier sensitivity at the SFP. In this case, the

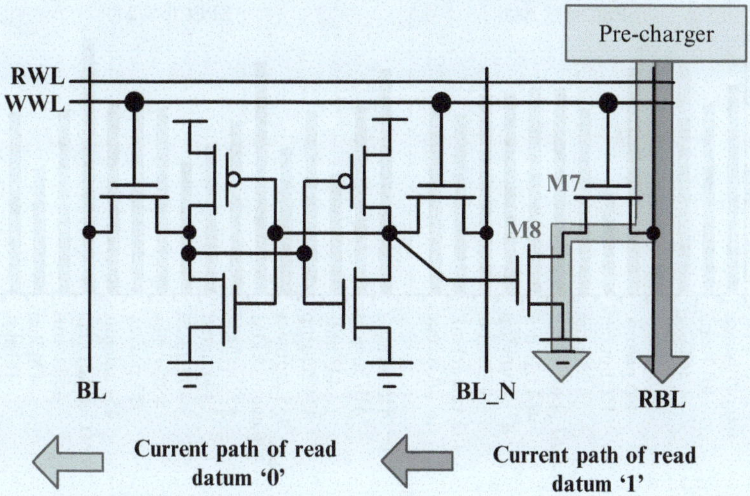

Fig. 3 Structure and reading path of 8T cells

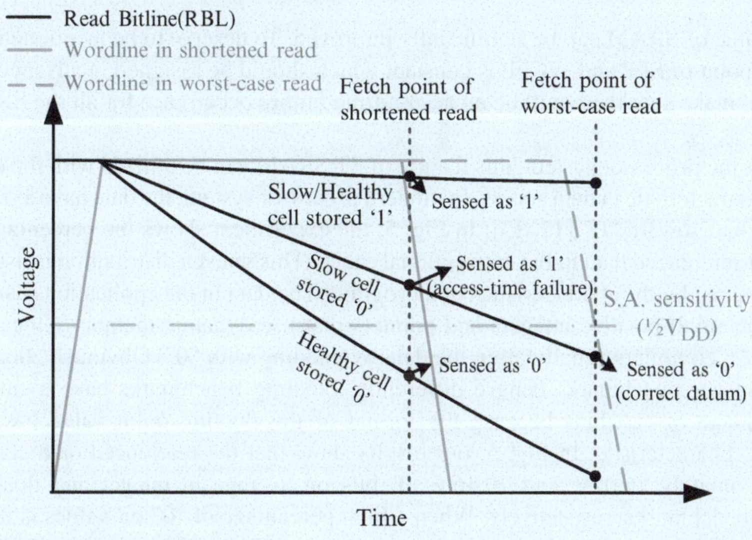

Fig. 4 Reading waveform of 8T cells

correct value "0" can be fetched. However, for a slow cell, the bitline discharges too slowly, which causes the bitline voltage at the SFP to remain greater than the sense amplifier sensitivity. Therefore, the value sensed by the sense amplifier is "1," which is incorrect. Fortunately, when there are enough read cycles, the bitline has sufficient time to discharge, and the voltage is less than the sense amplifier sensitivity at the worst-case fetch point (WCFP). In this case, the correct value of "0" is sensed by the sense amplifier. Therefore, if datum "0" can be stored without slow cells, the

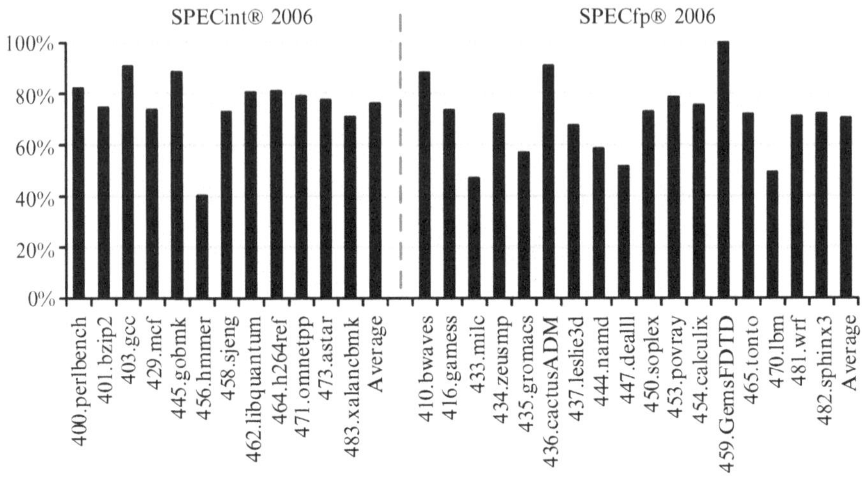

Fig. 5 Percentage of bit "0" of referenced data with SPEC 2006

read time of SRAM can be significantly improved. To deserve to be mentioned, the fetch point of SFP and WCFP is constant which should be assigned in advance and should make sure there will be no access-time failure occurrence for all the SRAM cells.

For the processor system, this feature of 8T SRAM can be utilized with the data-bias characteristic in data access. In modern processor system, the data have a strong value bias toward "0" [12, 13]. In Fig. 5, the experiment shows the percentage of "0" of referenced data in the conventional cache. This uneven distribution is usually determined by the characteristics of a program. The data in the applications usually contain small positive integers and pointers that use dynamic memory allocation. Besides, compilers usually align the data by padding with "0." Obviously, floating and integer benchmarks behave differently. Floating benchmarks have a smaller percentage of "0" bias because the format of the floating point value has less of this characteristic. In Fig. 5, our results show that the referenced data contain approximately 76.07% and 70.46% '0' bits on average in integer and floating-point benchmarks, respectively. When a high percentage of "0" bit values is in the referenced data and based on the characteristic of 8T SRAM cells that free from access-time failures when reading "1," some access-time failures can be avoided via inverting stored data.

3 Non-Capacity-Loss Fault-Tolerant Design to Reduce Timing Discrepancy in Local Memory

In the embedded processor system, local memory is usually used to provide faster accesses. Different with cache, local memory cannot have any capacity loss when applying the failure-tolerant designs. Based on our observations, the 8T SRAM read latency is significantly affected by the values that are stored in slow cells. We thus propose two designs that do not sacrifice any capacity for local memory. In these two designs, ZC-EDC provides the better reliability because the design not only can detect access-time failure but also can detect the other type of SRAM fault [2]. DTC-SRAM can provide the access-time failure tolerance without any access-time overhead. These designs are described in following sections in details.

3.1 Lightweight EDC with Zero Counting

Since access-time failures only occur when datum "0" is read, the ZC-EDC use a lightweight strategy of zero-bit counting to generate the error detection codes, then dynamically detect access-time failures with the generated codes. The access-time failures are detected with the shortened fetch point (SFP) which is explained in Sect. 2.3. If the detection result indicates data failures, the data fetch point is then extended to the worst-case fetch point (WCFP) to provide sufficient access time for failed access.

3.1.1 System Architecture and Execution Flow

Figure 6 illustrates the detailed architecture of the design of the ZC-EDC. In the ZC-EDC, there are three major parts. The first is the access-time failure detection mechanism, which is triggered by each cache read to determine the effects on the access time by any dynamic variations. This function is performed by an access-time failure detector, which includes a zero-bit counter and a comparator to check the "0" numbers of the read data and detect any access-time failures. Additionally, this zero-bit counter in the detector is used when the data are written. The second part of the ZC-EDC adjusts the access time for each access; this function is controlled by an access-time controller. Based on the result of the access-time failure checking procedure, the access-time controller gates the pre-charger to adjust the access with the assigned data fetch point. The third part of the ZC-EDC is to dynamically invert the data for decreasing the possibility of a datum "0" being stored on a cell that is experiencing an access-time failure.

Figure 7 illustrates the execution flow of the ZC-EDC. When the SRAM of the ZC-EDC writes the data into the SRAM, the zero-counting bits are calculated by a zero-bit counter and are stored into the SRAM bank. Conversely, when the data are read from the SRAM, they are checked by the corresponding zero-counting bits. Then, the ZC-EDC will modify the data fetch point to the WCFP if the number of "0" is unmatched.

To calculate the number of "0," a zero-bit counter [14, 15] is implemented in the ZC-EDC. As Fig. 7 indicates, the zero-counting process lies on the critical path of the cache read operation. Thus, the ZC-EDC must marginally increase the average access time.

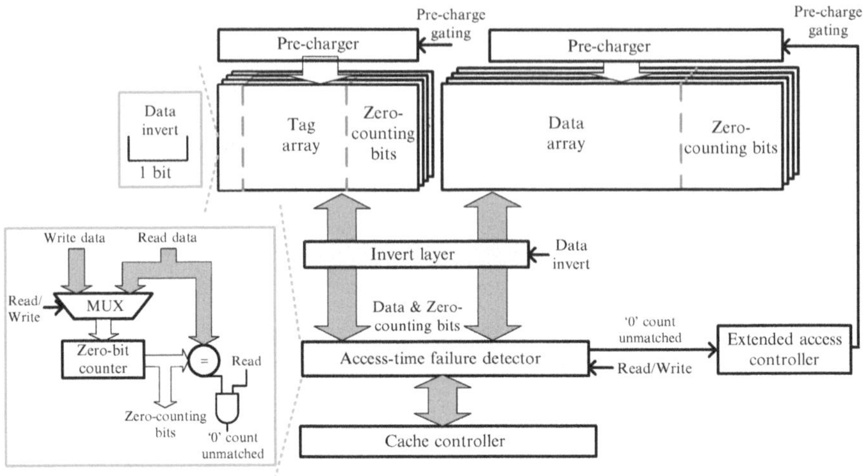

Fig. 6 Detailed ZCAL cache architecture

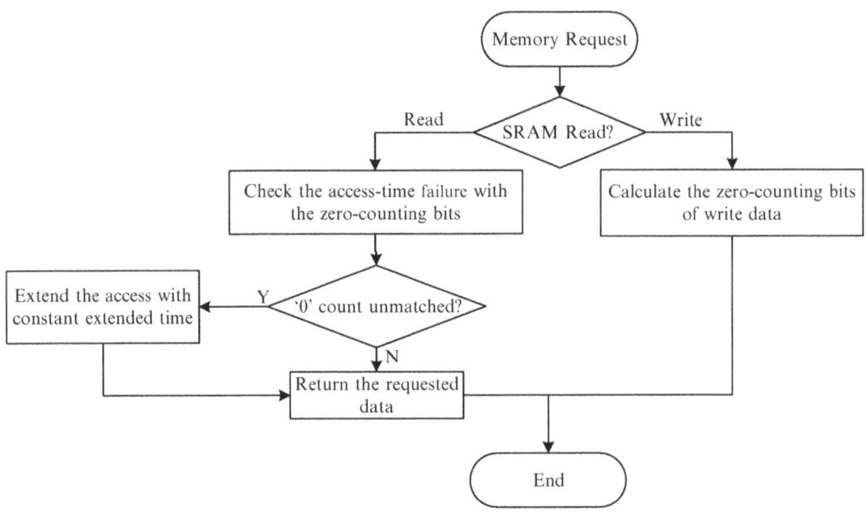

Fig. 7 Execution flow of ZCAL cache

3.1.2 Access-Time Failure Detection by "0" Counting

The "0" counting method should ensure that the access-time failures can be detected regardless of which access-time failures occur in the data section or the zero-counting bits. Table 1 presents four examples of different fault situations. From the examples described above, it is clear that if the data bits are faulty, then the "0" count of the data will decrease. Conversely, if the zero-counting bits fail, then the stored "0" count will increase. Thus, regardless of whether the access-time failures

Table 1 Example of different fault situations

Fault case	Data bits	Zero-counting bits
No fault	11110000 (4 zero bits)	0100(4)
Fault(s) in data bits	11110101 (2 zero bits)	0100(4)
Fault(s) in zero-counting bits	11110000 (4 zero bits)	1100(12)
Faults in both data bits and zero-counting bits	11111111 (0 zero bits)	1101(13)

occur in the data or the zero-counting bits and how many access-time failures occur, the access-time failure detector can always detect them.

3.1.3 Detection Granularity Trade-Off

The ZC-EDC can select different detection granularities (e.g., 8 zero-counting bits for a 128-bit cache line, 6 zero-counting bits for a 32-bit word). Finer-grained detection granularity can provide a better performance but will likely result in higher energy consumption. Similar to the ECC designs, the ZC-EDC has a decoding overhead when data are written to the cache. Every write operation requires counting the number of "0" and storing that number in another memory location. This operation will generate overhead in energy and access latency because the cache does not simultaneously write/read all of the data in a row.

3.2 Dynamic Timing Calibration SRAM

To avoid the latency overhead of local memory, we propose a dynamic timing calibration for 8T SRAM (DTC-SRAM) to detect the influence of the stored value at runtime. This method also adaptively adjusts the data fetch point in the read operation based on the type of stored data like ZC-EDC. In each write operation, the data fetch point of the updated row is set up to WCFP. When data are read, DTC-SRAM calibrates the read access time and records the suitable data fetch point for the read row. In this section, we describe the DTC-SRAM architecture and explain our method of timing calibration.

3.2.1 Architecture of DTC-SRAM

A dynamic timing calibrator can calibrate the appropriate data fetch point of referenced data. The calibrator compares the read data that are fetched at SFP and WCFP. If both data are equal, these data can be read within shortened read cycles. The details of this process will be introduced in next section.

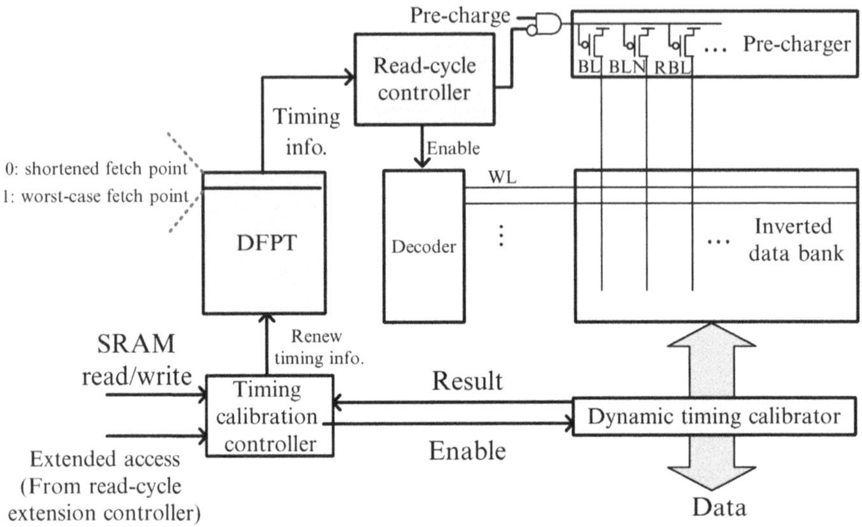

Fig. 8 Architecture of DTC-SRAM

A timing calibration controller updates the data fetch point of the current read row into the DFPT. In the read operation, if the data fetch point of read data is WCFP, the timing calibration controller will update the data fetch point based on the calibration result. In the write operation, the controller updates the data fetch point to the WCFP because the data are not yet checked by the calibrator. Obviously, the data fetch point of the updated data could be misjudged, but the appropriate data fetch point will be calibrated in the next read operation.

We use a read-cycle controller and the data fetch point stored in the DFPT to control the read cycle. The read-cycle controller obtains the data fetch point from DFPT and decides the count of read cycles. The controller disables the decoder to maintain the same active wordline and gate for all the bitlines pre-charge to control the read cycles.

Figure 8 shows the detailed architecture of our DTC-SRAM. We added four components to the original 8T SRAM: (1) a data fetch point table (DFPT) to record the appropriate fetch point of each row, (2) a dynamic timing calibrator to detect an appropriate data fetch point of the read row in the read operation, (3) a timing calibration controller to update the fetch point table, and (4) the read-cycle controller to control the read cycles by the pre-charge gating according to the recorded fetch point.

The DFPT is a small additional SRAM; it records the timing information for referenced data. Each block of timing information uses one bit to identify whether the referenced data belonged to the worst-case read or the shortened read. The read operation of DFPT must be completed before the next pre-charge of data array to indicate if the read cycle needs to be extended; therefore, the table must be designed

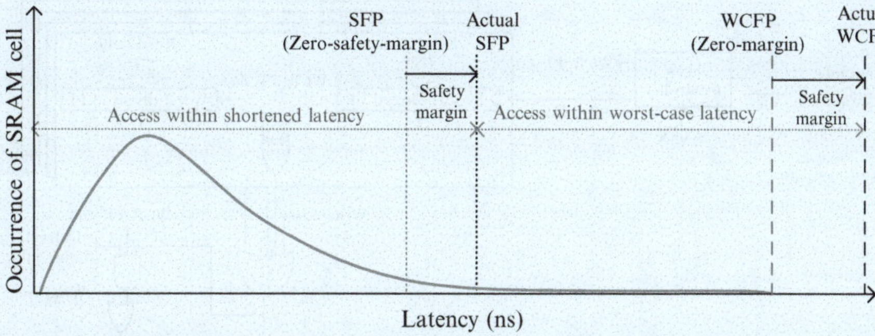

Fig. 9 Setup of safety margin against variations

with an optimal size or circuit technique to reduce its access latency and ensure that even the slowest cell can be read without any access-time failures within a certain time.

3.2.2 Dynamic Timing Calibration by Twice Data Fetch

To find the appropriate fetch point of each datum read, DTC-SRAM fetches the read data twice—at the fetch point of a shortened read and the fetch point of a worst-case read. Because the fetch data at WCFP are given sufficient time to read data, the data do not have any latency-related faults. DTC-SRAM uses the data of WCFP as the golden data and then compares data of two fetch points to check if the data of SFP are correct. If the data of SFP are correct, the read time can be shortened. Otherwise, the data should be read with worse-case latency.

The operating frequency of caches is usually decreased against the influence of process, voltage, and temperature (PVT) variations (or increases in the operating voltage). Figure 9 provides an example of this process. All of the SRAM cells can be read within the latency of the WCFP in a zero-safety margin. However, if a safety margin needs to be added, the safe read operations should be with the actual WCFP as the black broken line that is shown in Fig. 9. Similarly, for the secure dynamic timing calibration, SFP also needs to add a safety margin to assure that all of the cells determined to be healthy can be read with the shortened latency. The safety margin of SFP is narrower than the margin of WCFP because the dynamic timing calibration can detect the impact of latency from the process variation. Therefore, the safety margin of SFP only needs to consider the worse-case influence of dynamic variations (temperature and voltage).

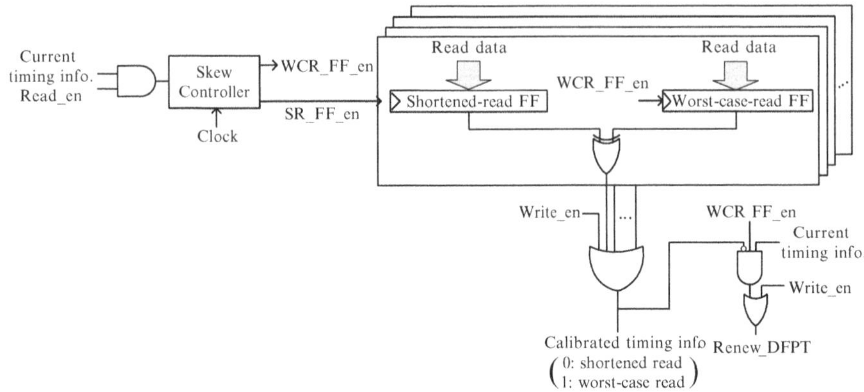

Fig. 10 Detailed architecture of the dynamic timing calibrator

3.2.3 Details of Dynamic Timing Calibrator

Figure 10 shows the detailed architecture of the dynamic timing calibrator. There are two types of flip-flops (FFs): shortened-read FFs and worst-case-read FFs, which fetch data at the shortened fetch point and the worst-case fetch point, respectively. These FFs are enabled by the data-reading-enable signal and the worst-case-read timing information. After the data are fetched, the calibrator compares the fetched data using XOR gates. If the data stored in the shortened-read FF and the worst-case-read FF are equal, these data can be read within shortened read cycles. Otherwise, these data must be read with worst-case cycles. Figure 10 shows an example waveform of the dynamic timing calibrator. In this example, we assume that a shortened read requires two cycles and a worst-case read requires three cycles. After two cycles and three cycles of read operations, the enable signal of the shortened-read FF (SR_FF_en) and enable signal of the worst-case-read FF (WCR_FF_en) are triggered, respectively. The first read is completed before SR_FF_en triggers; thus, the calibration result is "0" (i.e., the data can be read within shortened read cycles). In contrast, the second read cannot complete before SR_FF_en triggers; thus, the calibration result is "1." DFPT is updated with calibration results when the renew DFPT signal is triggered. A strict timing calibration is necessary to ensure correct timing information under all possible variations. Thus, by sending early skewed SR_FF_en signals, a safety margin that is used to fetch the shortened read cycle data against the worst-case combination of variations is added.

4 Flexible Space Management Strategies for L1 Cache to Reduce Aggressive Timing Discrepancy

Caches are usually used in the modern processor system. For the latency-sensitive L1 cache, the error-tolerant designs are required to avoid increasing the access latency, especially read latency. Therefore, previous error-tolerant designs such as ECC are not suitable to be used in L1 cache. We thus propose three cache designs that do not increase significant latency overhead of L1 caches: timing-aware LRU policy, bit-level failure-mask management strategy, and data allying management with a special wordline alliable SRAM. These designs are described in following sections in details.

4.1 Timing-Aware LRU Policy

As previously observed [16], the most recently used (MRU) line per set captures approximately 90% of the cache hits. However, conventional LRU policy will not be able to consider the occurrence of access-time failures and could thus potentially cause important data to be placed in slow blocks.

Therefore, if the MRU data are unfortunately stored in the access-time failure cache line, the frequent access of the MRU data will cause a significant loss in performance. To address this issue, caches can apply a dynamic access-time failure map that uses 1 bit to label the cache line if any access-time failure occurred before on a certain cache line. Once the access-time failure cache lines are labeled, the label will not be erased afterward. Therefore, by referring to the access-time failure map, the caches know if the faultless cache line exists in the referenced cache set and may be able to move the MRU data to the cache line that is access-time faultless, and most data can be fetched without extension.

Intuitively, when the traditional LRU policy changes the sequence for a cache hit or cache miss, data can be swapped to a faultless cache line. However, if a program involves a large amount of streaming data, data swap becomes unnecessary because cached streaming data will not be used again and streaming data will always result in cache misses. To avoid additional data swapping, we propose a latency-aware LRU policy. Figure 11 illustrates the behavior of the latency-aware LRU policy. Streaming data will not occupy normal cache lines in this method, and the method can resist streaming data. Therefore, there will be no need to swap data, which will allow additional slow-cache-line access to be avoided. For better tolerant ability of slow cell, we combine the ZC-EDC and this strategy to build a zero-counting and adaptive-latency cache (ZCAL cache).

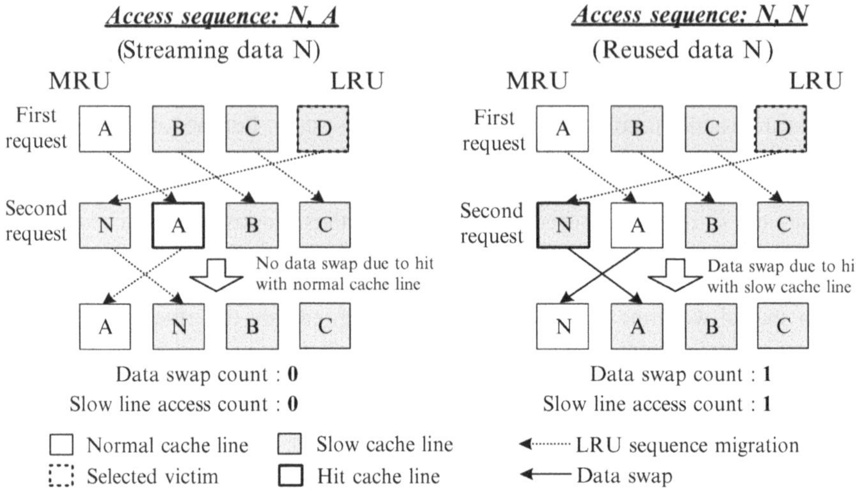

Fig. 11 Example of latency-aware LRU policy with hit swap

4.2 *Bit-Level Failure-Mask Management Strategy*

In the previous section, we described how the proposed DTC-SRAM calibrates the actual read cycle by dynamically considering the factor of stored data. However, expecting a certain value to cover slow cells is impractical when there are large numbers of slow cells. Therefore, based on DTC-SRAM, we propose a bit-level timing-failure-mask cache management strategy that exploits two cache characteristics, value bias and temporary locality, then build a cross-matching cache (CM cache) to enhance the ability to tolerate slow cells. The characteristics of value bias and temporary locality have been descripted in Sects. 2.3 and 4.1.

4.2.1 Access-Time Failure Masking via Data Mirroring

To improve the tolerance ability of massive access-time failures caused by slow cells, we propose a failure masking method that employs data mirroring. Figure 12 shows the concept for failure masking. Through the inverted data bank, the datum "1" may cause access-time failures. Assume that the original data and mirrored data are both "01110111." Slow cells (gray square bits) exist in both the referenced data and the mirrored data; thus, the datum "1" will be referenced as "0." In this situation, the data that are read out can pass through the OR layer to correct the referenced data. As Fig. 12 shows, if the value of the referenced datum differs in the original and mirrored data, then the read value is always correct if it is "1." The gray zeros in Fig. 12 are false zeros due to access-time failures. This simple method can mask most access-time failures.

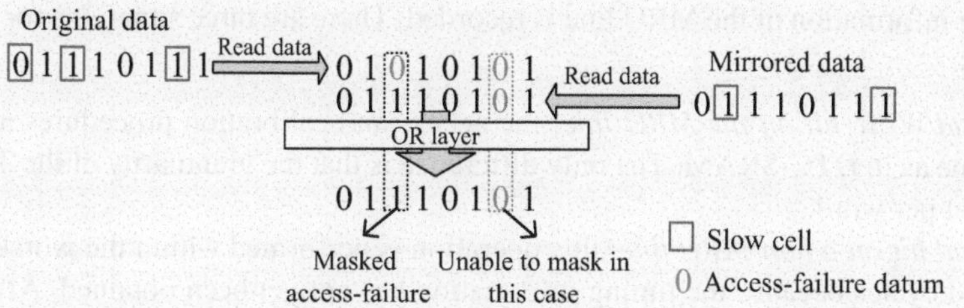

Fig. 12 Example of access-time failure masking

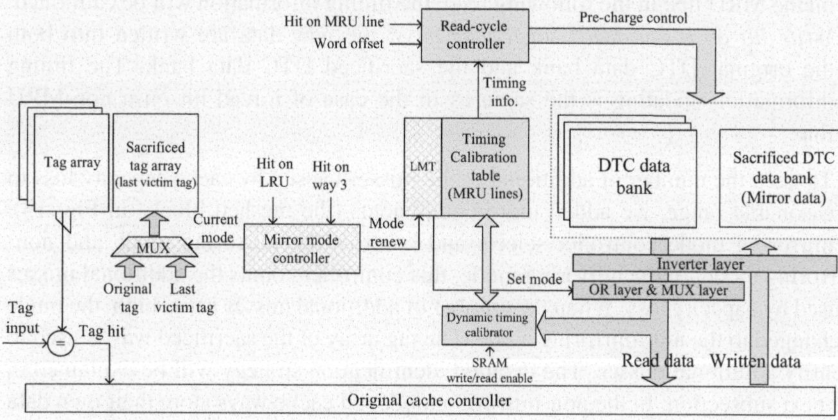

Fig. 13 Architecture of the cross-matching cache

Nevertheless, the seventh bit cannot be corrected because slow cells exist at the same position. In this rare situation, the access-time failure cannot be masked, but it can be detected by the dynamic calibrator and read with a worst-case read. With this strategy, we can typically read data within shortened-read cycles, even "0" data that are stored in slow cells.

4.2.2 Architecture of the Cross-Matching Cache

In addition to the DTC-SRAM components, two types of additional components (the dark gray blocks and meshed blocks shown in Fig. 13) are added to build the CM cache. The dark gray components are for the bit-level timing-failure mask, and the meshed blocks are for the mirror-mode changing, which will be explained later in the text.

In the CM cache, we sacrifice one cache way bank for the mirrored data of the MRU line per cache set for timing-failure masking. In contrast with DTC-SRAM, the TCT records the timing information of words in the MRU line after failure masking. The cache access flows are different with DTC-SRAM because only the timing information of the MRU line is recorded. There are three situations for cache access:

- *Read/Write hit on the MRU line*: the access and calibration procedures are the same as in DTC-SRAM. The only difference is that the granularity of the TCT is 1 bit per word.
- *Read hit on a non-MRU line*: this operation is performed within the worst-case-read cycles because the timing information has not yet been obtained. After the read, the new MRU line data are written to the sacrificed DTC data bank. Then, all words in the new MRU line are labeled as worst-case reads. If these data hit on the MRU line in the following read, the timing information will be calibrated.
- *Write hit on a non-MRU line/cache miss*: the new data are written into both the original DTC data bank and the sacrificed DTC data bank. The timing information updating is the same as in the case of a read hit on a non-MRU line.

To keep the number of additional cache misses caused by cache capacity loss in a reasonable range, we added more components (the meshed blocks in Fig. 13). A mirroring mode controller selects and controls the mirroring mode and non-mirroring mode. In the mirroring mode, this controller counts the additional misses caused by capacity loss. When the number of additional misses is too high, the mode is changed to the non-mirroring mode. The tag array of the sacrificed way is used to identify additional misses. The detailed identification strategy will be explained in the next subsection. In the non-mirroring mode, all cache ways store their own data and have no additional misses. In this mode, the TCT records the timing information of data that are not masked by the mirrored data.

When the mirroring mode controller changes the mode, each accessed set of cache changes its mode independently. A local mode table (LMT) is used to track the current mode of each set. Whenever a set of cache is accessed, the current mode information stored in the LMT is used to decide whether the mode needs to be changed. If the mode given by the mirroring mode controller is not matched with the current mode in the LMT, then the mode will be changed according to the mode given by the mirroring mode controller. The current mode information in the LMT is then updated; otherwise, the mode is not changed.

4.2.3 Additional Miss Detection and Prediction

In the non-mirroring mode, it is necessary to predict the number of additional misses that will be incurred by using the mirroring mode. A simple and effective method to predict additional misses is to count the hit rate of the least recently used (LRU) lines per cache set. When one cache way is lost, the data stored in the LRU lines are also lost; therefore, additional misses will occur. The operating mode is changed to the mirroring mode when the predicted number of additional misses is within a reasonable range.

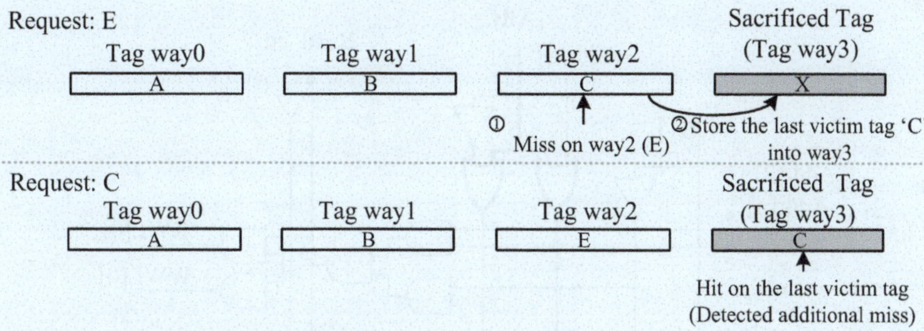

Fig. 14 Example of additional miss detection

Figure 14 shows an example of additional miss detection. There are two addresses requested: address "E," followed by address "C." When address "E" is requested, a cache miss occurs and way-2 is considered as a victim way. The tag of the last victim (tag way-2) is stored in the sacrificed tag space (tag way-3). Next, the second address, "C," is requested. The cache hits on the last victim tag; however, the corresponding data no longer exist. Hence, the hit on the last victim tag is detected and counted as an additional miss by the mirroring mode controller. Similarly, the mirroring mode is changed to the non-mirroring mode when the additional miss rate is too high.

4.3 Data Allying Management Strategy

Different with access-time failure designs that are with access-time adjustment, we propose Turbo cache that is based on an 8T SRAM cell with alliable wordlines. The alliable wordlines mean two wordlines are triggered while accessing the SRAM to speed up the bitline discharging time. With the read wordlines allying, 8T cell SRAM is able to perform with better reliability in an ultralow-voltage environment and decrease the read latency. Moreover, we propose specific cache management strategies to decrease the unnecessary boost penalty. With a Turbo cache, the system is able to instantaneously speed up the core and then is able to execute more applications.

4.3.1 8T SRAM with Alliable Read Wordline

In this work, an 8T SRAM with selectively allying read wordline circuitry is proposed to increase the read delay times with slight penalty. By inserting just two logic gates into the row decoder of 8T SRAM to achieve selective read wordline allying, there is no need of modifying cell structure, and the area penalty is slight as Fig. 15 shows. The proposed technique can double the read current and increase the

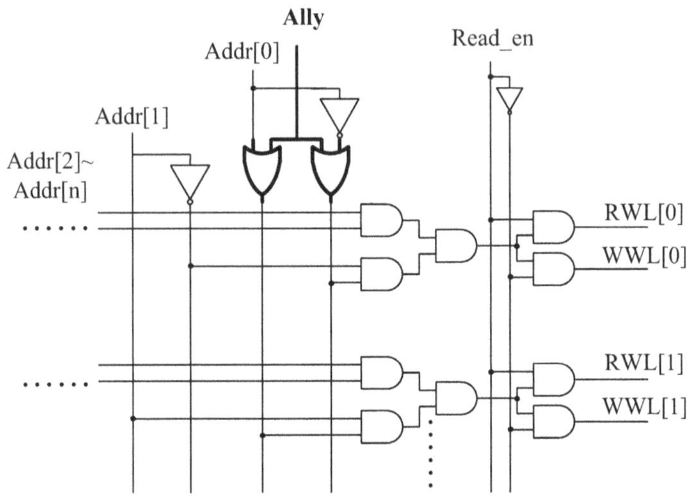

Fig. 15 The proposed row decoder for selective wordline allying

speed of 8T at read operation. In addition, selectively allying technique has more feasibility to cache designs because it will not sacrifice half of the cache capacity like the cache designs with 7T/14T SRAM [17].

Read wordlines allying technique can speed up the read speed of 8T SRAM. The allying technique forms an extra discharging path and doubles read current during read-0 operation. We simulate a single 8T SRAM column (128 × 1) at the worst case; the slowest process corner SNSP and the low temperature 0 °C reveal that the allying technique can largely speed up read-bitline discharging time to 49%.

For L1 cache, the benefit of decreasing the read latency of SRAM is able to decrease the read cycles of L1 caches. The proposed alliable wordline technique is simulated through a 1 KByte (256 × 32) single-ended 8T SRAM array, whose local bitline is 128 bits deep. The portion of read-bitline delay in the total array delay is approximately 54%, so the actual speedup of the total array delay after merging is 20% at 0.5 V, as shown in Fig. 16.

Figure 17 shows the SRAM array access latency that is normalized to the core cycle time at 0.9 V. When decreasing the voltage, the access latency of the SRAM and cycle time of the core is increased. Without our proposed alliable wordlines technique, one more cycle will be needed to access the SRAM when the voltage is below 0.7 V. With the technique, access latency can be shortened on average; the access latency of allied 8T SRAM can be reduced by more than 19% compared to conventional 8T SRAM in different voltages.

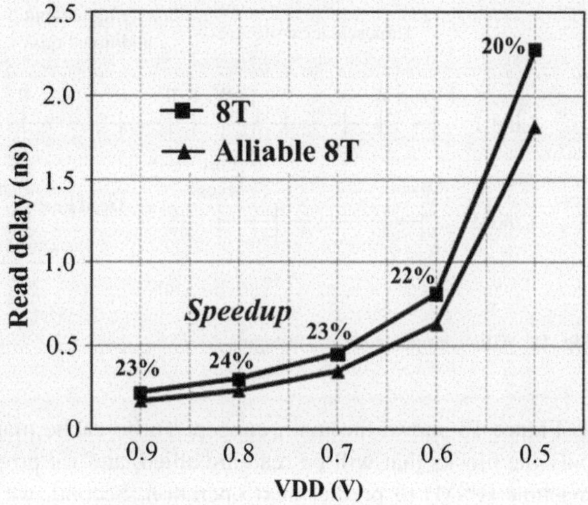

Fig. 16 Read delay evaluation results across 0.9–0.5 V of an 1 kB 8T SRAM macro

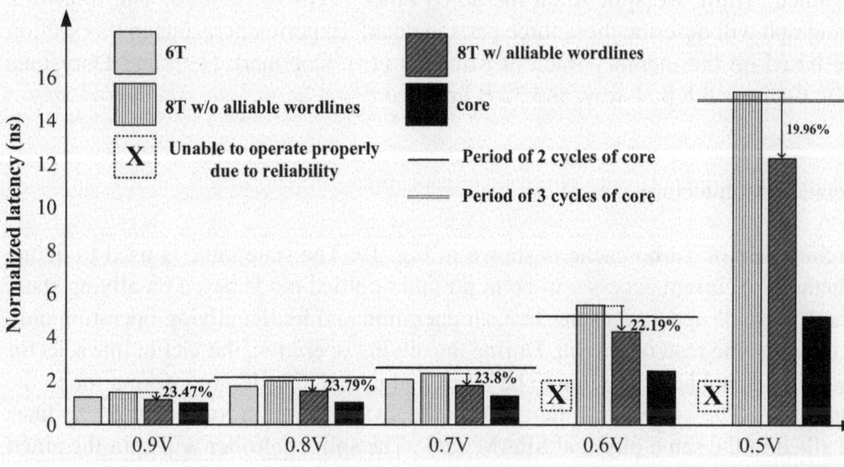

Fig. 17 SRAM access latency normalized to cycle time of core at 0.9 V

4.3.2 Turbo Cache Management Strategies to Reduce the Unnecessary Penalty

The pervious design like 7T/14T cache [17] is useful to decrease the read latency in every cache read operation; however, it may cause a high miss penalty in memory-intensive benchmarks due to the large capacity loss and cause additional energy consumption on cache write hit. Therefore, we propose the Turbo cache management strategies that are able to accelerate most of the read operations and effectively reduce the unnecessary penalty which includes miss penalty and allied write energy.

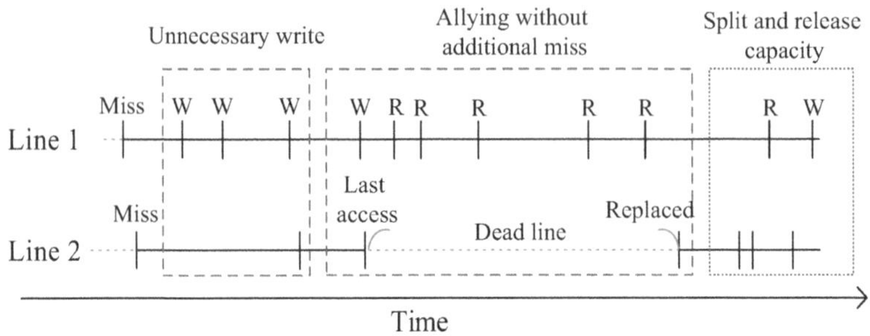

Fig. 18 Alliance management strategies

Figure 18 shows the main concept Turbo cache management strategies. First, only the blocks that will be read are allied, and we propose a low-cost finite-state machine (FSM) to predict next operation. Second, we choose the dead block to be allied. Third, we split when the allied block needs to be used. The following paragraph will describe these three parts in detail. Experiment results in this section are based on the memory trace of Mibench [18], Coremark [19], and Dhrystone [20]. Cache is 8 KB, 4-way, and 32 B line size.

Detailed Architecture

Architecture of Turbo cache is shown in Fig. 19. The state table is used to define whether the current access will be in normal or allied mode based on allying state. The FSM will update its state in each operation and let the allying operation only occur upon the read operation. During the allying operation, the victim line selector chooses a suitable victim line to be allied and update the allying information to the state table. The swap controller will do the swap operation to let two cache lines be allied in the same physical SRAM bank. The split controller will split the allied cache lines by updating the allying information if the FSM state indicates split mode. Although some hardware components are added on the path, these components do not affect the latency because they are not on the critical path. To make an alliance with a cache line, the valid bit of the victim line will be unset and updated to the tag section. The specific decoder restricts only the adjacent cache sets that can be allied; therefore we used remapping layer to remap the suitable cache set to be adjacent. When the core issues a cache request, the request will be sent to the remapping layer and the state table to trigger the wordlines simultaneously. Because every allying operation needs to know the LRU position of the set that is to be allied, the LRU information of each set is also kept in the state table instead of the tag section. In the state table, LRU information accounts for 2 bits, and the state information accounts for another 2 bits. All the strategies will be discussed in the following paragraph.

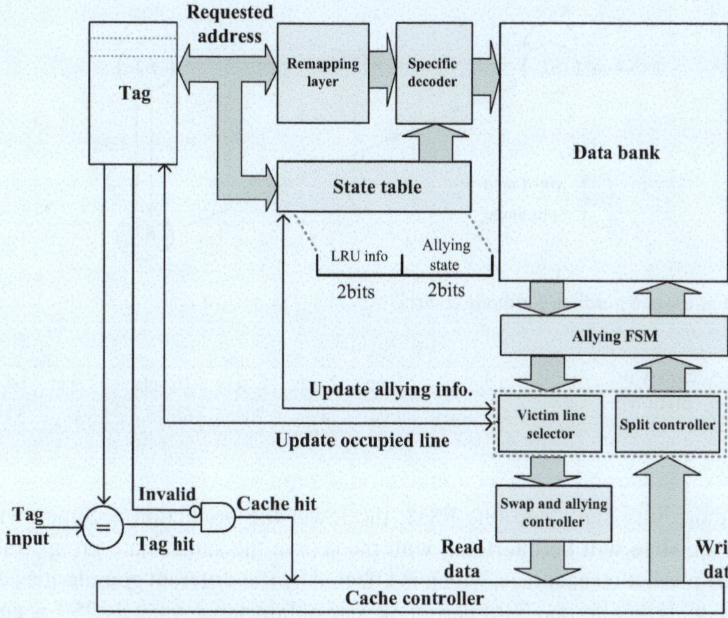

Fig. 19 The architecture of the Turbo cache

Strategy 1: Next Operation Prediction

The simplest way to maintain the read operation under allied mode is to keep the referenced line allied whether in a read or write operation. However, this causes large energy consumption when the line being written is under allied mode. Therefore, if the next access operation is a read operation, allied mode is worthwhile to decrease read latency. However, if the next operation is a cache write, we should not let the cache line being allied to prevent unnecessary overhead from writing on allied blocks. In Fig. 20, we use a 2-bit finite-state machine (FSM) to control changes in the mode. When encountering a read operation, the cache line will be allied for the next read operation. While encountering two consecutive write operations, the allied line will be split. The key observation behind this strategy is that there is a high continuity in the operation type. In MiBench, there is about 70% probability that the next operation type will be the same as the current type. Because the allying operation costs an additional write operation in the partner set, we split the allied pair if two write operations occur consecutively instead of as a single instance. This strategy prevents excess allying overhead in situations in which a read operation and write operation occur alternately. When the incoming operation type is a read operation, we predict that the next operation is also a cache read. If two consecutive write operations occur, we predict that the next operation is a cache write.

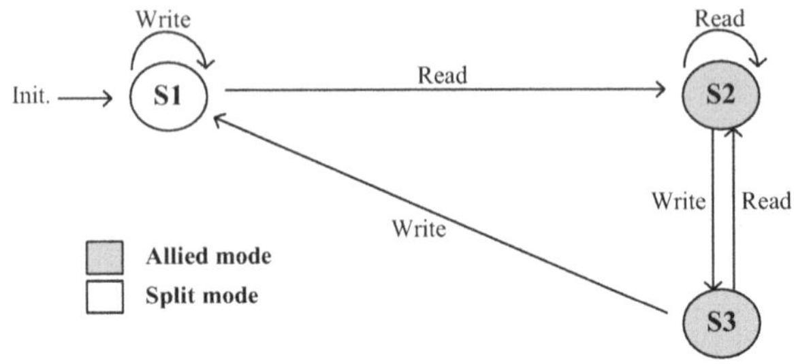

Fig. 20 Finite-state machine for mode control

Table 2 Accuracy of operation prediction

	1 set	2 sets	8 sets	16 sets	32 sets	64 sets
Accuracy	75%	74%	70%	66%	56%	53%

The more sets that share one FSM, the lower the prediction accuracy will be because the state will be interfered with the sets in the same share group. Table 2 shows the prediction accuracy, which is affected by the different granularities of the share group. If there is one FSM per set, we can obtain approximately 75% accuracy. If the whole cache (assume a total of 64 sets) shares one FSM, the accuracy will decrease to 53%, which is almost the same as guessing arbitrarily. However, the more FSM we use, the higher the area and energy consumption are. In this paper, we let two sets share one FSM, which is the most efficient.

Strategy 2: Victim Cache Line Selection

To reduce the miss rate, we have to choose a suitable line to be allied. In the ideal situation, if the occupied line is dead, which means that it will no longer be accessed before it becomes a victim line, it will not cause any additional misses. These kinds of cache lines are suitable to be allied. Many cache dead line prediction strategies have been proposed [21, 22]; however, these strategies are not suitable in the L1 cache because of additional latency or area overhead to keep the access counter or reference history table, and they have tremendous energy consumption. In the Turbo cache, we need a low-cost and low-overhead strategy for dead line prediction. Our first observation is that because of spatial locality, the line in the next set has a lower probability of being dead. We analyzed the probability of being dead by choosing the occupied line from the neighborhood to the line in the longer distance. According to the analysis, increasing distance is helpful for finding a dead line. The specific decoder we proposed could only trigger an adjacent wordline. To let a cache line allied with its ally in a longer distance, we should remap the address before it is sent to the decoder. For example, we can change the LSB and the third LSB before sending the address to the decoder to make a static address remapping with distance 8. As a result, the allied pair with distance 8 will be in the adjacent wordline.

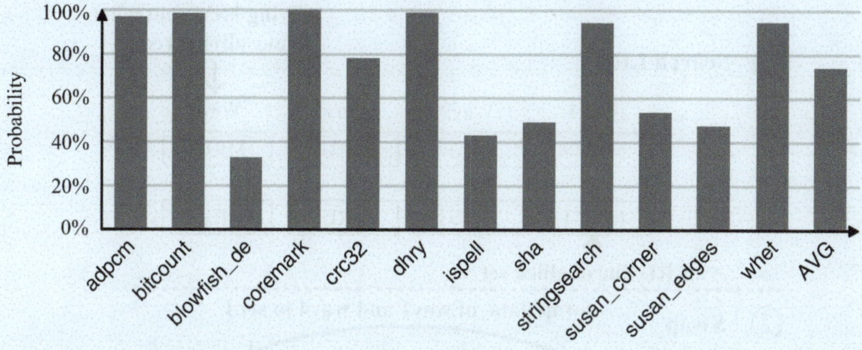

Fig. 21 The probability of a LRU line being a dead line

Another observation is that if the cache line becomes the least recently used (LRU) line, there is a high probability that it will become a dead line, which makes it a suitable victim to be allied. Figure 21 shows the probabilities of being dead if the cache line becomes LRU. Most parts of benchmarks show high probabilities of being dead when falling into LRU. On average, there is approximately a 75% probability of LRU being a dead line in MiBench benchmarks.

Strategy 3: Data Swap

Although we can find an LRU line via this low-cost method, with the restriction of the physical SRAM bank, the pair lines cannot be allied directly. If the way position of the pair is not the same, they are in different physical banks. In this situation, we have to swap cache line data before allying. Figure 22 shows the execution flow of the occupied line selection. At the first step, when way four of set 1 issues an allying request, we search the LRU line in set 2, which is in way one. In the second step, we swap the data in way one and way four in set 1 and then issue an allying request on way one. In the last step, way one can be allied with the LRU line (way two) in set 2. Data swapping will not block the CPU because the swap operation is not on the critical path. Swapping the way may require writing in two cache lines, and this will cause energy overhead. We will evaluate the swap overhead in Sect. 6.

When choosing a suitable victim line in the partner set, the miss rate is not the only consideration. If the target line being chosen is dirty, we need to write back before doing an allying operation. If so, energy consumption will be increased greatly as a result of accessing the lower level cache or outer memory. With these considerations, we set a priority to search the victim line. First, we search the line that has been allied. Second, we choose the LRU line. With the swapping scheme, we can choose the victim line from different physical bank. This strategy also has the restriction that only one line in each set is allied simultaneously to minimize the miss rate and the energy consumption of the write back operation.

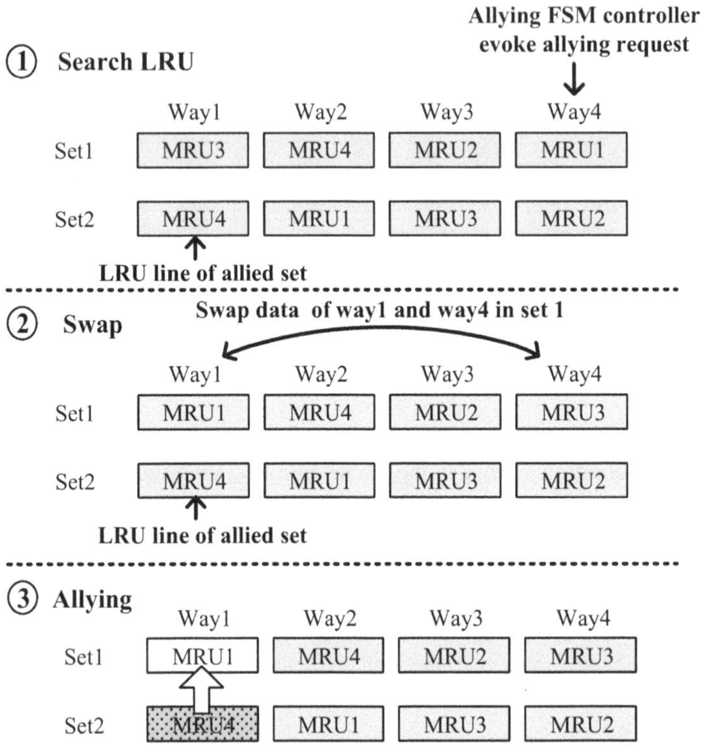

Fig. 22 Example of a swap operation

In order to keep the miss rate not being increased greatly, the allied pair will be split in the following situation. First, if the set of the allied line is accessed and causes a miss, the allied pair will be split. Second, if the referenced set is accessed with two consecutive write operations, the allied pair will be split also.

5 Evaluation

5.1 Experimental Environment

Table 3 shows our experimental environment. We performed our experiments using the MARSSx86 [23] full-system simulator, which supports x86 in-order and out-of-order (OoO) core simulations. We got the memory trace with MARSSx86 and simulate with a modified DineroIV [24]. The configuration of simulated local memory is 32 KB, 64-bits data width. The configuration of simulated L1 cache is a size of 32 KB, four ways, and a 64B line size (eight words). The L2 cache has a size of 4 MB, eight ways, and a 64B line size. We forwarded 50 billion instructions

Table 3 The simulator configuration

Full-system simulator	MARSSx86 [23]
L1 cache model	DineroIV [24], CACTI, Chip Measurement
Workload	SPEC2006 [25]
Processor	4 issue width out-of-order core
ISA	x86
Memory access cycles	Shortened read: 2 cycles
	Worst-case read: 3 cycles
Local memory configuration	32 KB, 64-bits data width
L1 cache configuration	32 KB, 4-way, 64B block size
L2 cache	4 MB, 8-way, 64B block size

and simulated 100 million instructions from SPEC2006 [25] for each simulation. We observed the simulation results for the 65-nm process SRAM under 0.5 V using Monte Carlo simulations and compared our findings with the simulation results of Chen et al. [9].

The different methods are used for obtaining the energy consumption and latency of the SRAM and the logic components. The access energy and the latency of SRAM memory were from a HSPICE simulation with a 65-nm process under 0.5 V. The overhead of additional SRAM bits are calculated based on the simulation result. For the evaluation of the logic components, we implemented all of the components and synthesize them under 0.5 V to get the average power result.

In the experiments, we compared several designs. All of the read/write operations of baseline local memory and cache take three cycles. For other designs, the shortened-read cycle is set to two cycles, whereas the worst-case read cycles are three. In these evaluations, we did not consider the improvement of write operations; therefore, all of the write operations take three cycles. In the timing table cache designs such as VL-cache [26], the timing table records access cycles in the cache line granularity. In the separated Vdd design [26], the Vdd is higher than other designs by 0.1 V (0.6 V); thus, the read/write cycle is set to two cycles without access-time failures.

We assume that the tag arrays of those cache designs have no access-time failures, and there are many proposed methods that can be applied [28, 29]. In this paper, we apply a higher Vdd to tag array and additional bits except zero-counting bits. The operating voltage of the data array is 0.5 V, whereas the higher operating voltage of the tag array is 0.6 V. This overhead is estimated in the following experiment.

5.2 Comparison of Slow Cell Tolerance Ability

Figure 23 shows the probability of worst-case read in different proposed designs for local memory. ZC-EDC performs better slow cell tolerance ability than

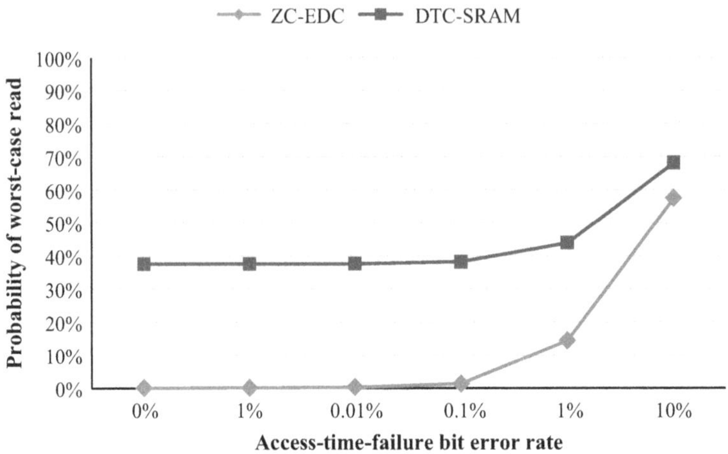

Fig. 23 Probability of worst-case read with local memory designs

DTC-SRAM. ZC-EDC can keep approximately 40% shortened read operations even operating in the environment with 10% access-time failure ratio. However, ZC-EDC takes more area, energy, and latency overhead because of the zero-counting bits and the calculation of zero-counting bits. On the other hand, although DTC-SRAM performs worse slow cell tolerance ability, it has no significant overhead. Nevertheless, DTC-SRAM has the misjudging problem to make additional worst-case read. That is the reason why DTC-SRAM has some worst-case read with error-free environment.

Figure 24 shows the probability of worst-case read in different access-time failure designs for L1 caches. VL Cache [26] (timing table designs) has coarse timing information granularity (line level). Their probability of a worst-case read is increased dramatically as the slow cell ratio is greater than 0.01%. ZCAL cache also has coarse timing information granularity, so the probability of a worst-case read is also increased dramatically. Although the ZCAL cache has insignificant tolerant ability, it can detect the other kind of SRAM failure and provide better reliability. CM cache records the timing information of MRU lines in the word granularity. With the bit-level masking, CM cache can operate very well even with 10% slow cells. In this experiment, we did not show the result of Turbo cache because the probability of a worst-case read of Turbo cache is variable depending on the strategies and is constant regardless of the access-time failure ratio.

5.3 Performance Analysis

Figure 25 shows the average memory access time (AMAT) in each proposed access-time failure-tolerant designs for local memory. The AMAT is normalized

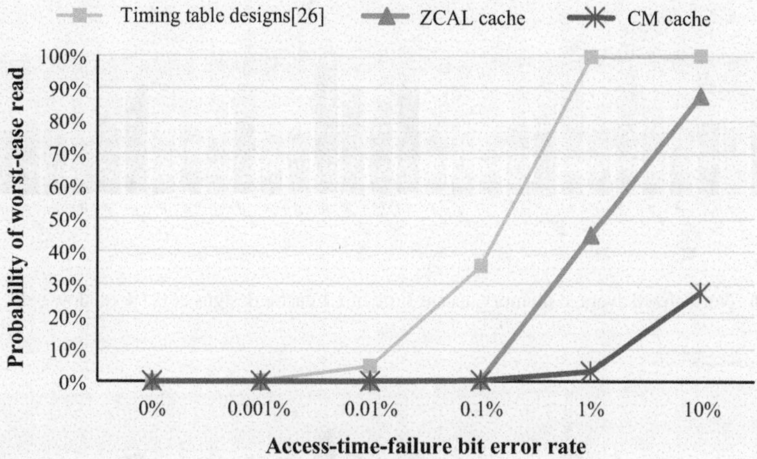

Fig. 24 Probability of worst-case read with L1 cache designs

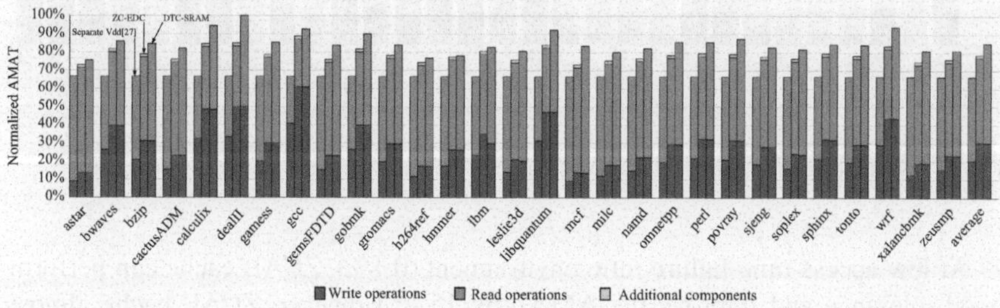

Fig. 25 Normalized average memory access time of local memory designs at 0.1% of slow cells

to the worst-case design. In local memory, the latency increment directly affects AMAT. The ZC-EDC has an increased AMAT of approximately 1.7% due to the bit counting procedure; the DTC-SRAM only increases an inverter layer with approximately 0.05% latency increment. On average, ZC-EDC and DTC-SRAM perform approximately 21% and 15% AMAT improvement with 0.1% slow cells. The separate Vdd [27] design has no worst-case accesses, but it has large energy overhead due to applying high Vdd.

Figures 26 and 27 show the average memory access time (AMAT) of each cache designs. The AMAT is normalized to the worst-case cache design. Figures 26 and 27 present the results for the cases in which the slow cell ratios are 0.1% and 10%, respectively.

Those cache designs have several additional logics on the critical path (path for cache read). Those logics slightly increase the latency (ZCAL cache, 2.2%; CM cache, 0.94%). To deserve to be mentioned, there is no additional logic that lays on the critical path. We included the increased latency in our performance analysis.

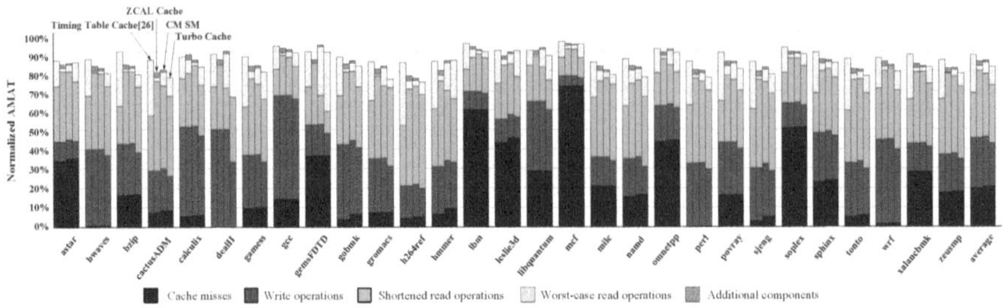

Fig. 26 Normalized average memory access time of L1 cache designs at 0.1% of slow cells

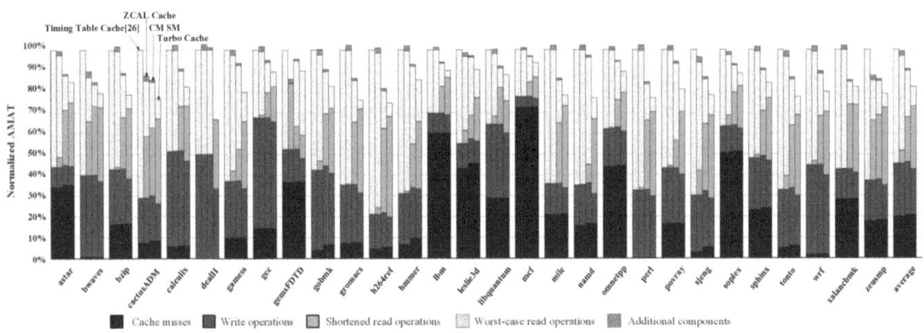

Fig. 27 Normalized average memory access time of L1 cache designs at 10% of slow cells

At low access-time failure ratio environment (0.1%), ZCAL cache can perform good tolerance and improve the AMAT by 15%. However, ZCAL cache almost becomes useless at the high access-time failure ratio environment (10%). CM cache has better tolerance ability at high access-time failure ratio environment. Therefore, CM cache can improve the AMAT by 15% and 9% at 0.1% and 10% of slow cell ratio, respectively. If the workloads tend to have many write operations, such as with GemsFDTD, then the AMAT improvement will be reduced because the read latency tends to be misjudged. Although we sacrificed a cache way, the additional cache miss impact on the AMAT is increased only by an average of 1%. Our proposed strategy for Turbo cache can minimize the penalty from losing capacity; therefore, Turbo cache can reduce the average memory access time by 18% on average compared to the baseline cache. The AMAT result of turbo-cache is regardless with access-time failure ratio.

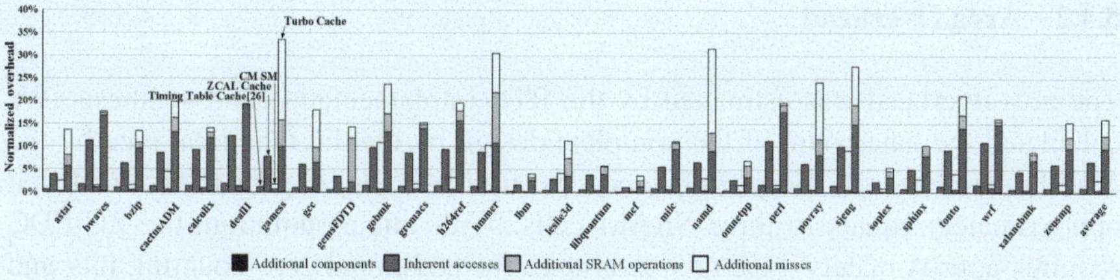

Fig. 28 Energy overhead of different cache designs

5.4 Design Complexity

5.4.1 Energy Overhead

The access-time failure-tolerant designs of local memory do not sacrifice any capacity; thus, the only cause of energy overhead comes from additional components. ZC-EDC needs to add a lot of zero-counting bits; it takes 11.4% energy overhead. On the other hand, DTC-SRAM has very small energy overhead due to the simple architecture (1.8%).

Figure 28 shows the energy overhead of each cache designs. The item of inherent cache access includes the energy overhead from additional logics, additional SRAM bits, and higher operating voltage. Separated Vdd designs use higher Vdd; thus, the energy consumption of inherent cache access is greater than in other designs (44% on average). The timing table designs have very small energy overhead from inherent cache access due to their simplicity. They have approximately 1% of energy overhead on average. The ZCAL cache exhibits approximately 6.67% energy overhead. This overhead estimation includes the SRAM and logic overhead. Although CM cache sacrifices the cache capacity to tolerate slow cell, the management strategy of CM cache can avoid the additional miss energy overhead through selective mirroring and has approximately an average of 2.6% energy overhead. The Turbo cache consumes 16% more energy on average compared to the baseline. The energy overhead is mostly used to read on the allied cache line. Reading on the allied cache line is 20% of energy overhead (compared to the energy of the read on a non-allied cache line). The percentage of read on an allied cache line is 59% of total read operation on average using our proposed strategy. On the average, 17% of write operations write on an allied cache line. The swap operation does not consume large amounts of energy because only 9% of the total cache operation needs an allying operation, and the energy consumed by a miss event due to capacity loss is minimized by using our strategy.

5.4.2 Area Overhead

The area overhead was estimated by the RTL implementation and synthesis. We calculated the gate count of the complete design by dividing the total area by the NAND gate area. In the process node we used, one logic gate is assumed that approximately equals to three SRAM cells. With this assumption, the ZC-EDC requires approximately 14% area overhead for additional zero-counting bits and additional logic, and the DTC-SRAM requires approximately 5% area overhead.

In the cache designs, the ZCAL cache is approximately 7% area overhead with cache-line-detection granularity. CM-FM and CM-SM designs have approximately 11% of area overhead due to a complex controller and the need to duplicate data into the sacrificed data array for data mirroring, whereas CM-SP has approximately 7% of area overhead compared to conventional cache design. The area overhead of Turbo cache from the logic of management component and the specific decoder is approximately 4.5%.

5.4.3 Consideration for Out-of-Order Processors

In current processor systems, hit under miss is a common design in which hit or miss information should be sent to the core after the reading tag. Referenced data should be accompanied by hit information and be sent to the core within fixed cycles. However, in these types of variable-latency cache designs, data may be sent with some extra cycles delayed. The extra cycles create unfixed latency for the core to receive data even in cache hit operations. Out-of-order (OoO) cores need a more complicated scheduler to handle variable receiving latencies.

6 Related Works

6.1 Reliable Low-Voltage Cache Designs

Wilkerson et al. [1] proposed two architectural methods to tolerate defective bits for low voltage. The Word-Disable (WDIS) scheme combines two defective consecutive lines in the same set into one faultless line by indicating the defective blocks in the defect map. For more reliable operation at low voltage, this method decreases the cache capacity and associativity by 50%. The second scheme, Bit-fix, uses one cache way to patch the defective bit in the other three ways in a 4-way set. Therefore, Bit-fix decreases the cache capacity and associativity by 25%. The ZerehCache (ZC) [2] scheme uses spare cache lines as redundant space. It groups faulty lines with no conflicts as a functional line and imports a complicated interconnection network between the row decoder and the cache sets for data remapping. Instead of spare cache lines, the Archipelago solution [3] selects sacrificial lines from different banks to patch defective bits and merges collision-free lines as a logical line. These solutions sacrifice capacity or exploit complicated data remapping to gain reliability at low voltage. These methods result in large performance losses and are thus not suitable for latency-sensitive L1 caches.

6.2 Error Correction Code Designs

By encoding the original data in a redundancy for check bits and decoding together when the data are read, ECC can detect and correct a limited number of errors that may occur at any time. Therefore, ECC designs can increase the reliability of SRAM. Recently, some ECC designs have been proposed for low-voltage caches to address a large number of faults under low-operating voltage.

Zeshan Chishti et al. proposed a multi-bit segmented ECC (MS-ECC) in [4]. The MS-ECC focuses on tolerating SRAM faults in low-voltage caches. MS-ECC supports both a high-voltage mode and a low-voltage mode. In the high-voltage mode, the entire cache capacity is available for high performance. In the low-voltage mode, MS-ECC trades off cache capacity for reliability at low voltage. A portion of the cache is used to store additional ECC information, thereby enabling more errors to be fixed. Instead of using BCH-based code, which has high complexity and latency, MS-ECC is equipped with an orthogonal Latin square code (OLSC), which has a faster coding time and more simply eliminates the impact of access latency. However, OLSC requires a large number of check bits. Therefore, MS-ECC sacrifices at most half of the cache capacity to store check bits and increased cache miss.

Alaa R. Alameldeen et al. [5] proposed a variable-strength ECC (VS-ECC). This design also focuses on low-voltage caches. Instead of employing full multi-bit correction codes, VS-ECC uses both strong and weak ECCs. In typical cases, VS-ECC employs a fast and simple ECC such as SECDED in lines with less than one fault. In addition, VS-ECC is equipped with a strong multi-bit ECC (e.g., 4EC5ED), which needs additional area and access latency for the small number of lines with multi-bit faults. VS-ECC may also disable some cache lines if the numbers of defective bits cannot be tolerated by both weak and strong ECCs. By leveraging weak ECC with strong ECC, VS-ECC requires fewer check bits and access latency than full multi-bit correction codes.

6.3 Robust Circuit Designs

Robust SRAM cells, 8T [6], and 10T [7] are also used to increase the reliability of SRAM without significant performance losses. These robust SRAM cells can maintain a better safety noise margin (SNM) in low-voltage conditions. Single-read bitline (SRBL) 8T SRAM increases the read stability by separating the read port.

Thus, the supply voltage can be scaled down lower than for 6T cells. However, the area above these robust cells must be considered, particularly the differential-10T (bit interleaving) cells that incur large area overhead. Recently, robust 8T SRAM has been widely used in modern L1 caches [8] without fault-tolerance mechanisms to provide reliable operation at low voltage. However, modern 8T SRAM L1 caches still suffer from the long-latency problem, and the problem is a critical issue in processor systems.

Hidehiro Fujiwara et al. proposed a dependable SRAM with 7T/14T cells [20] that can dynamically control its reliability. This design adds two transistors in two neighbor 6T cells. On average, each memory cell has seven transistors. The proposed SRAM cell design has normal mode, high-speed mode, and dependable mode. In normal mode, a one-bit datum is stored in one 7T memory cell, which is the most area efficient. In high-speed mode, the datum is stored in the 14T memory cell. The high speed is achieved when both wordlines of the 14T cell are driven, which enables a faster readout. In dependable mode, the datum is also stored in the 14T memory cell, but one wordline is asserted. Thus, this design can reduce both the reliability barrier and performance barrier effect.

6.4 Tolerating Access-Time Failure Designs

Mutyam [26] proposed a VL cache that uses a timing table created by the manufacturer during the testing process to record different access cycles of each cache set and a set predictor to predict the number of cycles that will be necessary for the next access. The cache access is replayed when the prediction is wrong. Zhai et al. [27] studied the activity factors of cores and caches and tuned them independently to determine the best operating voltage that addresses the reliability concerns and offers better performance. They found that when co-optimizing with the cores for the best overall performance, the optimal method used higher voltage for the cache than the core. However, speeding up mostly healthy cells leads to unnecessarily higher energy consumption.

6.5 Timing Speculation of the Pipeline

Razor [30, 31] is the work of circuit-level speculation to eliminate the worst-case safety guard band of the pipeline. Razor installs the timing-error-tolerant flip-flops on critical paths and scales the supply voltage of the pipeline adaptively. When timing errors occur because of the overly low voltage or the dynamic variation, Razor detects errors and recovers data to maintain functional work. Razor also calculates the error rate to scale the supply voltage properly. Thus, Razor can eliminate the worst-case safety margin and work at a lower voltage without being restricted by the delay of the longest path, resulting in significant energy savings.

7 Conclusion

In this chapter, we described the problem of the timing discrepancy between cores and caches and proposed several designs for local memory and L1 caches. The proposed designs consider on the characteristics of 8T SRAM that is the impact of the stored data. For local memory designs, we proposed ZC-EDC and DTC-SRAM to reduce the worst-case-read count. ZC-EDC can reduce the AMAT of local memories by 21% on average at 0.1% of slow cell ratios. And DTC-SRAM can reduce the AMAT of memory by 15% on average at 0.1% of slow cell ratios.

On the other hand, we proposed ZCAL cache, CM cache, and Turbo cache for access-time failure-tolerant cache. ZCAL cache uses ZC-EDC and timing-aware LRU policy. CM cache masks the slow cells in bit level and reduces the worst-case-read count. Turbo cache is based on an alliable 8T SRAM that is able to perform reliable ultralow-voltage operations and provide the alliable wordline function. Moreover, we also propose specific cache management strategies for decreasing unnecessary energy penalties. The ZCAL cache can reduce the AMAT of L1 caches by 15% on average at 0.1% of slow cell ratios, respectively. The CM cache can reduce the AMAT of L1 caches by 15% and 9% on average at 0.1% and 10% of slow cell ratios, respectively. The Turbo cache can reduce the AMAT of L1 caches by 18% on average.

References

1. Wilkerson, C.: Trading off cache capacity for reliability to enable low voltage operation. 35th International Symposium on Computer Architecture. IEEE (2008)
2. Ansari, A.: Zerehcache: armoring cache architectures in high defect density technologies. 42nd Annual IEEE/ACM International Symposium on Microarchitecture. IEEE (2009)
3. Ansari, A.: Archipelago: a polymorphic cache design for enabling robust near-threshold operation. 17th International Symposium on High Performance Computer Architecture. IEEE (2011)
4. Chishti, Z.: Improving cache lifetime reliability at ultra-low voltages. 42nd Annual IEEE/ACM International Symposium on Microarchitecture. ACM (2009)
5. Alameldeen, A.R.: Energy-efficient cache design using variable-strength error-correcting codes. 38th International Symposium on Computer Architecture. IEEE (2011)
6. Chang, L.: Stable SRAM cell design for the 32 nm node and beyond. In: Digest of Technical Papers. Symposium on Very Large Scale Integration Technology. IEEE (2005)
7. Chang, I.J.: A 32 kb 10T sub-threshold SRAM array with bit-interleaving and differential read scheme in 90 nm CMOS. IEEE J. Solid State Circuits. **44**, 650–658 (2009)
8. Gerosa, G.: A sub-2 W low power IA processor for mobile internet devices in 45 nm high-k metal gate CMOS. IEEE J. Solid State Circuits. **44**, 73–82 (2009)
9. Chen, G.: Yield-driven near-threshold SRAM design. IEEE Trans. Very Large Scale Integr. Syst. **18**, 1590–1598 (2010)

10. Mukhopadhyay, S.: Modeling of failure probability and statistical design of SRAM array for yield enhancement in nanoscaled CMOS. IEEE Trans. Comput. Aided Des. Integr. Circ. Syst. **24**, 1859–1880 (2005)
11. Humenay, E.: Impact of parameter variations on multi-core chips. In: In Workshop on Architectural Support for Gigascale Integration (2006)
12. Moshovos, A.: A case for asymmetric-cell cache memories. IEEE Trans. Very Large Scale Integr. Syst. **13**, 877–881 (2005)
13. Mazreah, A.: A novel zero-aware four-transistor SRAM cell for high density and low power cache application. In: International Conference on Advanced Computer Theory and Engineering, pp. 571–575. IEEE (2007)
14. Hossain, R.: Circuit for determining the number of logical one values on a data bus. Patent No. 6, 729, 168 (2004)
15. Dalalah, A.: New hardware architecture for bit-counting. In: 5th WSEAS International Conference on Applied Computer Science, pp. 118–128 (2006)
16. Petit, S.: Exploiting temporal locality in drowsy cache policies. In: Proceedings of the 2nd conference on Computing frontiers, pp. 371–377. ACM (2005)
17. Fujiwara, H.: A 7T/14T dependable SRAM and its array structure to avoid half selection. In: 22nd International Conference on Very Large Scale Integration Design, pp. 295–300. IEEE (2009)
18. Guthaus, M.R.: MiBench: a free, commercially representative embedded benchmark suite. International Workshop on Workload Characterization, pp. 3–14. IEEE (2001)
19. Gal-On, S.: Exploring CoreMark™–a benchmark maximizing simplicity and efficacy. The Embedded Microprocessor Benchmark Consortium (2012)
20. Weicker, R.P.: Dhrystone: a synthetic systems programming benchmark. Commun. ACM. **27**, 1013–1030 (1984)
21. Kharbutli, M.: Counter-based cache replacement algorithms. International Conference on Computer Design, pp. 61–68. IEEE (2005)
22. Khan, S. M.: Sampling dead block prediction for last-level caches. 43rd Annual IEEE/ACM International Symposium on Microarchitecture, pp. 175–186. IEEE (2010)
23. Marss-x86. Available. http://marss86.org/~marss86/index.php/Home
24. Edler, J.: Dinero IV trace-driven uniprocessor cache simulator. Available. http://pages.cs.wisc.edu/~markhill/DineroIV/ (1998)
25. SPEC CPU2006 Benchmarks. Available. http://www.spec.org/cpu2006/
26. Mutyam, M.: Process-variation-aware adaptive cache architecture and management. IEEE Trans. Comput. **58**, 865–877 (2009)
27. Zhai, B.: Energy efficient near-threshold chip multi-processing. In: Proceedings of the 2007 international symposium on Low power electronics and design, pp. 32–37. ACM (2007)
28. Ganapathy, S.: Effectiveness of hybrid recovery techniques on parametric failures. International Symposium on Quality Electronic Design. IEEE (2013)
29. Agarwal, A.: Exploring high bandwidth pipelined cache architecture for scaled technology. In: Design, Automation and Test in Europe Conference and Exhibition, 2003, pp. 778–783. IEEE (2003)
30. Ernst, D.: Razor: A low-power pipeline based on circuit-level timing speculation. Proceedings. 36th Annual IEEE/ACM International Symposium on Microarchitecture, pp. 7–18. IEEE (2003)
31. Das, S.: RazorII: in situ error detection and correction for PVT and SER tolerance. IEEE J. Solid State Circuits. **44**, 32–48 (2009)

Redesigning Software and Systems for Nonvolatile Processors on Self-Powered Devices

Chun Jason Xue

1 Introduction

Wearable devices are attracting increasing attention from both research and industry. Wearable technology enables the devices, such as smart watches, multifunction shoes, and intelligent glasses, to keep close contact with users in order to monitor the well-being status and respond to users' requirements and queries. As a traditional power source of embedded systems, battery is no longer a favorable choice for wearable devices due to (1) large size and weight, (2) safety and health concerns, and (3) frequent recharges. Therefore, researchers are actively pursuing power alternatives. Out of all possible solutions, energy harvesting is proposed to be one of the most promising techniques to meet both the size and power requirements of wearable devices.

Energy harvesting devices generate electric energy from its surroundings using direct energy conversion techniques [1]. Examples of power sources include but are not limited to solar [2–4], wind [5], vibration [6], electromagnetic radiation including light and RF [7–9], and piezo [10, 11]. It is also possible to harvest energy simultaneously from multiple sources in a system [12, 13]. The obtained energy can be used to recharge a capacitor or, in some cases, to directly power the electronics [1]. However, there is an intrinsic challenge with harvested energy. They are all *unstable* [14]. Figure 1 shows power traces collected from several representative ambient energy sources, including TV RF, piezo, and thermal and solar power, confirming the instability [15].

With an unstable power supply, the processor execution will be interrupted frequently. It is reported that the interval between adjacent power failures of computational RFIDs (CRFIDs) is less than one second [16, 17]. Frequent turning

C.J. Xue (✉)
City University of Hong Kong, Kowloon Tong, Hong Kong
e-mail: jasonxue@cityu.edu.hk

H. Yasuura et al. (eds.), *Smart Sensors at the IoT Frontier*,
DOI 10.1007/978-3-319-55345-0_5

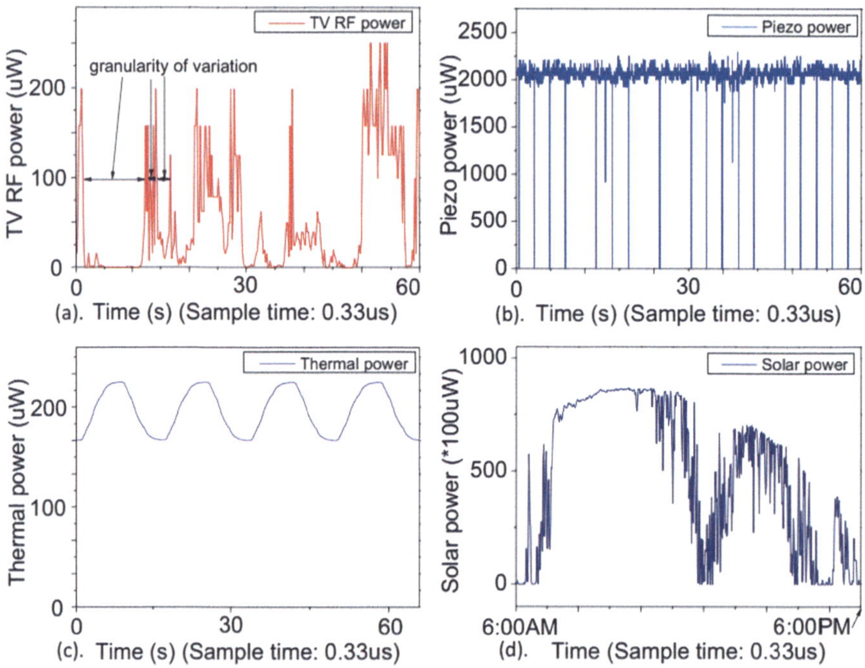

Fig. 1 Power traces [15]: (**a**) TV RF (**b**), piezo (**c**), thermal (**d**), solar

off and rebooting will impose extra burden on limited power budget. The load system would be forced to shut down if there is not enough energy available. In traditional CMOS-based processors, all the logic would be lost after shutdown and reboot, resulting in program re-execution from the very beginning. What is worse, in some cases, large tasks can never finish the execution since the intermediate results cannot be saved. To address this problem, nonvolatile processor (NVP) has been designed to enable instant on/off execution and keep accumulative progress for these devices [18, 19]. In the NVP, a nonvolatile memory (NVM) is attached to the processor. Every time there is a power outage, the processor's volatile state, will be saved into the NVM. When the next time power comes back on, the processor's state is copied back and the program execution can be resumed, as illustrated in Fig. 2. After the resumption, the program can continue the execution from the position being interrupted before power outage instead of starting over from the very beginning. Specific circuit can be designed to detect the power drop, which indicates the coming power outage, and when power runs out, a charge reserve in a small capacitor can be used to back up volatile contents to NVM [20].

Flash has been adopted as the NV memory for backup [17, 21]. A more popular choice is FRAM, which has comparable access efficiency to SRAM and the superior endurance as long as 10^{14} write cycles [22–25]. Zwerg et al. [20] presented an ultralow-power microcontroller unit which embedded FRAM as on-

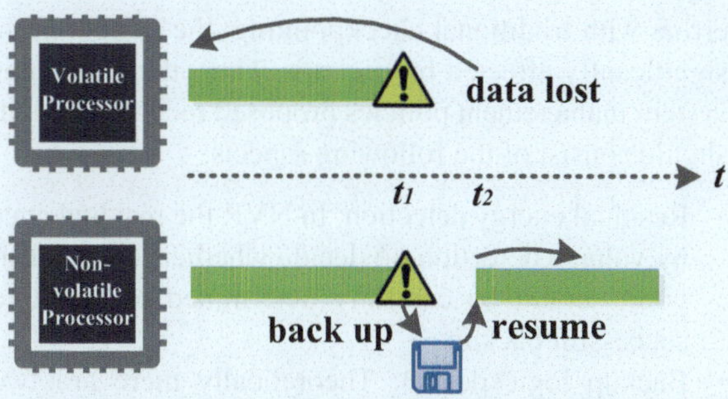

Fig. 2 Illustration of NVP behavior. Assuming a power failure at time t_1 and the recovery at t_2, the volatile processor needs to re-execute the program at t_2 due to the data loss, while the nonvolatile processor is able to continue the execution by memorizing status in NVM

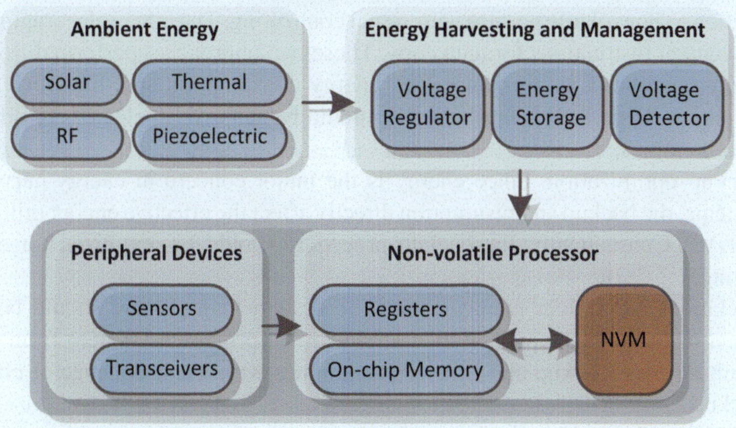

Fig. 3 A system architecture with energy harvesting system powered nonvolatile processors. This work aims to reduce the on-chip memory content to back up upon power failures

chip memory for fast write capability. When power runs out, a charge reserve in a 2nF capacitor is used to complete memory access to FRAM. Liu et al. proposed a ReRAM-based NVP with faster resumption and higher clock frequency [26]. Yu et al. [27] proposed a nonvolatile processor architecture which integrates nonvolatile elements into volatile memory at bit granularity. Wang et al. [28] developed a novel compare-and-write ferroelectric nonvolatile flip-flop which can be used in the checkpoint processor for energy harvesting applications. By copying volatile logic into nonvolatile memory, NVP is able to record the execution status and resume the execution from the exact place it was interrupted.

Figure 3 shows a general system architecture for NVP systems. Energy harvested from ambient environment is used to power the whole system. There is an energy storage, e.g., capacitor, to store a certain amount of energy. Upon a power failure, energy stored in the capacitor will be used to back up the volatile state into nonvolatile memories. Both the registers and volatile on-chip memory should be backed up. Due to the occurrence of backup, the NVP behaves quite differently from traditional volatile processors, necessitating backup-aware techniques in NVP systems. For example, the backup procedure induces potential consistency errors with traditional checkpointing; the system performance and energy cost are significantly affected by backups. Thus, there are adaptive architecture design and system management policies proposed recently. Specifically, the NVP development should consist of the following aspects:

- Residual energy detection. In NVP, the residual energy should be sensed, usually by voltage detection, to decide whether to trigger the backup or not. The trigger point should be carefully determined to guarantee sufficient energy left for successful backup;
- Backup logic design. Theoretically there are two ways to achieve the data backup/resumption. One is designing circuits for copying data from volatile portion to nonvolatile portion with signal controlling. The other is leveraging data movement instructions for data copy. These two approaches perform differently in area overhead, performance, and energy consumption and, thus, fit various scenarios. Thus, the backup schemes should be adaptively selected for different volatile logics;
- Backup optimization. Since energy is the major concern in energy harvesting systems, the backup and resumption directly affect the effective energy utilization in NVP. Consequently, the backup procedure should be optimized for energy saving;
- Backup-aware system management. The system management should be fine-tuned to fit the backup, such as mechanisms to protect the system from errors resulting from backup and software techniques with a consequence of efficient backup.

NVP-related work can be categorized according to the design levels [29], as summarized in Fig. 4. On the hardware techniques, there are existing work on NV flip-flop design [28, 30–32], processor logic exploration [33–36], NVP architecture design [18, 26, 27, 37–41], as well as NVP controller design [42]. These researches explore the fundamental design of NVP, confirming the feasibility of usage of NVP in reality. There are also researches on hardware-level optimizations for NVP such as maximum power point tracking [43] and compression-based backup [44, 45], proposing strategies to improve the energy utilization in NVP systems.

In this chapter, we summarize the software- and system-level design and optimization techniques proposed for NVP systems, covering on-chip memory management, software design and optimizations, and prototypes and tools for NVP. Specifically, there are research topics of backup-aware checkpoint locating, backup content reduction, register allocation, instruction scheduling, task scheduling, error correction, and so on. The goal of this chapter is to summarize and compare related works and give an overview of current status of software development for NVP on self-powered devices.

The remainder of this chapter is organized as follows. Section 2 presents the consistency issue in NVP and corresponding solutions. Section 3 summarizes the software-level design and optimization techniques for NVP, including checkpoint locating, optimizations for register and on-chip memory, as well as prototype and simulation tools. Section 4 concludes this paper.

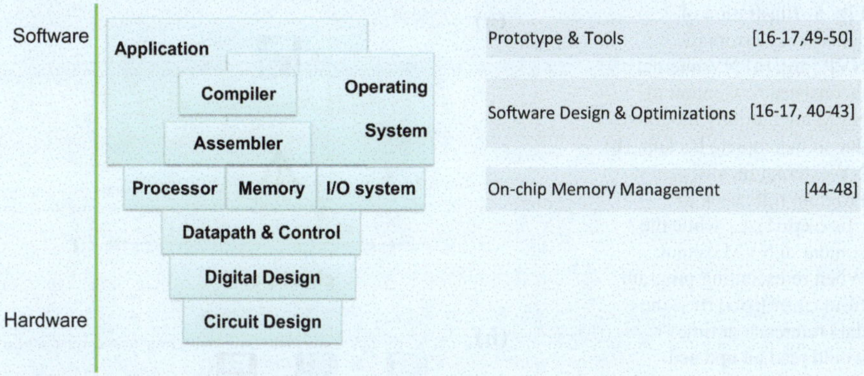

Fig. 4 Overview of the software-level design and optimizations in NVP

2 Software Techniques for System Consistency

It is important that the software running on NVP be error-free. Ransford et al. [46] summarize the consistency errors when using NVM to back up. Errors are categorized into NV-internal inconsistency and NV-external inconsistency, both could incur errors in NVP. NV-internal inconsistency happens if data are not fully updated to NVM before power depletion. System status cannot correctly resume due to the incomplete version stored in NVM. NV-external inconsistency happens when the NVM is updated after one checkpoint, and the energy is depleted before next checkpoint. In this case, after power resumes, the program will roll back to the last checkpoint, while the content in NVM cannot roll back. If the updated data in NVM is used during re-execution from last checkpoint, an error will occur due to wrong data references. Figure 5 illustrates these two kinds of errors. The existence of consistency errors greatly threatens the feasibility of NVP and, thus, should be carefully handled. In this section, solutions to eliminate these consistency errors are presented.

Xie et al. [47] discuss the consistency errors in NVP and propose a consistency-aware checkpointing solution to eliminate errors. The targeted architecture includes volatile registers and nonvolatile main memory, and the discussed errors belong to NV-external errors categorized in [46]. The proposed solution is to guarantee that there is a checkpoint between each load-store pair (such as "r" and "w" in Fig. 5). The rationality of eliminating errors is to guarantee not to use the updated data in NVM in program re-executions after rolling back. The authors then develop a set of algorithms to locate the potential errors and determine the checkpoint locations. To sum up, the principles to determine the checkpoints are as follows: first, there should be at least one checkpoint between each load-store pair; second, the maximum distance between two adjacent checkpoints should be limited within a threshold to avoid a large rollback overhead; third, since the system backs up at each

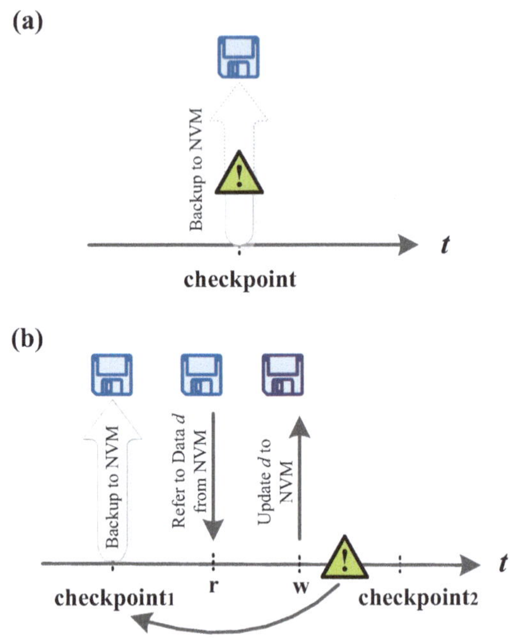

Fig. 5 Illustration of consistency errors in NVP [46]. (**a**) NV-internal inconsistency. Content in NVM are partially modified due to incomplete backup. (**b**) NV-external inconsistency. Program rolls back to `checkpoint`$_1$ while the content in NVM cannot. When re-executing program from `checkpoint`$_1$, the data reference at time "r" would read an updated version of data from NVM, inducing an error

checkpoint, the number of checkpoints should be minimized to reduce the backup cost. According to these principles, offline-based analysis is conducted to determine checkpoints for error-free NVP.

Lucia et al. [48] propose a programming and execution model called DINO to eliminate the NV-external errors. Different from [47], the authors claim that, for intermittent systems with hybrid memory architecture, checkpointing for volatile data is not sufficient. As a complementary, they propose to do data versioning for nonvolatile data. Specifically, programmers need to carefully define task boundaries to divide the whole program into atomic tasks. Then DINO nodes are inserted at the task boundaries, which will execute the checkpointing for volatile data and data versioning for nonvolatile data. The main idea is to capture all program status for both volatile and nonvolatile parts and to guarantee that they can be successfully resumed after reboots.

These techniques eliminate potential consistency errors in NVP, to guarantee the correct execution even with frequent backup and resumptions.

3 Software Design and Optimizations for Nonvolatile Processor

Once we guarantee the correctness of programs running on NVP, the software should be redesigned to be more efficient. Frequent backup/resumptions make NVP significantly different from traditional processors. Due to the criticality of energy in NVP, the cost of backup/resumption should be minimized to improve the effective usage of energy. Besides, software management schemes can be adaptively improved for NVP by incorporating effects of backup/resumptions. In this section, design and optimization techniques in software level are summarized.

3.1 Checkpoint Locating

Traditionally, checkpoints are periodically injected into programs. In systems with frequent power-offs, checkpoint locations need further consideration. In NVP, periodical checkpoints may waste energy during phases with comparatively sufficient and stable energy supply. Thus, power failure-driven backups are more suitable for NVP systems. However, trade-offs should be explored. On one hand, if the checkpoints are densely inserted, there will be much energy wasted on unnecessary checkpointing. On the other hand, if the distance between adjacent checkpoints is long but the power-off frequency is high, the overhead of rollbacks will be huge. There are several research groups focusing on the checkpoint locating.

Ransford et al. [16, 17] develop a software system called Mementos, which enables long-running computations to span power loss events. Mementos design consists of compile-time instrumentation and runtime energy-aware state checkpointing. At compile time, Mementos inserts trigger points, which are calls to a Mementos library function that estimates available energy, at control points in the program. The inserted positions include loop-latch, function return, and positions having a predetermined distance with last trigger point. At runtime, Mementos detects the capacitor voltage and triggers the checkpointing.

Sharing the similar idea of inserting checkpoints offline and flexibly enabling them on the fly, Mirhoseini et al. [21] embed checkpoints during the behavioral synthesis process. Specifically, they propose to identify the optimal checkpoints during high-level synthesis (HLS) procedure at design time and adaptively trigger some of them at runtime. Determination of optimal checkpoints is based on control data flow graph (CDFG). In the constructed graph, nodes represent basic blocks and edges represent data flows. On the basis of CDFG, a finite-state machine (FSM) is constructed by merging concurrency nodes in CDFG into the same state. Given the knowledge of program execution, a state order can be derived. Then the optimal checkpoints can be derived by examining all possible checkpoints. The backup logic is then designed and integrated into the hardware design. At runtime, the capacitor voltage is detected to decide whether to trigger a checkpoint or not. This scheme only suits the input-independent applications. The authors also observe that the structure and number of different states in design's FSM is independent of the program input, while the state order would be different. Consequently, the authors propose another strategy to locate the optimal checkpoints at design time. The two principles are, first, inserting checkpoints at the end of each feedback loop and, second, setting a limit on the distance between two consecutive checkpoints.

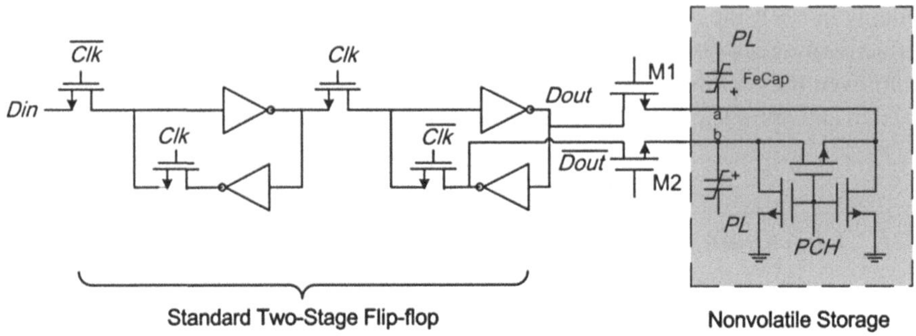

Fig. 6 The structure of nonvolatile flip-flop [28]

Mirhoseini et al. [49] proposed to include checkpoints in high-level synthesis during design time and activate checkpointing at runtime. One checkpoint is activated if it is estimated to be the end of energy supply. The estimation is based on the average consumption of instruction execution and thus may not guarantee the backup success.

3.2 *Register-Oriented Optimizations*

There are two ways to back up registers according to the size of register files. For processors with small number of registers, the registers can be all backed up upon each power failure since they are usually frequently updated and the backup procedures induce small overhead. For systems with large register files such as ultralow-power processors or graphics processing units (GPUs), registers can be selectively backed up to reduce the backup overhead while guaranteeing successful resumptions.

3.2.1 Backup for Small Register Files

For a small register file, all the registers can be simultaneously backed up. To accomplish this, the memory cell can be redesigned to consist of two portions: traditional volatile part and nonvolatile part for backup, as Fig. 6 shows. Upon power failures, the data in the standard two-stage flip-flop can be copied to the nonvolatile storage. This design is called nonvolatile flip-flop (NVFF). There are also other NVFF designs such as a magnetic flip-flop proposed in [20]. Even though NVFF achieves efficient backup, it is not suitable for large register files since it would induce large area overhead to attach nonvolatile storage into each cell.

3.2.2 Backup for Large Register Files

Instead of applying NVFF, researchers have been explored better ways for register backup in systems with large register files. Wang et al. [50] suggest a hybrid register architecture for NVPs with large register file, where the register file contains both volatile and nonvolatile registers. In this work, the authors propose to assign critical data into nonvolatile registers as many as possible to prevent critical data loss, so that the program can be resumed correctly after power on. In order to do this, critical data overflow-aware register allocation strategies are developed to minimize the possibility of critical data being spilled to volatile registers so that the failure rate of register backup can be reduced. The main idea is to map the life intervals of critical variables to free segments of nonvolatile registers so that they can have the longest overlap time.

Instead of register allocation, Xie et al. [51] propose a checkpoint-aware instruction scheduling algorithm to reduce writes to NV registers. This is motivated by the observation that the number of registers to back up at each instruction varies. Under a fixed checkpoint frequency, the authors propose to schedule instructions over multiple function units without violating the original interdependencies, so that the number of registers to back up can be reduced. In this work, the authors first analyze the minimum set of registers to back up at each checkpoint, based on which instructions are rescheduled with the objective of reducing the number to back up at checkpoints.

3.3 *On-Chip Memory Optimizations*

Backup for on-chip memory is quite different from registers due to the much larger size. In this subsection, we will discuss the backup for main memory and cache, respectively.

3.3.1 Backup for Main Memory

In a NVP system with volatile main memory, all the data in main memory should be backed up to guarantee successful resumption. Strategies are proposed to reduce the backup cost of main memory.

Zhao et al. [52] propose an optimization strategy to reduce stack size to backup upon power failures. Motivated by the observation that the size of stack to back up varies along program execution, the authors propose to flexibly reallocate the checkpoints to positions with less stack content to back up. This scheme works with the assumption that all other contents in main memory are fully backed up upon power failures. Figure 7 shows an example. Assuming an energy warning is received at time t_1, four frames should be backed up with the instant backup strategy; if the program continues the execution to t_2, there is only one frame left for

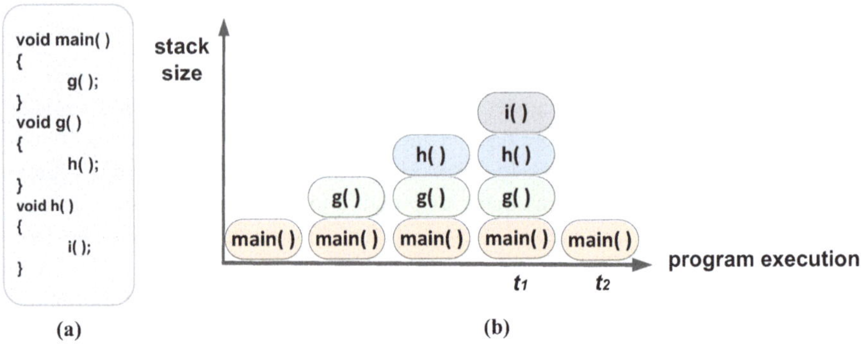

Fig. 7 Backup location can be flexibly determined considering stack size [52]

backup, indicating a more energy-efficient backup choice. So the main idea is that, when receiving power failure signals, instead of instant backup, the program has the flexibility to execute further steps to look for a better location for backup, with the objective of minimizing the stack size to back up while guaranteeing successful backup with limited available energy. The backup location is determined based on offline analysis. The challenge is to accurately model the stack size at each instruction and search for the feasible backup locations within the range of available energy.

Li et al. [53] also target optimization of stack backup while from a different angle. The authors assume fixed checkpoint locations and propose to trim the stack space by address sharing among objects and functions with disjoint live ranges. In this case, the stack content to be backed up can be effectively reduced. The stack allocation and management policies are modified to achieve this goal. A heuristic graph coloring algorithm is proposed for allocation of data and function call sites, with the objective of sharing addresses among all objects and call sites to the greatest possible extent. After trimming, the backup cost can be reduced with smaller stack size.

3.3.2 Backup for Cache

Not all the contents in cache need to be backed up since some of them also reside in main memory and thus can leverage the backup of main memory. The data must be backed up are the dirty blocks that have not been written back to main memory. There are two possible architectures to support the cache backup. One is to attach NVM at the cache level to back up dirty blocks, and the other is to write back dirty blocks to main memory before main memory's backup.

Li et al. [54] propose a backup flow consisting of a partial backup process and a runtime prewrite back scheme to reduce the cache content to write back upon power failures. The main idea of partial backup process is to predict dead blocks in cache

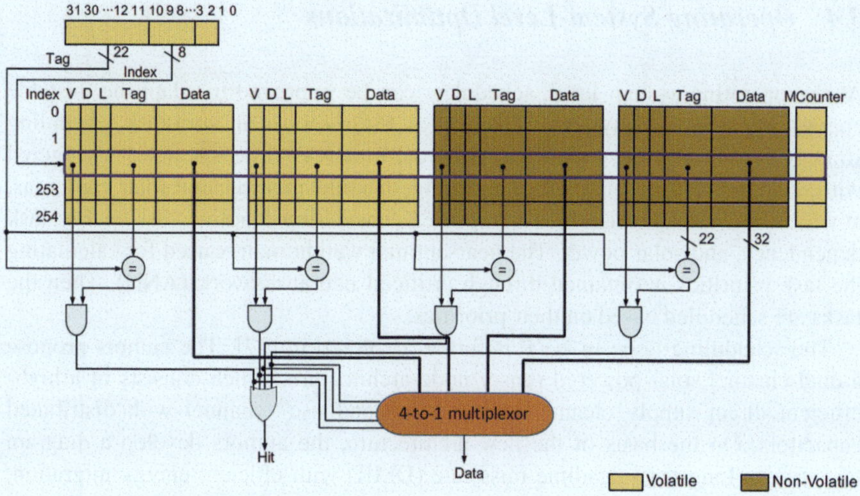

Fig. 8 Hybrid cache architecture in NVP [55]

and exclude them of writing back. The recently used bits (RUB) are exploited for classification of dead/live blocks, based on which a dead block prediction scheme is constructed. A threshold is set to limit the number of dirty blocks within cache, and some dirty blocks with large RUBs are prewritten back to the nonvolatile parts when the number of dirty blocks exceeds the threshold.

Xie et al. [55] explore the cache architecture in NVP and corresponding backup strategies. They analyze the hybrid cache, where there are both volatile and nonvolatile blocks in each set, as shown in Fig. 8. The nonvolatile blocks can be used either for caching data or backing up data upon power failures. The authors propose to reserve sufficient nonvolatile cache blocks to back up dirty ones, so that the cache content can be correctly resumed. In order to achieve this, for each set, the number of dirty blocks in volatile part is counted, and the corresponding number of nonvolatile blocks is reserved. Other nonvolatile cache blocks are normally used for caching. The block placement directly affects the performance of program execution due to the different access costs of volatile and nonvolatile material. Besides, only dirty blocks in volatile part need to be backed up, so the placement also has impact on the backup cost. On the basis of these two considerations, block placement and migration policies between volatile and nonvolatile portions are proposed. Proactive write back policy is also designed to avoid too many dirty volatile blocks being backed up upon power failures. This work provides a guideline for cache management in NVP, with the objective of successful and efficient cache checkpointing.

3.4 Operating System-Level Optimizations

At the operating system level, schedulers can be improved to adapt the unstable energy supply in NVP systems. Zhang et al. [56] propose an intra-task scheduling strategy to minimize the deadline miss in real-time NVP. The scheduler is triggered with scenario changes such as task finishing, deadline missing, and solar variations, at which time the task priorities are updated considering deadline, task energy, task dependency, and solar power. The near-optimal weight matrix used for calculating the task priorities is obtained through artificial neural network (ANN). Then the tasks are scheduled based on their priorities.

The scheduling issue in NVP is further explored in [57]. The authors propose a dual-channel solar-powered sensor node architecture, which consists of a high-efficient direct supply channel and a "store and use" channel with distributed capacitors. On the basis of the new architecture, the authors develop a diagram to optimize long-term deadline miss rate (DMR) with efficient energy migration, where energy can be migrated among distributed capacitors. The proposed diagram contains offline and online parts. The former determines the optimal capacitor sizes and DMR training samples for artificial ANN training. The latter adopts the ANN to determine the real-time optimal capacitor size, scheduling pattern, and task queue, followed by an algorithm for better DMR.

3.5 Prototype and Tools

NVP prototypes have been developed by different research groups. Mementos [16] is constructed for computational RFIDs, integrating the checkpointing schemes for the maximum forward progress. Jayakumar et al. [58] propose a lightweight, in situ checkpointing technique called QUICKRECALL where the Ferroelectric RAM (FRAM) is used for status backup. Both systems can protect the system from frequent power losses by state checkpointing and are implemented and verified in the Texas Instruments, MSP430 family of microcontrollers.

Heidari et al. [59] propose a multisource energy harvesting system to combine multiple harvesting sources to provide a more stable power supply. Taking indoor photovoltaics (PV), piezoelectric (PZ), and thermoelectric generator (TEG) as examples, the authors discussed issues including maximum power extraction and converter parameter optimization in NVP systems.

Simulation tools are able to provide efficient way for NVP verification and evaluations, in absence of real NVP systems. Gu et al. [60] develop a simulator for nonvolatile processors named NVPsim based on gem5 [61]. NVPsim involves modeling on voltage detector, backup/restore controller, and NVP state machine and is able to report, for various NVP architectures, the breakdown of energy consumption of hardware modules as well as statistic file for performance and energy analysis. NVPsim enables emulations of NVP behaviors as a verification platform to assess efficacy of newly proposed strategies.

3.6 Discussions

It can be observed that techniques across various design levels have been explored, where cross-layer strategies can be applied in combination for error-free, high-performance, and energy-efficient NVP. It needs cross-layer schemes since various levels may affect each other, and optimizations should be done by simultaneously considering combinational behaviors. For example, cache backup in NVP is closely related to main memory backup. Writing back dirty blocks from cache can release the backup burden of cache while may affect the backup procedure of main memory. Thus, optimizations of NVP should globally consider all components to achieve the best system design. Besides, the hardware-software co-design should be further explored for efficient backup. For example, NVFF designed for register backup in hardware is performance and energy efficient but with comparatively large area overhead; software-directed backup is slow while with no extra circuits. The trade-off should be investigated for NVP system.

Operating system-level management can be potentially further studied to develop backup-aware schedulers, memory management and optimizations, file systems, and so on, to integrate more NVP-adaptive strategies.

4 Conclusion

Due to the backup and resumption procedures, the NVP system has potential consistency errors, and the backup/resumption significantly affects the correctness, performance, and energy efficiency of NVP systems. Recently, there are researches proposing solutions to pursue correct and efficient NVP design from software and system's perspective. This paper provides an overview of the software technique for NVP design and optimizations in self-powered devices, including consistency error categorization, error correction, checkpoint locating, backup content reducing, adaptive compiler design, scheduler design, NVP prototype, and simulation tool development. This chapter gives an overview of the current status of software development in NVP and also a guideline of future work in NVP systems.

References

1. Sudevalayam, S., Kulkarni, P.: Energy harvesting sensor nodes: survey and implications. IEEE Commun. Surv. Tutorials **1**(3), 443–461 (2011)
2. Raghunathan, V., Kansal, A., Hsu, J., Friedman, J., Srivastava, M.: Design considerations for solar energy harvesting wireless embedded systems. In: International Symposium on Information Processing in Sensor Networks (IPSN). IEEE Press, Piscataway (2005)
3. Taneja, J., Jeong, J., Culler, D.: Design, modeling, and capacity planning for micro-solar power sensor networks. In: International Conference on Information Processing in Sensor Networks (IPSN), pp. 407–418 (2008)
4. Zhang, D., Liu, Y., Li, J., Xue, C.J., Li, X., Wang, Y., Yang, H.: Solar power prediction assisted intra-task scheduling for nonvolatile sensor nodes. IEEE Trans. Comput. Aided Des. Integr. Circuits Syst. (TCAD) **1**(5), 724–737 (2016)
5. Weimer, M.A., Paing, T.S., Zane, R.A.: Remote area wind energy harvesting for low-power autonomous sensors. In: IEEE Power Electronics Specialists Conference (PESC), pp. 1–5 (2006)
6. Kulah, H., Najafi, K.: Energy scavenging from low-frequency vibrations by using frequency up-conversion for wireless sensor applications. IEEE Sens. J. **1**(3), 261–268 (2008)
7. Naderiparizi, S., Parks, A.N., Kapetanovic, Z., Ransford, B., Smith, J.R.: WISPCam: a battery-free RFID camera. In: 2015 IEEE International Conference on RFID (RFID), pp. 166–173 (2015)
8. Talla, V., Kellogg, B, Ransford, B., Naderiparizi, S., Gollakota, S., Smith, J.R.: Powering the Next Billion Devices with Wi-Fi (2015). ArXiv e-prints
9. Sample, A.P., Yeager, D.J., Powledge, P.S., Mamishev, A.V., Smith, J.R.: Design of an RFID-based battery-free programmable sensing platform. IEEE Trans. Instrum. Meas. **1**(11), 2608–2615 (2008)
10. Shenck, N.S., Paradiso, J.A.: Energy scavenging with shoe-mounted piezoelectrics. IEEE Micro **1**(3), 30–42 (2001)
11. Kymissis, J., Kendall, C., Paradiso, J., Gershenfeld, N.: Parasitic power harvesting in shoes. In: Second International Symposium on Wearable Computers, Digest of Papers, pp. 132–139 (1998)
12. Park, C., Chou, P.H.: Ambimax: autonomous energy harvesting platform for multi-supply wireless sensor nodes. In: Annual IEEE Communications Society on Sensor and Ad Hoc Communications and Networks, pp. 168–177 (2006)
13. Mirhoseini, A., Koushanfar, F.: Learning to manage combined energy supply systems. In: IEEE/ACM International Symposium on Low-power Electronics and Design (ISLPED), pp. 229–234 (2011)
14. Kansal, A., Hsu, J., Zahedi, S., Srivastava, M.B.: Power management in energy harvesting sensor networks. ACM Trans. Embed. Comput. Syst. **6**(4) (2007)
15. Ma, K., Zheng, Y., Li, S., Swaminathan, K., Li, X., Liu, Y., Sampson, J., Xie, Y., Narayanan, V.: Architecture exploration for ambient energy harvesting nonvolatile processors. In: International Symposium on High Performance Computer Architecture (HPCA), pp. 526–537 (2015)
16. Ransford, B., Sorber, J., Fu, K.: Mementos: system support for long-running computation on RFID-scale devices. In: International Conference on Architectural Support for Programming Languages and Operating Systems (ASPLOS), pp. 159–170 (2011)
17. Ransford, B., Clark, S.S., Salajegheh, M., Fu, K.: Getting things done on computational RFIDs with energy-aware checkpointing and voltage-aware scheduling. In: HotPower (2008)
18. Wang, Y., Liu, Y., Li, S., Zhang, D., Zhao, B., Chiang, M.-F., Yan, Y., Sai, B., Yang, H.: A 3us wake-up time nonvolatile processor based on ferroelectric flip-flops. In: European Solid-State Circuits Conference (ESSCIRC), pp. 149–152 (2012)
19. Sheng, X., Wang, Y., Liu, Y., Yang, H.: SPaC: a segment-based parallel compression for backup acceleration in nonvolatile processors. In: Design, Automation & Test in Europe Conference & Exhibition (DATE), pp. 865–868 (2013)

20. Zwerg, M., Baumann, A., Kuhn, R., Arnold, M., Nerlich, R., Herzog, M., Ledwa, R., Sichert, C., Rzehak, V., Thanigai, P., Eversmann, B.: An 82 uA/MHz microcontroller with embedded FeRAM for energy-harvesting applications. In: International Solid-State Circuits Conference (ISSCC), pp. 334–336 (2011)
21. Mirhoseini, A., Songhori, E.M., Koushanfar, F.: Idetic: a high-level synthesis approach for enabling long computations on transiently-powered ASICs. In: Pervasive Computing and Communication Conference (PerCom), pp. 19–31 (2013)
22. Ducharme, S., Reece, T.J., Othon, C., Rannow, R.K.: Ferroelectric polymer Langmuir-Blodgett films for nonvolatile memory applications. IEEE Trans. Device Mater. Reliab. **1**(4), 720–735 (2005)
23. Horii, Y., Hikosaka, Y., Itoh, A., Matsuura, K., Kurasawa, M., Komuro, G., Maruyama, K., Eshita, T., Kashiwagi, S.: 4 Mbit embedded FRAM for high performance system on chip (SoC) with large switching charge, reliable retention and high imprint resistance. In: International Electron Devices Meeting, pp. 539–542 (2002)
24. Nakamoto, H., Yamazaki, D., Yamamoto, T., Kurata, H., Yamada, S., Mukaida, K., Ninomiya, T., Ohkawa, T., Masui, S., Gotoh, K.: A passive UHF RF identification CMOS Tag IC using ferroelectric RAM in 0.35-um technology. IEEE J. Solid State Circuits **1**(1), 101–110 (2007)
25. Shiga, H., Takashima, D., Shiratake, S., Hoya, K., Miyakawa, T., Ogiwara, R., Fukuda, R., Takizawa, R., Hatsuda, K., Matsuoka, F., Nagadomi, Y., Hashimoto, D., Nishimura, H., Hioka, T., Doumae, S., Shimizu, S., Kawano, M., Taguchi, T., Watanabe, Y., Fujii, S., Ozaki, T., Kanaya, H., Kumura, Y., Shimojo, Y., Yamada, Y., Minami, Y., Shuto, S., Yamakawa, K., Yamazaki, S., Kunishima, I., Hamamoto, T., Nitayama, A., Furuyama, T.: A 1.6 GB/s DDR2 128 Mb chain FeRAM with scalable octal bitline and sensing schemes. IEEE J. Solid State Circuits **1**(1), 142–152 (2010)
26. Liu, Y., Wang, Z., Lee, A., Su, F., Lo, C.P., Yuan, Z., Lin, C.C., Wei, Q., Wang, Y., King, Y.C., Lin, C.J., Khalili, P., Wang, K.L., Chang, M.F., Yang, H.: 4.7 a 65nm ReRAM-enabled nonvolatile processor with 6× reduction in restore time and 4× higher clock frequency using adaptive data retention and self-write-termination nonvolatile logic. In: 2016 IEEE International Solid-State Circuits Conference (ISSCC), pp. 84–86 (2016)
27. Yu, W.k., Rajwade, S., Wang, S.E., Lian, B., Suh, G.E., Kan, E.: A non-volatile microcontroller with integrated floating-gate transistors. In: International Conference on Dependable Systems and Networks Workshops (DSN-W), pp. 75–80 (2011)
28. Wang, J., Liu, Y., Yang, H., Wang, H.: A compare-and-write ferroelectric nonvolatile flip-flop for energy-harvesting applications. In: International Conference on Green Circuits and Systems (ICGCS), pp. 646–650 (2010)
29. Liu, Y., Li, Z., Li, H., Wang, Y., Li, X., Ma, K., Li, S., Chang, M.-F., John, S., Xie, Y., Shu, J., Yang, H.: Ambient energy harvesting nonvolatile processors: from circuit to system. In: Design Automation Conference (DAC), pp. 150:1–150:6 (2015)
30. Zhao, W., Belhaire, E., Javerliac, V., Chappert, C., Dieny, B.: A non-volatile flip-flop in magnetic FPGA chip. In: International Conference on Design and Test of Integrated Systems in Nanoscale Technology (DTIS), pp. 323–326 (2006)
31. Zhao, W., Moreau, M., Deng, E., Zhang, Y., Portal, J.M., Klein, J.O., Bocquet, M., Aziza, H., Deleruyelle, D., Muller, C., Querlioz, D., Romdhane, N.B., Ravelosona, D., Chappert, C.: Synchronous non-volatile logic gate design based on resistive switching memories. IEEE Trans. Circuits Syst. Regul. Pap. **1**(2), 443–454 (2014)
32. Sakimura, N., Sugibayashi, T., Nebashi, R., Kasai, N.: Nonvolatile magnetic flip-flop for standby-power-free SoCs. IEEE J. Solid State Circuits **1**(8), 2244–2250 (2009)
33. Kim, M.S., Liu, H., Swaminathan, K., Li, X., Datta, S., Narayanan, V.: Enabling power-efficient designs with III–V tunnel FETS. In: IEEE Compound Semiconductor Integrated Circuit Symposium (CSICs), vol. 10 (2014)
34. Swaminathan, K., Liu, H., Li, X., Kim, M.S., Sampson, J., Narayanan, V.: Steep slope devices: enabling new architectural paradigms. In: Proceedings of the 51st Annual Design Automation Conference (DAC), pp. 1–6. ACM (2014)

35. Liu, H., Li, X., Vaddi, R., Ma, K., Datta, S., Narayanan, V.: Tunnel FET RF rectifier design for energy harvesting applications. IEEE J. Emerging Sel. Top. Circuits Syst. **1**(4), 400–411 (2014)
36. Heo, U., Li, X., Liu, H., Gupta, S., Datta, S., Narayanan, V.: A high-efficiency switched-capacitance HTFET charge pump for low-input-voltage applications. In: International Conference on VLSI Design, pp. 304–309. IEEE (2015)
37. George, S., Ma, K., Aziz, A., Li, X., Khan, A., Salahuddin, S., Chang, M.-F., Datta, S., Sampson, J., Gupta, S., Narayanan, V.: Nonvolatile memory design based on ferroelectric FETs. In: Proceedings of the 53rd Annual Design Automation Conference (DAC), pp. 118:1–118:6 (2016)
38. Ma, K., Li, X., Li, S., Liu, Y., Sampson, J.J., Xie, Y., Narayanan, V.: Nonvolatile processor architecture exploration for energy-harvesting applications. IEEE Micro **1**(5), 32–40 (2015)
39. Ma, K., Li, X., Liu, Y., Sampson, J., Xie, Y., Narayanan, V.: Dynamic machine learning based matching of nonvolatile processor microarchitecture to harvested energy profile. In: Proceedings of the IEEE/ACM International Conference on Computer-Aided Design (ICCAD), pp. 670–675 (2015)
40. Bartling, S.C., Khanna, S., Clinton, M.P., Summerfelt, S.R., Rodriguez, J.A., McAdams, H.P.: An 8MHz 75 μA/MHz zero-leakage non-volatile logic-based cortex-m0 MCU SoC exhibiting 100% digital state retention at V_{DD}=0v with <400ns wakeup and sleep transitions. In: 2013 IEEE International Solid-State Circuits Conference Digest of Technical Papers, pp. 432–433 (2013)
41. Sakimura, N., Tsuji, Y., Nebashi, R., Honjo, H., Morioka, A., Ishihara, K., Kinoshita, K., Fukami, S., Miura, S., Kasai, N., Endoh, T., Ohno, H., Hanyu, T., Sugibayashi, T.: 10.5 A 90nm 20MHz fully nonvolatile microcontroller for standby-power-critical applications. In: International Solid-State Circuits Conference Digest of Technical Papers (ISSCC), pp. 184–185 (2014)
42. Liu, Y., Suy, F., Wangy, Z., Yang, H.: Design exploration of inrush current aware controller for nonvolatile processor. In: IEEE Non-Volatile Memory System and Applications Symposium (NVMSA), pp. 1–6 (2015)
43. Wang, C., Chang, N., Kim, Y., Park, S., Liu, Y., Lee, H.: Storage-less and converter-less maximum power point tracking of photovoltaic cells for a nonvolatile microprocessor. In: Asia and South Pacific Design Automation Conference (ASP-DAC), pp. 379–384 (2014)
44. Wang, Y., Liu, Y., Liu, Y., Zhang, D., Li, S., Sai, B., Chiang, M.-F., Yang, H.: A compression-based area-efficient recovery architecture for nonvolatile processors. In: Proceedings of the Conference on Design, Automation and Test in Europe (DATE), pp. 1519–1524 (2012)
45. Wang, Y., Liu, Y., Li, S., Sheng, X., Zhang, D., Chiang, M.-F., Sai, B., Hu, X., Yang, H.: PaCC: a parallel compare and compress codec for area reduction in nonvolatile processors. IEEE Trans. Very Large Scale Integr. VLSI Syst. PP(99), 1491–1505 (2013)
46. Ransford, B., Lucia, B.: Nonvolatile memory is a broken time machine. In: Proceedings of the workshop on Memory Systems Performance and Correctness (MSPC), pp. 1–3 (2014)
47. Xie, M., Zhao, M., Pan, C., Hu, J., Liu, Y., Xue, C.J.: Fixing the broken time machine: consistency-aware checkpointing for energy harvesting powered non-volatile processor. In: Design Automation Conference (DAC), pp. 184:1–184:6 (2015)
48. Lucia, B., Ransford, B.: A simpler, safer programming and execution model for intermittent systems. In: ACM SIGPLAN Conference on Programming Language Design and Implementation (PLDI), pp. 575–585 (2015)
49. Scott, J., Lee, L.H., Arends, J., Moyer, B.: Designing the low-power m*core architecture. In: IEEE Power Driven Microarchitecture Workshop, pp. 145–150 (1998)
50. Wang, Y., Jia, H., Liu, Y., Li, Q., Xue, C.J., Yang, H.: Register allocation for hybrid register architecture in nonvolatile processors. In: IEEE International Symposium on Circuits and Systems (ISCAS), pp. 1050–1053 (2014)
51. Xie, M., Pan, C., Hu, J., Yang, C., Chen, Y.: Checkpoint-aware instruction scheduling for nonvolatile processor with multiple functional units. In: Asia and South Pacific Design Automation Conference (ASPDAC), pp. 316–321 (2015)

52. Zhao, M., Li, Q., Xie, M., Liu, Y., Hu, J., Xue, C.J.: Software assisted non-volatile register reduction for energy harvesting based cyber-physical system. In: Design, Automation & Test in Europe Conference & Exhibition (DATE), pp. 567–572 (2015)
53. Li, Q., Zhao, M., Hu, J., Liu, Y., He, Y., Xue, C.J.: Compiler directed automatic stack trimming for efficient non-volatile processors. In: Design Automation Conference (DAC), pp. 183:1–183:6 (2015)
54. Li, H., Liu, Y., Zhao, Q., Gu, Y., Sheng, X., Sun, G., Zhang, C., Chang, M.-F., Luo, R., Yang, H.: An energy efficient backup scheme with low inrush current for nonvolatile SRAM in energy harvesting sensor nodes. In: Proceedings of the 2015 Design, Automation & Test in Europe Conference & Exhibition (DATE), pp. 7–12 (2015)
55. Xie, M., Zhao, M., Li, H., Pan, C., Zhang, Y., Liu, Y., Xue, C.J., Hu, J.: Checkpoint aware hybrid cache architecture for NV processor in energy harvesting powered systems. In: International Conference on Hardware/Software Codesign and System Synthesis (CODES+ISSS) (2016, to appear)
56. Zhang, D., Li, S., Li, A., Liu, Y., Hu, X.S., Yang, H.: Intra-task scheduling for storage-less and converter-less solar-powered nonvolatile sensor nodes. In: International Conference on Computer Design (ICCD), pp. 348–354 (2014)
57. Zhang, D., Liu, Y., Sheng, X., Li, J., Wu, T., Xue, C.J., Yang, H.: Deadline-aware task scheduling for solar-powered nonvolatile sensor nodes with global energy migration. In: Design Automation Conference (DAC), pp. 1–6 (2015)
58. Jayakumar, H., Raha, A., Raghunathan, V.: QUICKRECALL: a low overhead HW/SW approach for enabling computations across power cycles in transiently powered computers. In: International Conference on VLSI Design, pp. 330–335 (2014)
59. Heidari, S., Ding, C., Liu, Y., Wang, Y., Hu, J.: Multi-source energy harvesting management and optimization for non-volatile processors. In: Sixth International Green Computing Conference and Sustainable Computing Conference (IGSC), pp. 1–2 (2015)
60. Gu, Y., Liu, Y., Wang, Y., Li, H., Yang, H.: NVPsim: a simulator for architecture explorations of nonvolatile processors. In: Asia and South Pacific Design Automation Conference (ASP-DAC), pp. 147–152 (2016)
61. Binkert, N., Beckmann, B., Black, G., Reinhardt, S.K., Saidi, A., Basu, A., Hestness, J., Hower, D.R., Krishna, T., Sardashti, S., Sen, R., Sewell, K., Shoaib, M., Vaish, N., Hill, M.D., Wood, D.A.: The gem5 simulator. SIGARCH Comput. Archit. News **1**(2), 1–7 (2011)

Part II
Sensing Technology for IoT

OEICs for High-Speed Data Links and Tympanic Membrane Transducer of Hearing Aid Device

Wei-Zen Chen, Shih-Hao Huang, and Jhong-Ting Jian

1 OEICs for Intensive Data Link

Nowadays, the applications of cloud service become pervasive in daily life. High-density and energy-efficient data links bridging computing and storage devices are demanding to accommodate the explosive bandwidth of data center. In contrast to electrical cables, optical interconnects are lighter weighted with a wider channel bandwidth and lower electromagnetic interference (EMI).

A typical optical receiver front end employs a shunt-shunt feedback TIA and a post-limiting amplifier to convert the photocurrent received by the photodetector (PD) into voltage form for post-signal processing. To remedy PD's parasitic capacitance for a high-speed operation, the TIA requires a high-gain and wide-bandwidth core voltage amplifier. It may encounter severe design trade-offs among the input sensitivity, power dissipation, and operating speed. In this chapter, a nested-feedback broadband TIA is proposed without using shunt-peaking inductors. The gain-bandwidth product is improved by more than 3× compared to the prior art.

Conventionally, an optical receiver is composed of multi-chips implemented in different technologies [1–4]. For example, the TIA and post-limiting amplifier are fabricated using a Bipolar/CMOS or BiCMOS process, while the PD is implemented in a more expensive InGaAs or GaAs technology, as is shown in Fig. 1a. To realize a cost-effective and miniaturized receiver front end, monolithically integrated optoelectronic integrated circuits (OEICs) are preferable [5–14]. As shown in Fig. 1b, the OEIC integrates an on-chip PD, a TIA, and a post-limiting amplifier on

W.-Z. Chen (✉) • J.-T. Jian
National Chiao-Tung University, Hsinchu City, Taiwan
e-mail: wzchen@mail.nctu.edu.tw; jhongting.jian@gmail.com

S.-H. Huang
MediaTek Inc., Hsinchu City, Taiwan
e-mail: shih-hao.huang@mediatek.com

H. Yasuura et al. (eds.), *Smart Sensors at the IoT Frontier*,
DOI 10.1007/978-3-319-55345-0_6

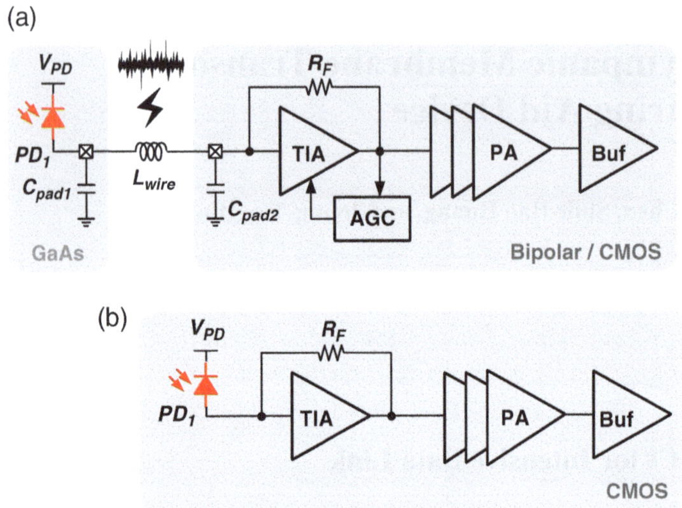

Fig. 1 (**a**) Hybrid and (**b**) monolithic optical receiver front end

a single chip. By eliminating the bonding wire inductor (L_{wire}), the signal integrity can be improved by reducing cross talk for multichannel integration. Besides, the signal bandwidth can be benefited by eliminating the parasitic capacitances (C_{pad1} and C_{pad2}) associated with the bonding pads and ESD devices at the input node of the TIA. To meet the speed requirement, spatially modulated photodetectors (SMPD) in CMOS technologies are proposed. They are able to operate at tens of Gb/s range to enable single chip OEICs for intensive data links in a computing platform.

1.1 CMOS Photodetectors

Light detection in a CMOS technology is performed by a reverse-biased P/N junction diode, which is mainly based on the substrate to well or diffusion to well junctions. Figure 2 shows the cross-sectional view of a photodetector comprised of N_{well} to $P_{substrate}$ junction. When the PD is reverse biased and illuminated, the photo-generated carriers in the depletion region quickly drift in opposite directions because of the presence of the electric field. Additionally, some of the photo-generated carriers in the neutral region diffuse randomly to the depletion region and contribute to the photocurrent. But most of them are recombined before reaching the depletion-neutral region boundary.

Given the incident photon flux density as

$$\Phi_O(t) = \frac{P_{in}(t)}{A}\frac{\lambda}{hc}(1-r) \quad (1)$$

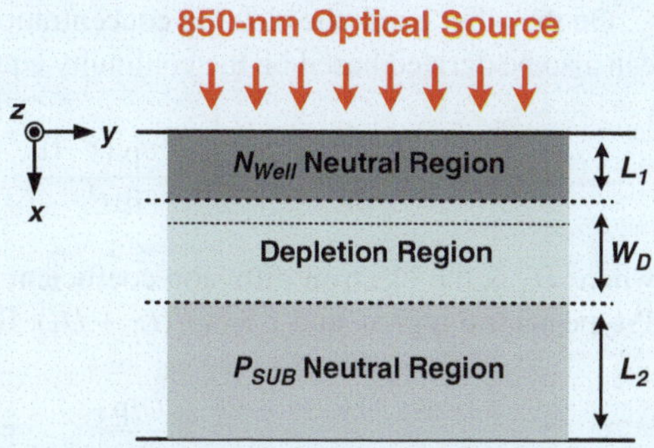

Fig. 2 Cross-sectional view of the P/N junction PD

where P_{in} is the input optical power, A is the active area of PD, λ is the optical wavelength, h is the Plank's constant, c is the speed of light, and r is the reflectivity of the photodetector. The generation rates of carriers per unit volume can be represented as

$$g(x,t) = \Phi_{\mathrm{O}}(t)\alpha e^{-\alpha x} \tag{2}$$

In the N-type neutral region, the hole concentration $p_n(x,t)$ can be derived based on the continuity equation:

$$\frac{\partial p_{\mathrm{n}}(x,t)}{\partial t} = D_{\mathrm{p}}\frac{\partial^2 p_{\mathrm{n}}(x,t)}{\partial x^2} - \frac{p_{\mathrm{n}}(x,t)}{\tau_{\mathrm{p}}} + g(x,t) \tag{3}$$

where D_{p} is the hole diffusion coefficient and τ_{p} is the hole diffusion time in the N-type neutral region. With the boundary conditions

$$\left.\frac{\partial p_{\mathrm{n}}}{\partial x}\right|_{x=0} = 0 \tag{4}$$

$$p_n|_{x=L1} = 0 \tag{5}$$

The hole diffusion current can be derived as [15]

$$I_{\mathrm{diff.N}}(s) = \sum_{m=1}^{\infty} \mathrm{Aq}D_p\alpha \frac{(2m-1)\pi(-1)^{m+1}e^{-\alpha L_1} + 2\alpha L_1}{(\alpha L_1)^2 + \left[\frac{(2m-1)\pi}{2}\right]^2}\frac{(2m-1)\pi}{2L_1}(-1)^{m+1} \times \frac{\Phi_{\mathrm{O}}(s)}{s + D_{\mathrm{P}}\left[\frac{(2m-1)\pi}{2L_1}\right]^2 + \frac{1}{\tau_{\mathrm{P}}}} \tag{6}$$

On the other hand, the electron concentration $n_p(x,t)$ in the P-type neutral region can also be derived based on the continuity equation:

$$\frac{\partial n_p\left(x',t\right)}{\partial t}=D_{\mathrm{n}}\frac{\partial^2 n_{\mathrm{p}}\left(x',t\right)}{\partial x'^2}-\frac{n_{\mathrm{p}}\left(x',t\right)}{\tau_{\mathrm{n}}}+g\left(x',t\right) \tag{7}$$

where D_n is the electron diffusion coefficient, τ_n is the electron diffusion time in P-type neutral region, and $x' = x-(L_1+D_1)$. By applying the boundary conditions

$$\left.\frac{\partial n_{\mathrm{p}}}{\partial x'}\right|_{x'=0}=0 \tag{8}$$

$$n_{\mathrm{p}}\big|_{x'=L2}=0 \tag{9}$$

The electron diffusion current in the P-type neutral region can be derived as

$$I_{\mathrm{diff.P}}(s)=\sum_{m=1}^{\infty}\mathrm{Aq}D_n\alpha e^{-\alpha(L_1+D_1)}\frac{2m\pi\left[1-(-1)^m e^{-\alpha L_2}\right]}{(\alpha L_2)^2+(m\pi)^2}\frac{m\pi}{2L_2}\times\frac{\Phi_o(s)}{s+D_{\mathrm{n}}\left(\frac{m\pi}{2L_1}\right)^2+\frac{1}{\tau_{\mathrm{n}}}} \tag{10}$$

In the depletion region, the fast drift current bandwidth is inversely proportional to its depletion region width (W_{D}). Given the electron saturation velocity as v_{s}, the 3-dB bandwidth of drift current can be expressed as [15]

$$f_{\mathrm{dr}}\approx\frac{0.4v_{\mathrm{S}}}{W_{\mathrm{D}}} \tag{11}$$

For a CMOS P/N junction PD, the depletion width is usually less than 1 μm, and the f_{dr} is more than 25 GHz. The drift current can be expressed as

$$I_{\mathrm{dr}}(s)=\mathrm{Aq}\Phi_{\mathrm{o}}(s)\left[e^{-\alpha L_1}-e^{-\alpha(L_1+D_1)}\right] \tag{12}$$

Combining Eqs. (9), (10), and (11), the PD current of a reverse-biased P/N junction can be expressed as

$$I_{\mathrm{PD}}=I_{\mathrm{diff.N}}(s)+I_{\mathrm{diff.P}}(s)+I_{\mathrm{dr}}(s) \tag{13}$$

The parasitic capacitance associated with the photodetector is

$$C_{\mathrm{PD}}=A\sqrt{\frac{q\varepsilon_{\mathrm{S}}N_{\mathrm{N}}N_{\mathrm{P}}}{2\left(V_{\mathrm{bi}}+V_{\mathrm{R}}\right)\left(N_{\mathrm{N}}+N_{\mathrm{P}}\right)}} \tag{14}$$

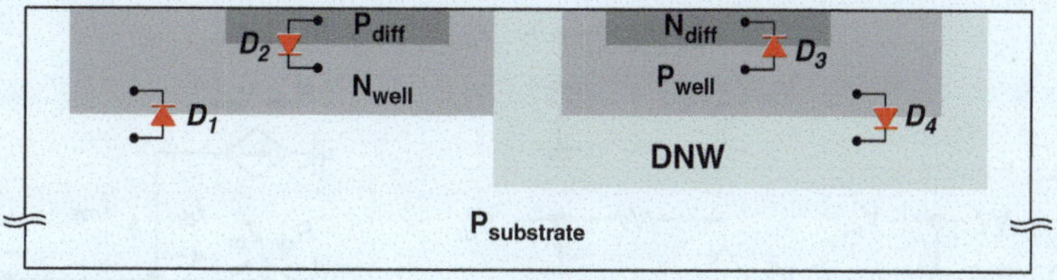

Fig. 3 P/N junction diodes in the generic CMOS technology

where N_N and N_P, respectively, denote the doping concentration of N-type and P-type region. According to Eq. (14), a PD with a lower doping concentration or a higher reverse-biased voltage (V_R) has a smaller PD capacitance, which is critical in designing a low noise receiver front end.

In a triple-well CMOS process as shown in Fig. 3, depletion regions for photo sensing can be composed of $P_{substrate}/N_{well}$, P_{diff}/N_{well}, P_{well}/N_{diff}, or P_{well}/deep N_{well} (DNW) junctions, which are represented as D_1, D_2, D_3, and D_4, respectively. When D_1 is illuminated, the depletion region generates fast drift current, whereas the neutral P and N regions generate slowly diffusive current. For a short distance optical link, an 850 nm VCSEL is popular as a light source at the transmitter side. As the penetration depth of the 850-nm light into silicon is about 18 μm, which is much deeper than the shallow well (<3 μm) or diffusion region. Thus there is only 15% of photons absorbed in the well region. Meanwhile, a large portion of photocurrent is generated from $P_{substrate}$ and leads to slowly diffusive currents, which limit the bandwidth of photodetectors to tens of MHz frequency range. By eliminating the slowly diffusive current generated in the $P_{substrate}$ with N_{well} or DNW, such as D_2, D_3, and D_4, a PD with a much wider bandwidth can be achieved.

Based on Eqs. (13) and (14), the equivalent circuit model of a photodetector composed of $P_{substrate}/N_{well}$ junction is depicted as shown in Fig. 4. Here the PD current is decomposed of $P_{substrate}$ diffusion current, N_{well} diffusion current, and fast drift current. Their bandwidths are about 5 MHz, 1 GHz, and 25 GHz, respectively. The corresponding responsivity of these three components can be modeled as R_{psub}, R_{nw}, and R_{dr}, which are about 319 (mA/W), 30 (mA/W), and 30 (mA/W). Figure 5 shows the $P_{substrate}/N_{well}$ junction PD frequency response, which is mainly dominated by the slowly diffusive current generated in the $P_{substrate}$.

The bandwidth of a single chip OEIC can be enhanced by getting rid of the slowly diffusive carriers in the photodetector [5–14]. From technology aspect, this approach can be achieved by using a SOI [6] or BiCMOS [7] process, but they are more costly compared to standard CMOS technologies [8] proposes a CMOS photodetector comprising of diffusion/well junctions and employing deep N_{well} for $P_{substrate}$ diffusive carrier isolation. However, the photodetector exhibits an 8× larger parasitic capacitance that impedes high-speed operation. To reduce the parasitic capacitance associated with photodetectors [9], employs a $P_{diffusion}/N_{well}$ PD with a smaller active region (16.54 μm by 16.54 μm), which is only 1/9 the cross-sectional area of a multimode fiber. But it also suffers from the degradation of receiver input

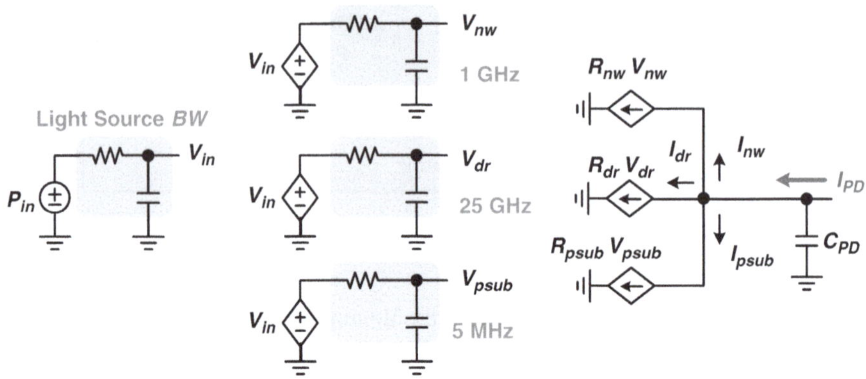

Fig. 4 Circuit schematic model of $P_{substrate}/N_{well}$ junction PD

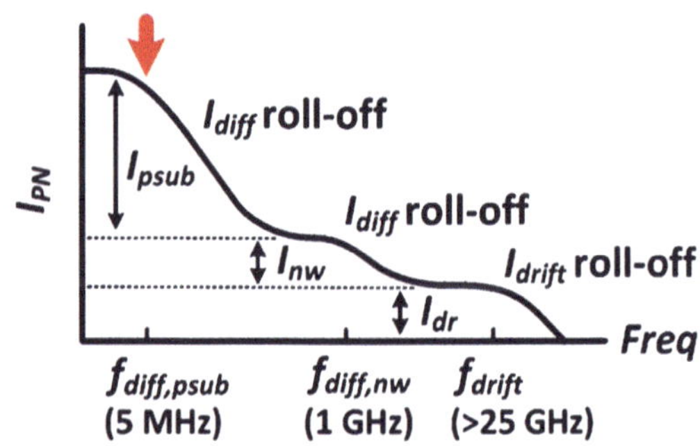

Fig. 5 Frequency response of $P_{substrate}/N_{well}$ junction PD

responsivity. On the other hand, the PD bandwidth can also be compensated with an equalizer from circuit design's perspective [7, 12–14]. However, due to the slowly roll-off characteristic of the CMOS PD's frequency response [14], a high-order and sophisticated equalizer is required under PVT variations.

Table 1 summarizes the performance benchmark of different PDs in CMOS technology. For a PD with size of $55 \times 55\ \mu m^2$ and fabricated in a 0.18-μm CMOS process, the parasitic capacitance of D_1 to D_4 are 353 fF, 2358 fF, 2142 fF, and 1480 fF, respectively, under a reverse-biased voltage of 1.2 V. Though the intrinsic bandwidth of D_2, D_3, and D_4 are increased compared to D_1, their responsivity are much lower and also have a much larger parasitic capacitance due to their heavier doping concentration in the P/N regions. It imposes design challenges to the broadband TIA design. On the contrary, D_1 has a better responsivity [14] and a smaller junction capacitance, but its intrinsic bandwidth is only at tens of MHz range.

Table 1 Comparison of the P/N junction PDs with reverse-biased voltage of 1.2 V

P/N junction PD type	R (mA/W)	$f_{\text{-3dB}}$ (MHz)	$C_{PD}/55^2$ μm² (fF)
$D1$: $P_{substrate}/N_{well}$ PD	379	10	353
$D2$: P_{diff}/N_{well} PD	30	1300	2358
$D3$: P_{well}/N_{diff} PD	30	1900	2142
$D4$: P_{well}/DNW PD	49	1200	1480

1.2 Spatially Modulated Photodetectors

To realize a fully integrated CMOS optical receiver capable of operating at Gb/s range, spatially modulated photodetectors (SMPD) are proposed [16, 17]. Figure 6 shows the layout and cross-sectional view of a strip-type SMPD, which is composed of a row of $P_{substrate}$ to N_{well} junction photodetectors alternately covered and uncovered by light-blocking materials, such as metal layers. The covered detectors are named as dark detectors, while the uncovered detectors are called light detectors. As the slowly diffusive carriers from $P_{substrate}$ region diffuse in all directions, the light detectors capture the fast drift carriers and slowly diffusive carriers, while the dark detectors capture only the slowly diffusive carriers. Figure 7 depicts the photo-generated current components of the light detectors and dark detectors. By applying the outputs of dark and light detectors to a differential TIA, the slowly diffusive carriers can be partially removed. Let I_{nL} and I_{nD}, respectively, represent the electron diffusion current of light and dark detectors in the $P_{substrate}$-neutral region, I_{pL} be the hole diffusion current in the N_{well}-neutral region, and I_{dr} be the drift current in the depletion region. The differential SMPD current is expressed as ($I_{light.} - I_{dark}$)

$$I_{light} - I_{dark} = \left(I_{nL} + I_{dr} + I_{pL}\right) - I_{nD} = \left(I_{nL} - I_{nD}\right) + I_{dr} + I_{pL} \tag{15}$$

Since I_{dr} is a high-speed component and I_{pL} also has a wide-bandwidth response thanks to a shallow diffusion depth, the remaining ($I_{nL} - I_{nD}$) would determine the effective bandwidth of SMPD according to Eq. (15). By applying this spatially modulated layout topology, the 3-dB bandwidth of a strip-type SMPD can be increased from about 10 to 850 MHz [14]. However, it is still insufficient for multi-Gb/s operations.

To more effectively cancel the slowly diffusive carriers, a two-dimensionally (2-D) meshed SMPD architecture is proposed [17]. The layout and cross-sectional view of the meshed SMPD are shown in Figs. 8a, b, respectively, which is laid out as a chessboard pattern. It consists of a PD array alternatively covered and uncovered by light-blocking metal layers. Compared to a strip-type SMPD, the slowly diffusive carriers generated from $P_{substrate}$ can be more equally captured by the neighbored dark detectors. Also, the meshed structure reduces the distance that PD carriers drift. Thus it benefits from a smaller R-C delay and a higher intrinsic bandwidth.

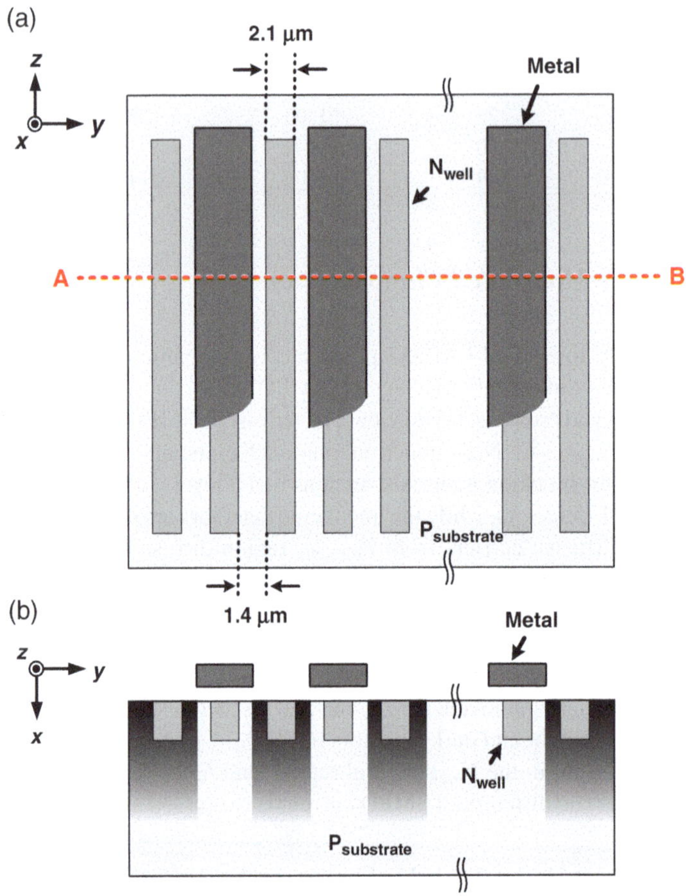

Fig. 6 (**a**) Top view and (**b**) cross-sectional view of the strip SMPD

By using the differential sensing scheme, a large portion of slowly diffusive carriers can be removed. Thus high-speed optical detection can be achieved by the proposed meshed SMPD but at the expense of a reduced responsivity.

The responsivity of a SMPD can also be increased by applying a higher reverse-biased voltage over it. As the depletion region at both the horizontal and vertical junctions is enlarged, more drift carriers can be generated. To investigate the effects of reverse-biased voltage (V_R) on the responsivity and bandwidth of CMOS PD, V_R of 1.2 V and 14.2 V are applied to characterize their performance. The PD is integrated with an optical receiver using chip-on-board assembly for performance characterization. The measured frequency responses of strip and meshed SMPDs are illustrated in Fig. 9. In the strip-type SMPD, the strip width of a unit detector is 2.1 μm, and the spacing is 1.4 μm. The detector is laid out in an octagon shape with

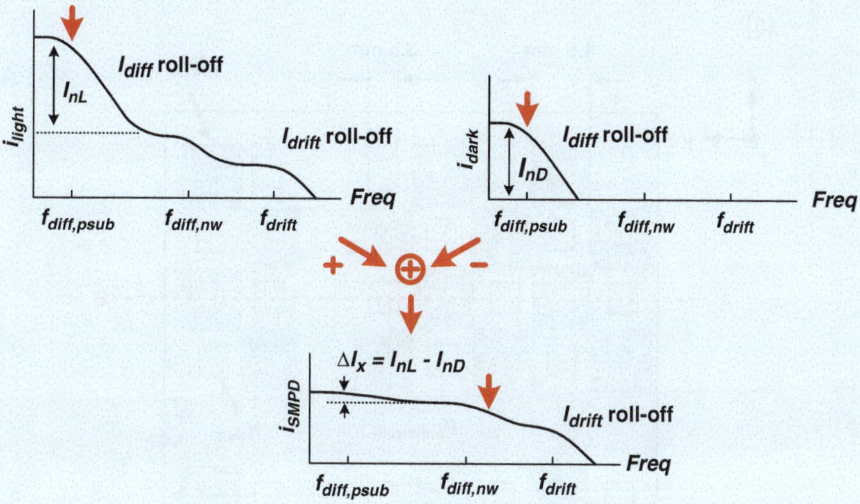

Fig. 7 Frequency responses of spatially modulated photodetector (SMPD)

an area of 55 μm × 55 μm to comply with the diameter of a multimode optical fiber. On the other hand, the meshed PD is composed of 3.5 μm × 3.5 μm unit detectors, and the pitch size is 1.4 μm, as is illustrated in Fig. 8.

The responsivity-bandwidth product of PDs is chosen as the figure of merit to evaluate their performance under different layout schemes. Compared to a strip-type SMPD, a meshed-type PD has a better responsivity-bandwidth product. It can be improved by 2.6× and 2.0× under a V_R of 1.2 V and 14.2 V, respectively. For the meshed-type SMPDs, their responsivity is 20 mA/W in the low-voltage mode and is boosted to 29 mA/W in the high-voltage mode. The responsivity can be improved by 1.5× by increasing V_R.

Table 2 summarizes the performance of SMPDs in a 0.18-μm CMOS technology. For CMOS PDs, the strip-type SMPDs have a larger responsivity though its bandwidth is much lower. On the contrary, the meshed-type SMPDs under a reverse-biased voltage of 14.2 V has a $f_{\text{-3dB}}$ of 6.9 GHz, which is suitable for 10-Gb/s operation without an equalizer.

The bandwidth of meshed-type SMPDs can be further improved by using a scaled CMOS technology. Figure 10 shows PD layout shrink from a 0.18 μm to a 40-nm technology node. By technology scaling, the diffusive carriers from $P_{substrate}$ region are more equally captured by the neighbored dark detectors owing to a shorter diffusion length [16]. The difference of the slowly diffusive currents in light and dark detectors can be expressed as

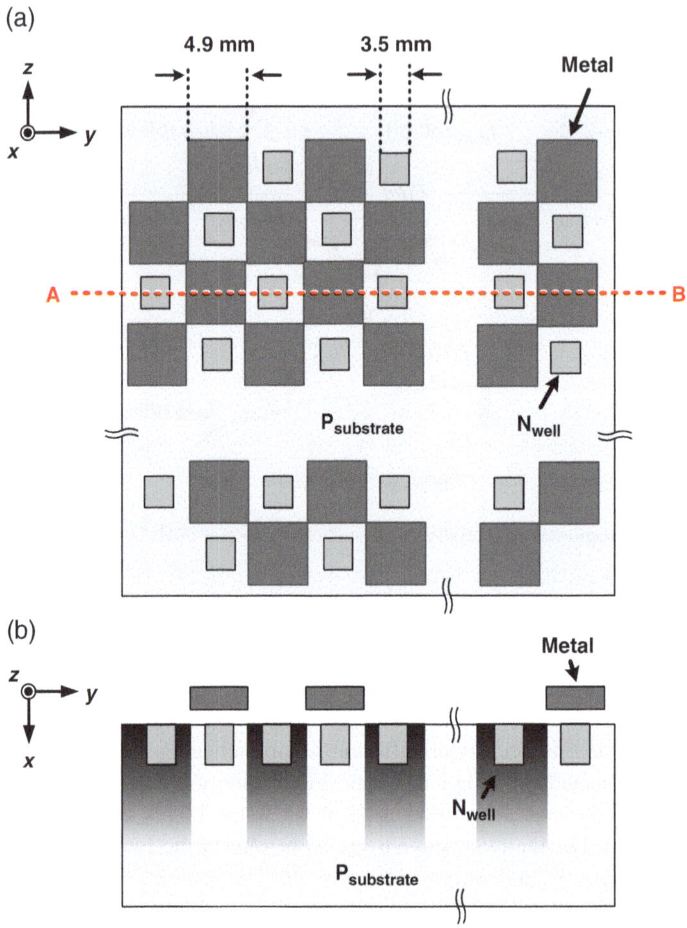

Fig. 8 (**a**) Top view and (**b**) cross-sectional view of proposed meshed SMPD

$$\Delta I_X(s) = I_{nL}(s) - I_{nD}(s)$$
$$= \sum_{m=1}^{\infty}\sum_{n=1}^{\infty} f(m,n) \frac{L_Y L_Z}{s + D_n \pi^2 \left[(2m-1)^2 L_Y^{-2} + (2n-1)^2 L_Z^{-2}\right] + \tau_n^{-1}} \tag{16}$$

where the L_Y and L_Z denote the width and length of a unit detector, D_n is the electron diffusion coefficient, τ_n is the electron diffusion time, and *f(m,n)* is a polynomial function independent of L_Y and L_Z. Limited by design rules, L_Y (=L_Z) is chosen

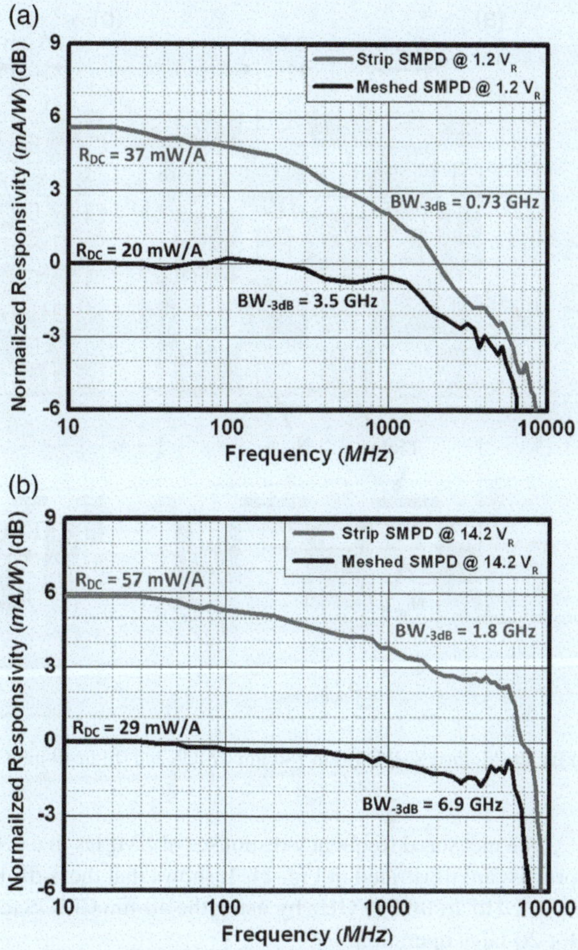

Fig. 9 Measured frequency responses of strip and meshed SMPDs in a 180-nm CMOS process under the reverse-biased voltage of (**a**) 1.2 V and (**b**) 14.2 V, respectively

Table 2 Comparison of SMPDs and commercial PD for $\lambda = 850$ nm

PD Type	Process	Responsivity (mA/W)	$f_{\text{-3dB}}$ (GHz)	C_{PD}[a] (fF)	V_R (volt)
Strip SMPD	0.18-μm CMOS	37	0.73	368	1.2
Strip SMPD	0.18-μm CMOS	57	1.8	213	14.2
Meshed SMPD	0.18-μm CMOS	20	3.5	354	1.2
Meshed SMPD	0.18-μm CMOS	29	6.9	206	14.2

[a]SMPD C_{PD} is simulated by HSPICE

as 4.9 μm in a 0.18-μm process and can be reduced to 1.6 μm in the 40-nm technology node. According to (16), ΔI_x can be further reduced thanks to the shrink of L_Y and L_Z. Thus the intrinsic bandwidth of CMOS SMPD can become wider by shrinking device pitch in a more advanced technology.

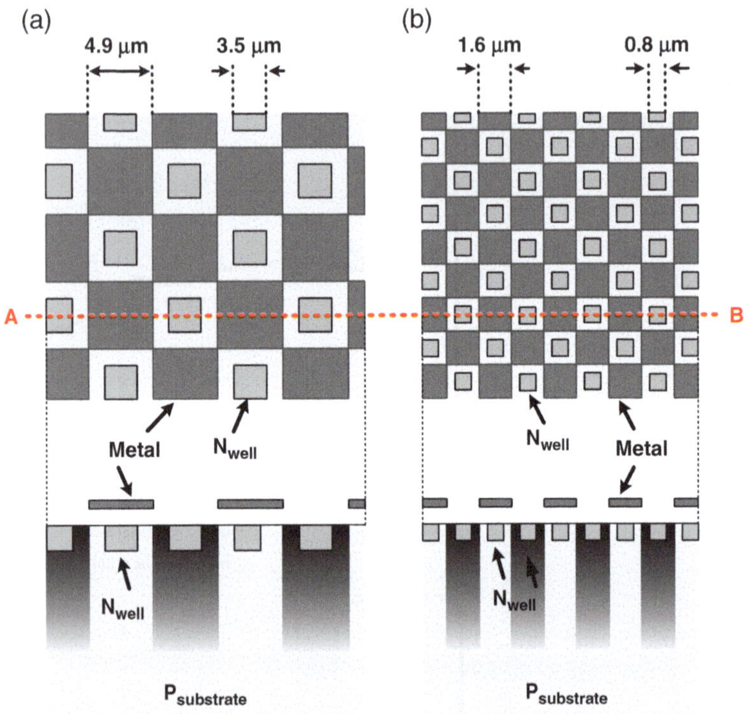

Fig. 10 Meshed SMPD (**a**) in 180-nm CMOS and (**b**) in 40-nm CMOS

The measured frequency responses of SMPDs in a 0.18-μm and a 40-nm CMOS process are illustrated in Fig. 11. It shows that the 3-dB bandwidth of SMPD can be boosted to around 14 GHz by using the 40-nm CMOS technology, which is suitable for 20-Gb/s operation.

Applying a higher reverse-biased voltage on PD not only improves its responsivity but also lowers its parasitic capacitance for a higher-speed operation. As the negative supply voltage on a $P_{substrate}$ may induce a higher substrate noise, to avoid interfering the receiver front end in a single chip integration, all the active circuits are surrounded by voltage islands isolated by N_{well} and deep N_{well}. Another concern arisen from a high reverse-biased voltage is the junction-breakdown issue. As shown in Fig. 12, the cathode of SMPD is connected to the TIA input. For a PD reverse-biased voltage of 14 V and $V_{in,TIA}$ of 0.8 V, the $P_{substrate}$ should be biased at −13.2 V . The maximum reverse-biased voltage happens at the $P_{substrate}/N_{well}$ (or $P_{substrate}$/DNW) junction, which is 14.2 V given the V_{DD} of 1 V. The applicable PD reverse-biased voltage is limited by the tolerable junction-breakdown voltage in the CMOS technology.

On the other hand, the heavier doping concentration in an advanced nanometer CMOS technology leads to a higher parasitic capacitance associated with it. For an input capacitance of less than 200 fF, the sensing region of the meshed SMPDs

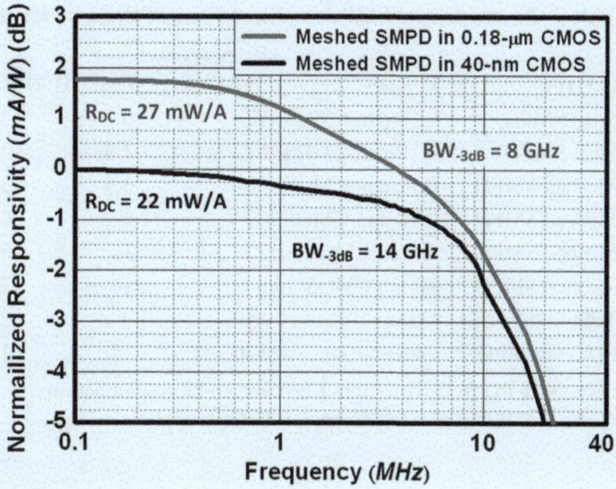

Fig. 11 Frequency responses of meshed SMPD in 0.18-μm and 40-nm CMOS

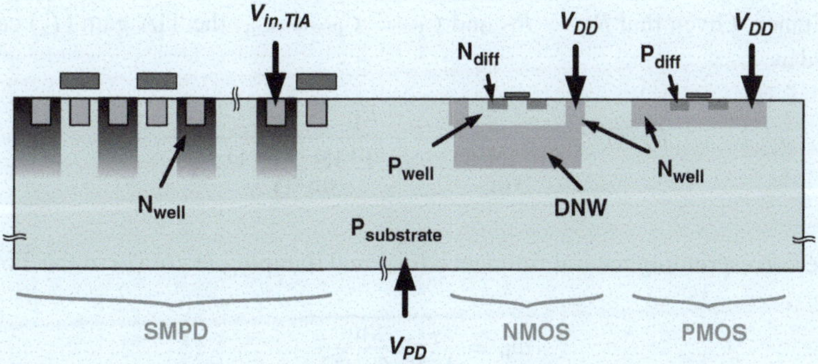

Fig. 12 Integration of SMPD and transistors (PMOS, NMOS) for CMOS OEIC

can be scaled down to facilitate the integration with a high-speed TIA but at the expense of degrading PD's responsivity. To sustain the receiver's input sensitivity, it demands more stringent noise requirement for the succeeding amplifier design.

Table 3 summarized the performance comparison of CMOS versus. GaAs PDs. Though the CMOS PDs can provide a pronounced bandwidth, their responsivity is about 20–80× lower compared to commercial GaAs counterparts. The corresponding sensitivity is therefore 13–19 dB worse. As a result, CMOS OEIC requires an ultralow noise receiver front end to resolve the incoming optical data.

1.3 Nested-Feedback TIA

Typically, a high-sensitivity TIA is based on a common-source amplifier with a large shunt-shunt feedback resistor. As shown in Fig. 13a, C_A and C_D denote the input

Table 3 Comparison of SMPDs and commercial PDs ($\lambda = 850$ nm)

PD type	Process	Responsivity (mA/W)	f_{-3dB} (GHz)	C_{PD} (fF)	V_R (Volt)	Area (μm^2)
Strip SMPD	0.18-μm CMOS	37	0.73	368	1.2	50 × 50
Strip SMPD	0.18-μm CMOS	57	1.8	213	14.2	50 × 50
Meshed SMPD	0.18-μm CMOS	20	3.5	354	1.2	50 × 50
Meshed SMPD	0.18-μm CMOS	29	6.9	206	14.2	50 × 50
Meshed SMPD	40-nm CMOS	22	14	556	14	50 × 50
Meshed SMPD	40-nm CMOS	8	14	200	14	30 × 30
TPD-8D12–052[a]	GaAs	650	9	220	3	75 × 75
PDCAxx-30-GS[b]	GaAs	530	17	100	2.5	30 × 30

[a]GaAs PIN PD from TrueLight Corp. for 10-Gb/s applications
[b]GaAs PIN PD from Albis Corp. for 25-Gb/s applications

and output capacitances of the core amplifier $A_C(s)$, and C_{PD} is the PD's parasitic capacitance. Given that $R_F >> R_D$ and $C_{IN} = C_{PD} + C_A$, the TIA gain (T_z) can be derived as

$$T_Z(s) \approx \frac{R_F}{\frac{R_F C_{IN} C_D}{g_m} s^2 + \frac{R_F C_{IN} + R_D C_D}{g_m R_D} s + 1} \tag{17}$$

The corresponding natural frequency (ω_n) and damping factor (ζ) of the TIA are

$$\omega_n = \sqrt{\frac{g_m}{R_F C_{IN} C_D}} \tag{18}$$

$$\varsigma = \frac{1}{2} \frac{R_F C_{IN} + R_D C_D}{\sqrt{g_m R_D^2 R_F C_{IN} C_D}} \tag{19}$$

For a maximally flat gain response ($\zeta = 0.707$), the TIA bandwidth (ω_{TIA}) can be derived as

$$\omega_{TIA} = \frac{\sqrt{2} g_m R_D}{R_F C_{IN}} \tag{20}$$

Besides, the core amplifier's 3-dB bandwidth, $\omega_p = (R_D C_D)^{-1}$, should be at least $\sqrt{2}\omega_{TIA}$.

Given C_{PD} and C_A of about 200 fF and 30 fF respectively, a 20-Gb/s, 500-Ω TIA demands a core amplifier with a 3-dB bandwidth of more than 20 GHz and a voltage gain of more than 17 dB. The corresponding gain bandwidth product is about 140 GHz, which is challenging to be implemented in a 40-nm CMOS technology. To overcome the bottleneck, a TIA incorporating both active and passive nested feedback is proposed, as is shown in Fig. 13b. The inner loop is a voltage

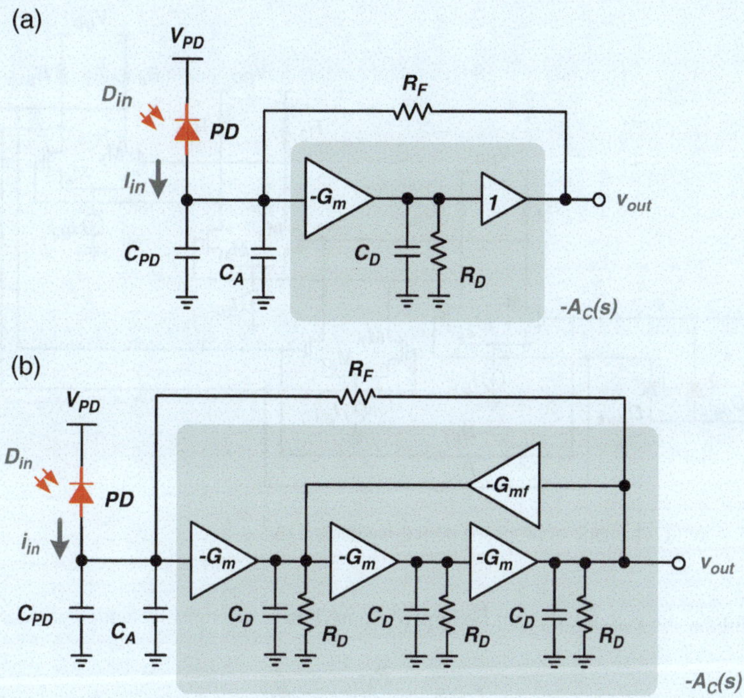

Fig. 13 (**a**) Shunt-shunt feedback TIA and (**b**) proposed nested-feedback TIA

amplifier composing of another transconductance gain stage (G_m) followed by a transimpedance amplifier with active feedback (G_{mf}). The outer loop comprising the passive feedback resistor (R_F) performs the current to voltage conversion.

Figure 14 shows the circuit schematic of the nested-feedback TIA. The output currents of SMPD are applied to a differential TIA to perform current subtraction. Let g_m denote the transconductance of M_1 to M_6, g_{mf} be the transconductance of M_7 and M_8, R_D be the resistance of R_1 to R_6, and C_D be the parasitic capacitance at the drain node of M_1 to M_6, the transfer function of core voltage amplifier can be derived as

$$A_C(s) = \frac{g_m}{g_{mf}} \left(\frac{g_{mf} g_m^2 R_D^3}{(1 + sC_D R_D)^3 + g_{mf} g_m^2 R_D^3} \right) \tag{21}$$

Besides, the conversion gain of the proposed TIA becomes

$$T_Z(s) = \frac{v_{out}(s)}{i_{in}(s)} = \frac{R_F}{1 + (1 + sC_{IN} R_F) A_C^{-1}(s)} \tag{22}$$

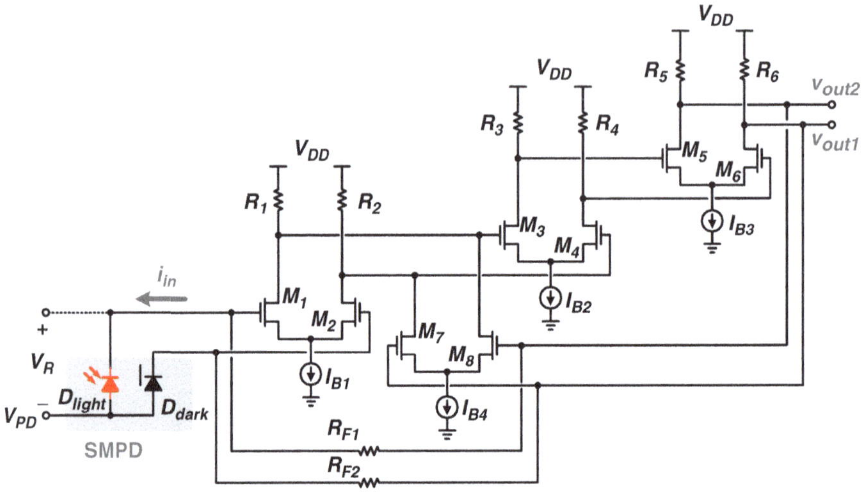

Fig. 14 Circuit schematic of proposed nested-feedback TIA

With $g_m >> g_{mf}$, the T_z can be approximated as

$$T_Z = \frac{R_F}{1 + A_C^{-1}} \approx R_F \tag{23}$$

For a maximally flat response, the frequency response of a high-order TIA can be approximated using a two-pole model without losing much design insight. We have

$$T_Z(s) = \frac{R_F}{1 + \frac{1.64}{g_m^3 R_D^3} + \left(\frac{1.64 C_{IN} R_F}{g_m^3 R_D^3} + \frac{1.59 C_D R_D}{g_m^3 R_D^3}\right)s + \frac{1.59 C_D R_D C_{IN} R_F}{g_m^3 R_D^3} s^2} \tag{24}$$

The corresponding natural frequency (ω_n) and damping factor (ζ) can be derived as

$$\omega_n \approx 0.8 g_m R_D \sqrt{\frac{g_m}{R_F C_{IN} C_D}} \tag{25}$$

$$\varsigma = \frac{C_{IN} R_F + 0.795 C_D R_D}{\sqrt{\left(g_m^3 R_D^3 + 1.64\right)\left(1.59 C_{IN} R_F C_D R_D\right)}} \tag{26}$$

Comparing Eqs. (26) to (18), the natural frequency of TIA can be improved by a factor of 0.8 $g_m R_D$ through nested feedback, which is about 2.5× in this design.

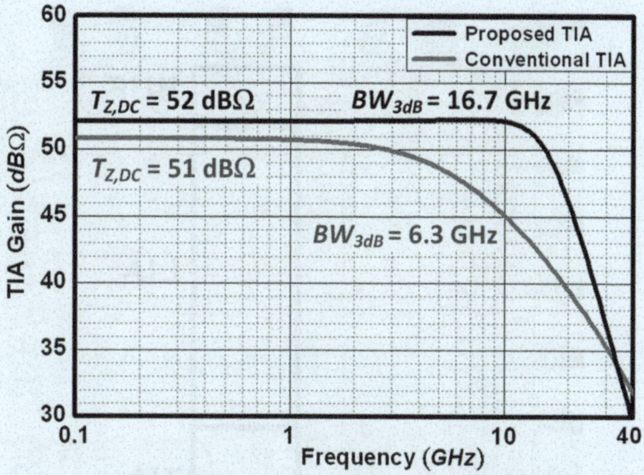

Fig. 15 Simulated magnitude response of proposed and conventional TIA

Figure 15 shows the performance comparisons between the nested-feedback TIA and the conventional TIA under the same power dissipation. By boosting up the gain of the core amplifier, the nested-feedback TIA provides a 2.5× bandwidth improvement and also a higher gain compared to the conventional architecture.

The input-referred noise current $\overline{I^2_{\text{n,TIA}}}$ of the TIA can be derived as

$$\overline{I^2_{\text{in,TIA}}} = \frac{8kT}{R_\text{F}} + \frac{1}{R_\text{F}^2}\left[\overline{V^2_{\text{n,SC}}} + \frac{1}{g_\text{m}^2}\left(g^2_\text{mf}\overline{V^2_{\text{n,SC}}} + g^2_\text{mf}\frac{\overline{V^2_{\text{n,SC}}}}{g^2_\text{m}R^2_\text{D}}\right)\right] \tag{27}$$

where $\overline{V^2_{\text{n,SC}}}$ represents the input-referred noise voltage of a single-stage source coupled pair amplifier and can be expressed as

$$\overline{V^2_{\text{n},sc}} = 2\left(\frac{4kT}{g_\text{m}} + \frac{4kT}{g^2_\text{m}R_\text{D}}\right) \tag{28}$$

According to Eq. (27), the input-referred noise current of the proposed TIA is dominated by the TIA's feedback resistor (R_F). A large R_F is preferred in order to improve receiver sensitivity. In this design, the integral input-referred noise current is 2.4 μA_{rms}. The corresponding input sensitivity is about 33.6 μA_{PP} ($14 \times 2.4\ \mu$) for BER of less than 10^{-12} [18].

1.4 A Multichannel OEIC with CMOS Photodetector

In an active optical cable, multichannel bidirectional link transceivers are assembled in a compact connector, such as SFP, QSFP, etc. Typically, PD array in a III-V technology is integrated with Si-based receiver through wire bonding on a PCB.

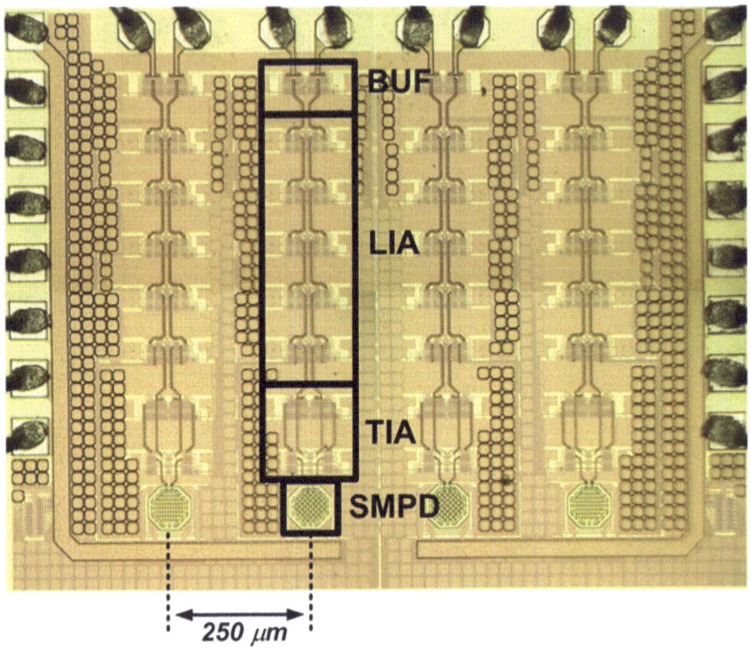

Fig. 16 Die photo of the four-channel 20-Gb/s OEIC

As the operating speed goes higher and higher, it encounters serious cross talk and signal integrity issue due to mutual coupling through bonding wire inductors. To alleviate this problem, an external PD array can be replaced by an on-chip spatially modulated photodetector (SMPD) array, which is then coupled to a four-channel parallel receiver. CMOS OEICs are low cost and bonding wire-free without resorting to flip chip bonding technology. To demonstrate the design concept, a 20-Gb/s (5Gb/s × 4), four-channel receiver array is implemented in a generic 0.18-μm CMOS technology. In this design, both strip- and meshed-type SMPDs are adopted in different channels to investigate their merits and demerits under different operation speeds.

Figure 16 shows the chip photograph. Here channels (#1, #4) are made up of strip-type and (#2, #3) of meshed-type SMPDs. Each channel is composed of a nested-feedback TIA followed by a limiting amplifier. The chip size is 1.3 × 1.0 mm, and the pitch size is 250 μm. The receiver IC is mounted on a printed circuit board for measurement. The SMPDs are powered with a negative supply voltage of −13 V and are surrounded by deep N_{well} to minimize mutual coupling and avoid interfering the body bias of receiver array at ground level. The receiver circuits are operated under a single 1.8 V supply. The overall conversion gain is 116 dBΩ, and the differential output swing is 820 mV_{pp}. The total power dissipation is 640 mW.

Figure 17 shows the measured bit error rate (BER) performance. The extinction ratio of VCSEL is 7.4 dB. All the channels are operating simultaneously for

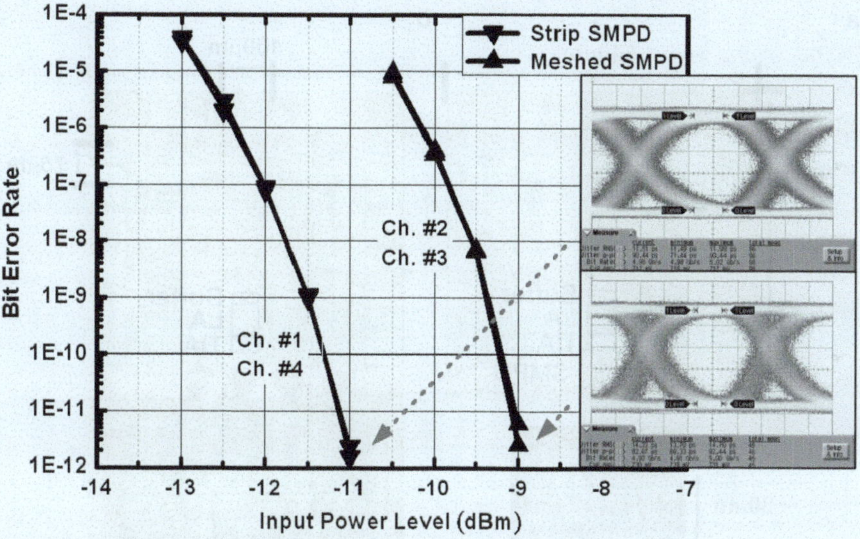

Fig. 17 BER curves and eye diagram for all four channels measured at 5 Gb/s

BER test. The meshed-type SMPD achieves a wider bandwidth at the cost of a lower responsivity compared to the strip-type architecture. Under a 5-Gb/s/channel operation speed (20-Gb/s throughput), it reveals that the input sensitivity of a receiver with a strip-type SMPD is better than that with a meshed-type SMPD by 2–3 dB. The input sensitivity of the two types are about −11 dBm and –9 dBm, respectively. The cross talk effect is evaluated by BER penalty. Measurement results show that the input sensitivity is degraded by less than 0.1 dB comparing multichannel to single channel operation. Thus the fully integrated multichannel OEIC demonstrates strong potential for future data-intensive (16×, 20×) optical links.

1.5 A 20-Gb/s OEIC with CMOS PD

A fully integrated CMOS OEIC capable of operating at tens of Gb/s is also demonstrated. Figure 18a showsthe chip micrograph of an optical receiver integrated with an on-chip CMOS meshed SMPD, and Fig. 18b shows the same CMOS receiver but integrated with a commercially available GaAs PD for performance comparison. The receiver chip is implemented in a generic 40-nm CMOS technology. As the heavier doping concentration in the advanced CMOS process leads to a higher parasitic capacitance associated with the SMPD, the optical sensing region is designed to be 30×30 μm to avoid severely deteriorating the receiver bandwidth.

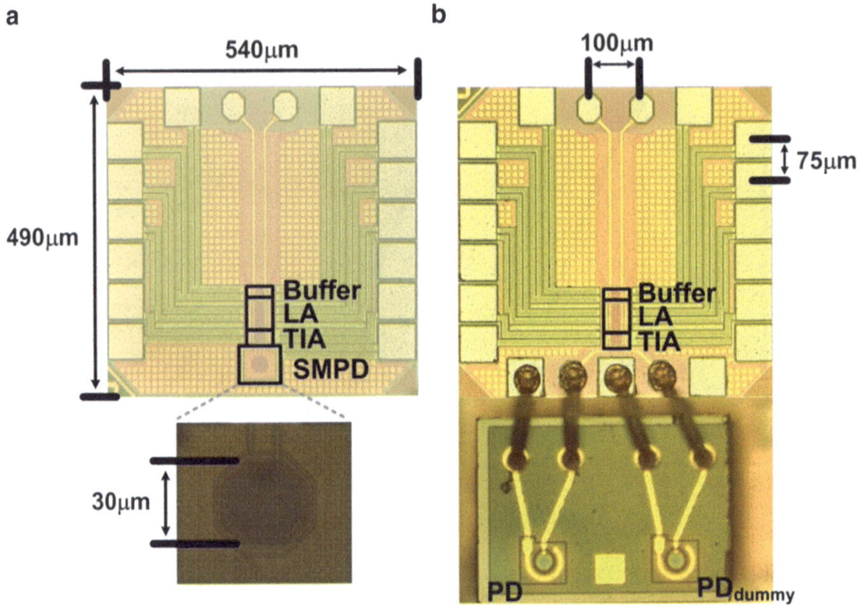

Fig. 18 Chip photos of 20-Gb/s receiver with (**a**) CMOS PD and (**b**) commercial GaAs PD

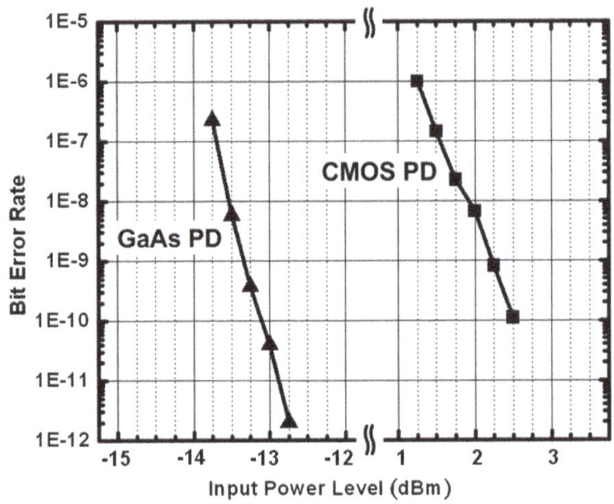

Fig. 19 Measured BER performances of optical receivers with CMOS PD and commercial GaAs PD at 20-Gb/s operating speed

Under a single 1-V supply voltage, the power dissipation of the Si-OEIC is 30 mW, among which 9 mW is consumed by the output buffer.

A PRBS-7 test pattern is utilized to modulate an 850-nm VCSEL (VI System V40-850 M) light source, which is coupled to the receiver chip for performance measurement. The eye diagrams are measured by Agilent 86100C, and the BER performance are characterized by using Anritsu MP1800A. Figure 19 summarizes

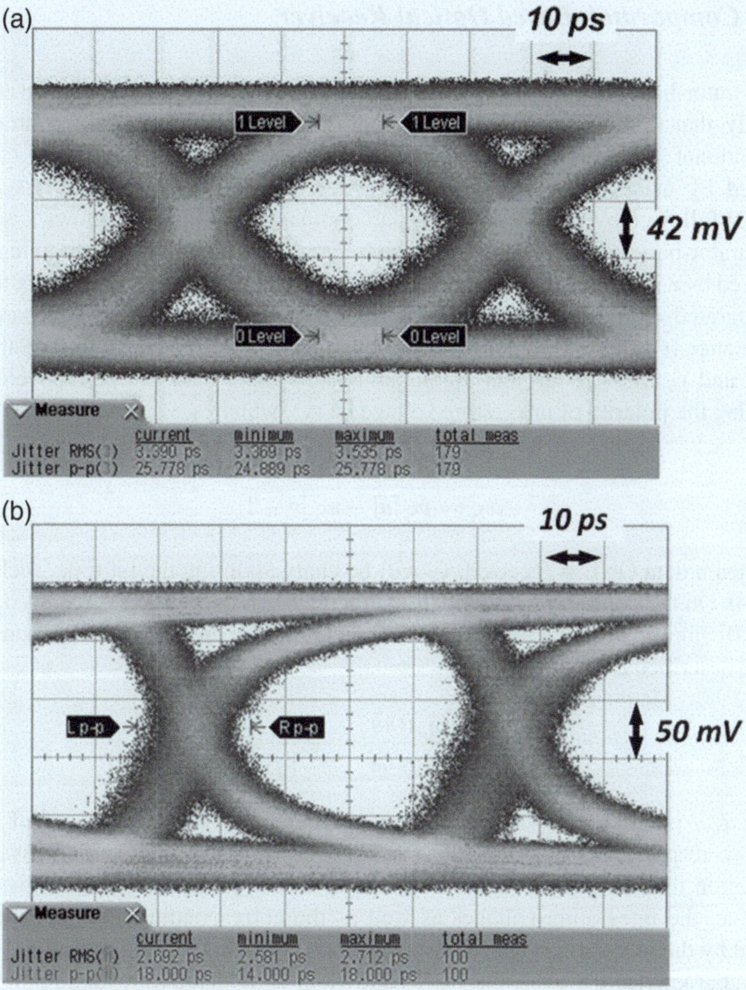

Fig. 20 Measured 20-Gb/s eye diagram at sensitivity level of optical receiver with (**a**) CMOS PD and (**b**) commercial GaAs PD

the BER performance of the optical receiver under a reverse-biased voltage of 14 V. For a BER of less than 10^{-10} at 20-Gb/s operation, the measured input sensitivity of the optical receivers with CMOS and GaAs PDs are 2.5 dBm and -13 dBm, respectively. The input sensitivity of the fully integrated CMOS OEIC is mainly limited by the responsivity of PD. Figure 20 shows the measured eye diagrams of the receiver with input power at sensitivity level. The corresponding data jitters are 3.39 ps_{rms} (25.78 ps_{pp}) and 2.69 ps_{rms} (18 ps_{pp}), respectively, with CMOS and GaAs PDs. This optical receiver provides a conversion gain of 80 dBΩ and is capable of delivering 450 mV_{pp} to 50-Ω output loads.

1.6 Comparator-Based Optical Receiver

Comparator-based optical receivers [16, 19–23] have attracted many research efforts recently that demonstrated promising energy and area efficiency in contrast to conventional TIA+LA-based counterparts. In the receiver front end, they can be realized by using either a photocurrent integrator [16, 19–21] or a TIA stage [22, 23] followed by voltage samplers and comparators. Figure 21 shows a typical comparator-based optical receiver, which is composed of a current integrator followed by a full-rate clocked comparator. As is proposed in [19], the photocurrent is integrated over sampling capacitor (C_S) in parallel with the PD's parasitic capacitance (C_{PD}), so as to convert it into voltage form directly. By differentiating $v_S\,[n]$ and $v_S\,[n-1]$ through a comparator, the input data can be recovered by detecting the polarity of integrating voltage (Δv_S), which is

$$\Delta v_S = v_S\,[n] - v_S\,[n-1] \tag{29}$$

When a data ONE is received, v_S will be charged during the bit time, such that $\Delta v_S > 0$. On the contrary, v_s will be discharged when a data ZERO is received, thus $\Delta v_S < 0$. Taking the PD's bandwidth into account, Δv_S in a 1-UI integration time can be described as

$$\Delta v_S = \frac{\int i_{PD}(t)dt}{C_{PD} + C_s} = \frac{\alpha_{INT} I_{PD} T_B}{C_{PD} + C_S} \tag{30}$$

where α_{INT} is a PD-bandwidth-dependent coefficient, T_B is the bit time, and $i_{PD}(t)$ and I_{PD}, respectively, represent the instantaneous and DC photocurrent over the integration time. As the integration time is inversely proportional to its operating data rate, the integrating voltages as well as the corresponding SNR are severely limited by the integrating capacitances ($C_S + C_{PD}$) at a high-speed operation.

To characterize the α_{INT}, the PD's bandwidth is modeled as a single-pole low-pass filter with a 3-dB bandwidth of ω_p for simplicity. The worst-case integrating

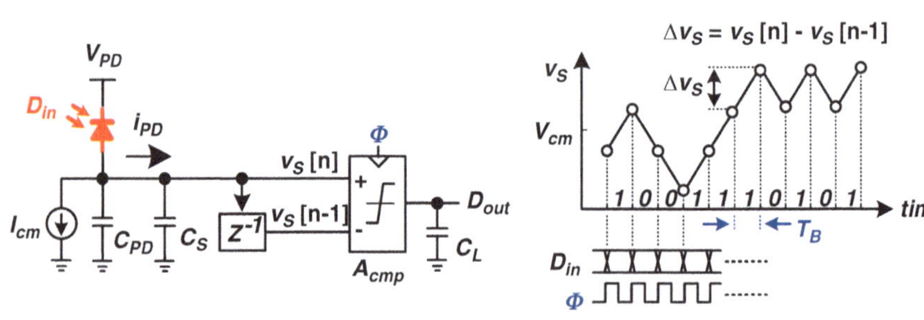

Fig. 21 Integrating-type receiver front end and its corresponding integrating voltage

voltage (Δv_s) occurs when a ZERO (or ONE) pulse is received after a long run of ONEs (or ZEROs). In the case of 1-*UI* integration time, the maximum integrating voltage can be derived as

$$\Delta v_{\mathrm{S}} = \frac{I_{\mathrm{PD}} T_{\mathrm{B}}}{C_{\mathrm{PD}} + C_{\mathrm{S}}} \left[1 - 2\frac{t_1}{T_{\mathrm{B}}} + \frac{2}{\omega_{\mathrm{p}} T_{\mathrm{B}}} \left(1 - 2e^{-\omega_{\mathrm{p}} t_1} + e^{-\omega_{\mathrm{p}} T_B} e^{-\omega_{\mathrm{p}} t_1} \right) \right] \tag{31}$$

where

$$t_1 = \frac{1}{\omega_{\mathrm{p}}} \ln \left(2 - e^{-\omega_{\mathrm{p}} T_{\mathrm{B}}} \right) \tag{32}$$

$$\alpha_{\mathrm{INT}} = 1 - 2\frac{t_1}{T_{\mathrm{B}}} + \frac{2}{\omega_{\mathrm{p}} T_{\mathrm{B}}} \left(1 - 2e^{-\omega_{\mathrm{p}} t_1} + e^{-\omega_{\mathrm{p}} T_B} e^{-\omega_{\mathrm{p}} t_1} \right) \tag{33}$$

Figure 22 illustrates α_{INT} under different PD bandwidth in terms of data rate (D_{R}). It shows that α_{INT} degrades from 78% to 15% if the PD bandwidth is reduced from 1 to 0.2 D_{R}.

The effective transimpedance gain (A_{RX}) combing with a current integrator and a dynamic comparator can be derived as

$$A_{\mathrm{RX}} = \frac{v_{\mathrm{out}}}{I_{\mathrm{PD}}} = \left(\frac{\alpha_{\mathrm{INT}} T_{\mathrm{B}}}{C_P + C_{\mathrm{S}}} \right) \exp \left(\frac{g_{\mathrm{m}} T_{\mathrm{R}}}{C_{\mathrm{out}}} \right) \tag{34}$$

where g_{m} is the transconductance gain of the dynamic comparator, T_{R} is the regeneration time of latch, and C_{out} denotes the parasitic capacitance at the comparator output. According to Eq. (34), A_{RX} can be boosted exponentially by increasing g_{m} and T_{R}. To achieve a higher input sensitivity and relax the speed requirement of a

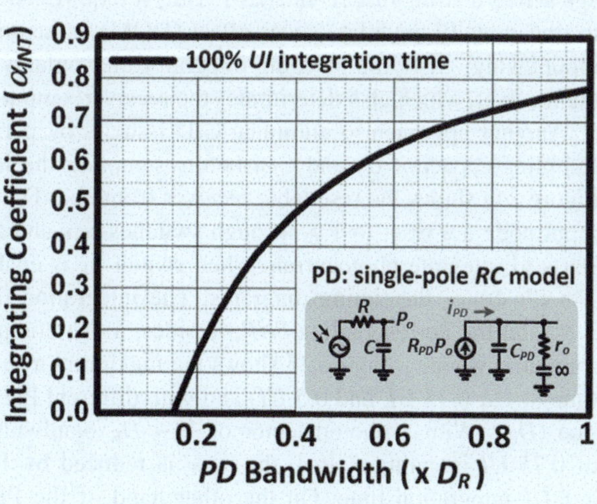

Fig. 22 α_{INT} versus i_{PD} bandwidth with single-pole RC model

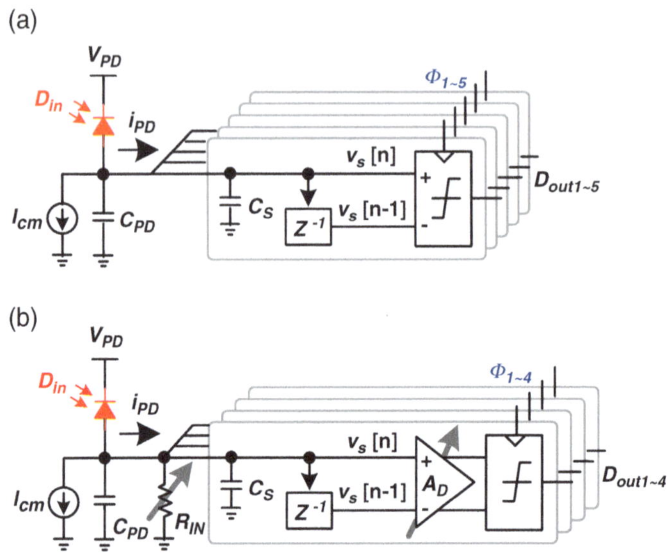

Fig. 23 Time-interleaved integrating-type optical receiver [19] (**a**) without and (**b**) with shunting an input resistor [20]

clocked comparator, a common approach is to extend T_R by using a time-interleaved architecture [16, 19–23], as is shown in Fig. 23a.

A major drawback of the integrating-type receivers is that they are less tolerant of consecutive identical digits (CID), which may overload the integrator. To circumvent the CID issue [19], shunts an input resistor, R_{IN}, in parallel with the photodetector, as is shown in Fig. 23b. It prevents voltage overload while receiving long-run CID bits. However, the receiver sacrifices data-dependent integrating voltage across a string of CID bits [20]. Thus it requires additional circuit functionality to dynamically modulate slicer offset (DOM), so as to maintain a constant slicer input swing. Also, the bleeding integrator unavoidably introduces excess noise to the receiver, which also deteriorates the receiver's input sensitivity.

Another approach to eliminate CID issue is by periodically resetting both the sampling capacitor C_S and C_{PD} before sampling the input photocurrent [16, 21]. Figure 24a shows the resettable receiver front end. During $\Phi = 0$, both $\mathrm{v_s}$ and the input node are reset to V_{cm} through switches $\mathrm{S_{R1}}$ and $\mathrm{S_{R2}}$. On the other hand, C_S starts to integrate photocurrent when switch $\mathrm{S_1}$ is enabled during $\Phi = 1$. Figure 24b illustrates the timing diagrams. The integration time is reduced by half, so is $\mathrm{v_s}$ and the corresponding *SNR*. Besides, a fast comparator is needed to resolve triangular wave v_s. Figure 25 shows the maximum α_{INT} under different integration time (1 *UI*, 0.75 *UI*, and 0.5 *UI*) and with different PD bandwidth in terms of data rate (D_R). With a photodetector of $1 - D_R$ bandwidth and a current integrator of 0.75-*UI* integration time, the α_{INT} is reduced by 10% compared to that with a 1-*UI* integration time. On the other hand, if the PD bandwidth is reduced to

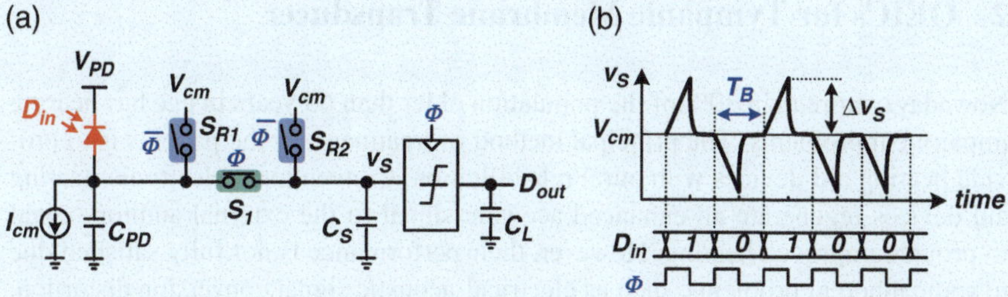

Fig. 24 Resettable integrating-type optical receiver front end and its corresponding integrating voltage

Fig. 25 α_{INT} versus i_{PD} bandwidth under different integration time

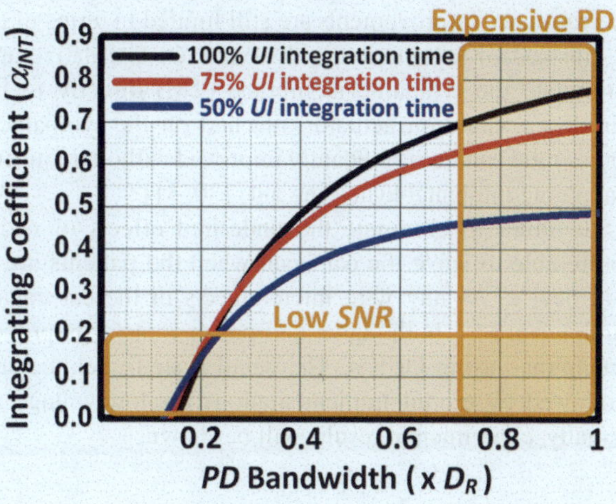

0.3 D_R, α_{INT} is reduced by half compared to that with a wide-bandwidth (1 D_R) PD. Meanwhile, the difference of α_{INT} for 1-*UI* and 0.75-*UI* integration time is negligible. If the PD bandwidth is reduced to be less than 0.2 D_R, the integration time becomes less relevant to α_{INT}, but a smaller value of α_{INT} would degrade the overall *SNR*. The shortcomings in the prior art [16, 19–23] can be remedied by using a current-boosting preamplifier in front of the integrating-type receiver, as is proposed in [24]. Compared to the prior art in [25–28] operating at 25 Gb/s, the input sensitivity can be potentially improved by 2X (or 3 dB) under a fixed power consumption. As the areas of analog circuits do not scale as aggressively as their digital counterparts when using more advanced process, it is expected that the area saving of a comparator-based receiver can be even more significant in a nanometer CMOS process. It demonstrates strong potential for high-density and short-reach optical interconnects.

2 OEICs for Tympanic Membrane Transducer

Nowadays, more than 30% of the population older than 65 years in age has hearing impairment problems. The principal method of treatment for the patients is to provide hearing aid devices with aural rehabilitation. Conventional electronic hearing aid devices regenerate an enhanced acoustic signal to the external auditory canal to provide improved hearing. However, their performance is not fully satisfied due to some inherent problems, such as electrical-acoustic signal conversion distortion, occlusion effects, and acoustic feedback. Some engineering strategies have been proposed to improve hearing in the presence of background noise, such as adopting directional microphones, array microphones, or noise reduction algorithms. The performance improvements are still limited in terms of comfort and clearness [29].

Instead of stimulation with sound, an alternative approach is to mechanically stimulate the tympanic membrane (TM) directly to improve sound quality. By driving a vibration actuator attached on the tympanic membrane, the signal-to-noise ratio can be significantly improved without requiring surgery, such as methods involving implantable hearing aids [30, 31].

In order to eliminate the undesired effects of occluding the ear canal, it is preferable to leave the ear open when the patients are equipped with the hearing aid device. As any wire interconnects in the ear canal is not desirable, a novel architecture for both signal and power transfer is proposed to stimulate tympanic membrane using OEICs. The detail system architecture will be firstly described, followed by circuit implementations for low-voltage and low-power operations. Finally, experimental results will be shown.

2.1 Tympanic Membrane Transducer with Optical Signal and Power Transmission

Figure 26 illustrates the system architecture of a hearing aid device with OEICs. It is composed of a carrier with sound processor, microphone, laser diode (LD) driver, the ear canal transceiver, and the tympanic membrane transducer with photodiode on top of it. The carrier is a wearable device, which delivers signal and power wirelessly to the tympanic membrane transducer through ear canal transceiver. Thus no battery is required at the transducer side for lightweight and small form factor. The transducer is driven by current to generate stable vibration and stimulate the membrane mechanically. By leaving the ear canal open, they eliminate the undesired effects of occluding the ear canal.

Wireless power transfer (WPT) technology dated back to 1899 and was firstly proposed by Nicola Tesla. It is now widely adopted in wireless chargers for consumer and implantable medical devices. Figure 27 shows the typical WPT architecture. At the transmitter side, the power source is converted to time-varying electromagnetic fields and emitted through an antenna or magnetic coils. The energy

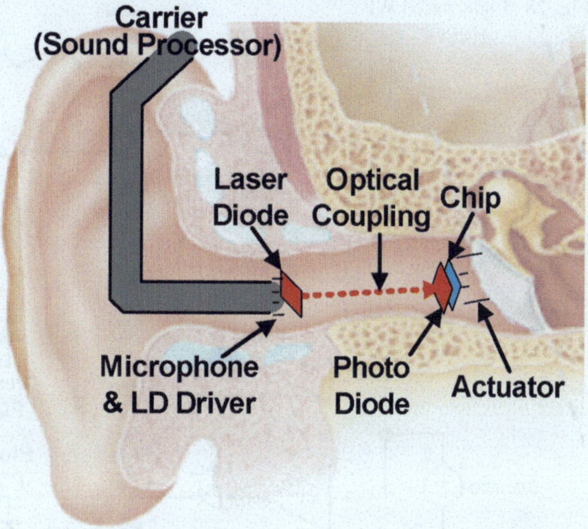

Fig. 26 Tympanic membrane transducer with optical signal and power transmission

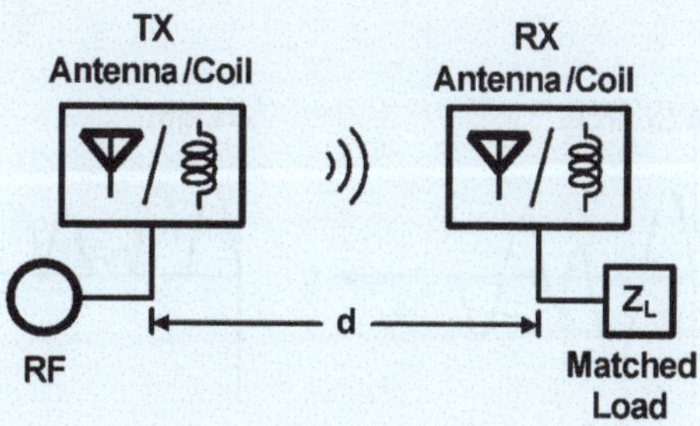

Fig. 27 WPT through coils and antenna

picked up at the receiver side is then converted back to current to drive the electrical loads. Typically, WPT through magnetic fields uses inductive coupling between coils. On the other hand, for far field power transferred, the power beams conveyed by radio-frequency (RF) carriers are radiated through antennas. As the dimension of the inner canal is small, the power conversion efficiency is deteriorated by either using inductive coupling through coils or RF coupling through antennas due to its limited size. Contrarily, by incorporating E/O and O/E conversion with solid-state devices, laser beams pave the way for wireless power transfer with a small form factor.

Figure 28 shows the light-based WPT architecture, which is similar to a typical optical transceiver. The E/O and O/E conversion are performed by a laser driver combining with laser diode at the carrier (sound processor) side and a photodetector at the transducer (membrane) side. In addition to data communication between the transmitter and the receiver, the optical energy received by the photodetector is converted to current to power the driver of transducer.

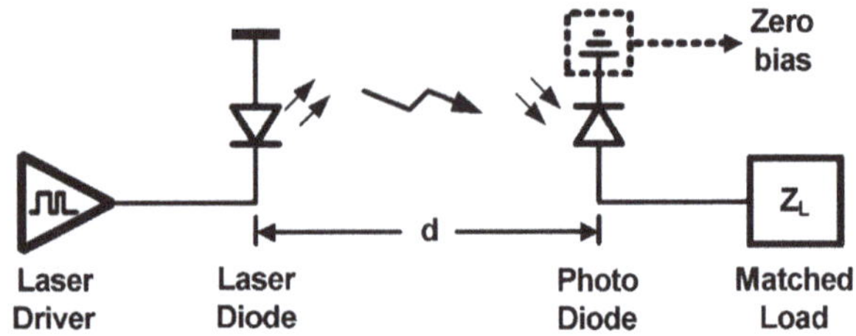

Fig. 28 Light-based WPT

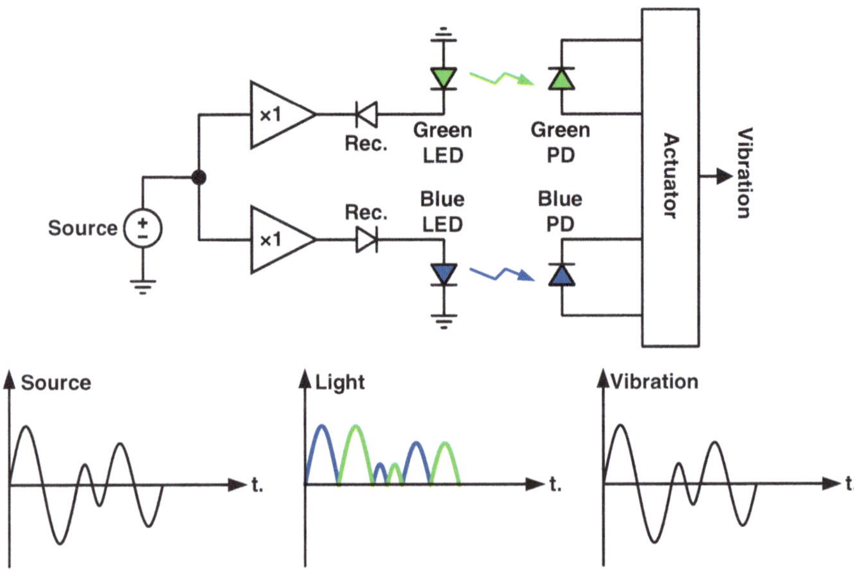

Fig. 29 Analog architecture

2.2 Light-Driven Transducer

Light-driven transducer is capable of delivering broadband audio signal. Figure 29 shows the transducer driver based on analog modulation [32]. At the transmitter side, the audio signal is firstly decomposed by half wave rectifiers, whose outputs are then utilized to modulate different color (green and blue) LEDs. At the actuator side, two PDs pick up the optical power, convert it into photocurrents, and then drive two opposite direction winding coils to restore the audio signal and provide stable oscillation. In this architecture, the finite turn on voltage of the rectifier may induce signal crossover distortion. Meanwhile, the sound quality is also limited by the nonlinearity in the O/E and E/O conversion.

To alleviate signal distortion caused by E/O and O/E conversion, a digitally modulated light-driven transducer is proposed as shown in Fig. 30 [32]. Here the audio signal is firstly pulse width modulated (PWM) to convert audio signal from voltage domain to phase domain. The PWM differential outputs modulate

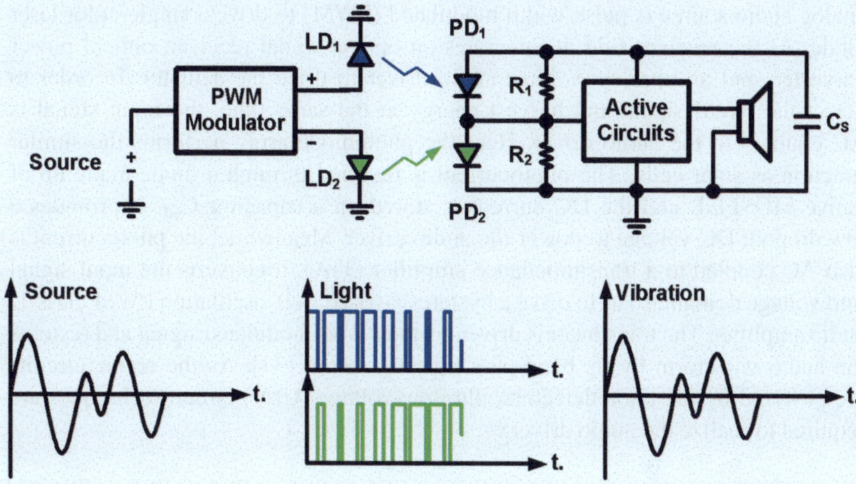

Fig. 30 PWM architecture

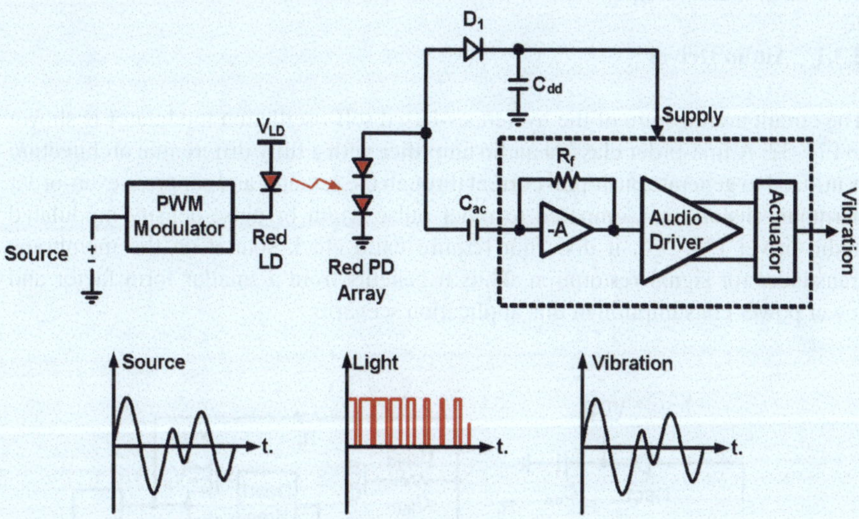

Fig. 31 A single wavelength light-driven transducer

different wavelength laser diodes (LD_1 and LD_2) in a binary mode. Thus the signal quality is less irrelevant to the linearity in E/O conversion. At the receiver side, the PWM-modulated photocurrent is DC rectified and restored on a capacitor to drive the actuator. For both cases in Figs. 29 and 30, two sets of LDs and PDs with different color filters are needed. It increases the form factor of the transducer and complicates the system integration.

To overcome the aforementioned shortcomings, a single wavelength light-driven transducer is proposed, as is shown in Fig. 31 [33]. At the transmitter side, the analog audio source is pulse width modulated (PWM) to drive a single-color laser diode. At the receiver side, it integrates an optical signal receiver, optical power harvester, and an ultralow-voltage audio driver to drive the actuator. In order to detect the PWM signal and harvest energy at the same time, the input signal is AC coupled to the audio driver. Here the photodiode array performs the similar function as solar cells. The photocurrent is rectified through a diode made up of native MOSFET, and the DC current is stored on a capacitor C_{dd}. It provides a low dropout DC voltage to power the audio driver. Meanwhile, the photocurrent is also AC coupled to a transimpedance amplifier (TIA). It converts the input signal into voltage domain, so as to drive a hysteresis-based, self-oscillation PWM class-D audio amplifier. The transducer is driven by the PWM-modulated signal and restores the audio waveform by its band-pass filtering nature [34]. As the active circuits are powered by the photodetectors, ultralow-voltage (ULV) circuit techniques are required to realize the audio driver.

2.3 Circuit Design

2.3.1 Audio Driver

The circuit architecture of the hysteresis-based self-oscillation modulator is shown in Fig. 32. A first-order class-D audio amplifier with a fully differential architecture is utilized to generate push-pull current through the actuator and suppress even-order harmonic distortion. Compared to other pulse width or pulse-density modulated audio driver [34, 35], it does not require extra clock source on the membrane transducer for signal restoration. Thus it benefits from a smaller form factor and lower power consumption in this application scenario.

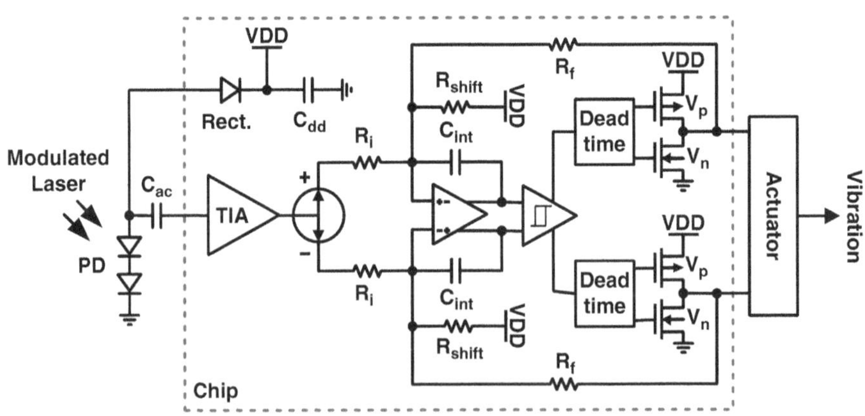

Fig. 32 Hysteresis-based self-oscillation modulation architecture

Let V_i be the input signal swing, V_h be the hysteresis window of the comparator, and V_p be the supply voltage. The modulation index (M) of the modulator is defined as

$$M = \frac{V_i}{V_p} \tag{35}$$

and

$$M = \frac{V_h}{V_p} \tag{36}$$

The oscillation frequency of the audio amplifier can be described as

$$f_{PWM} = \frac{1 - M^2}{4 \cdot H \cdot RC} \tag{37}$$

As the oscillation frequency depends on the signal amplitude, the power spectral density of the modulated signal will not concentrate on a single tone. It reduces EMI and also power dissipation [34, 35].

2.3.2 Transimpedance Amplifier

The transimpedance amplifier (TIA) is an inverter-based shunt-shunt feedback amplifier, which converts the photocurrent into differential output voltages (V_{i+}, V_{i-}), as is shown in Fig. 33. To enhance the transconductance of gain stage, back-gate bias circuit technique is adopted to adjust the threshold voltages of the active devices by V_p and V_n.

2.3.3 Bias Generator

Under a low supply voltage constraint, it becomes very challenging to implement an operational amplifier with sufficient gain under PVT variations. To overcome this problem, a back-gate bias generator is implemented to provide adequate bias voltages to the amplifier and maintain its operation at the high-gain region. Figure 34a illustrates an inverter-based gain stage, and Fig. 34b shows the corresponding back-gate bias generator. The bias generator is composed of a replica of the gain stage (M_{p1}, M_{n1}), followed by a two-stage error amplifier (M_{p2}, M_{n2}, M_{p3}, M_{n3}). The second-stage error amplifier is Miller compensated to maintain a sufficient phase margin. For a DC operating point of VDD/2 at V_1, the desired output voltage at V_2 is also targeted at VDD/2. If V_2 is lowered than the expected value, the error amplifier output V_4 becomes lower and so does V_{bp} and V_{bn}. It turns out that the PMOS threshold voltage $|V_{TP}|$ is reduced, while NMOS threshold voltage

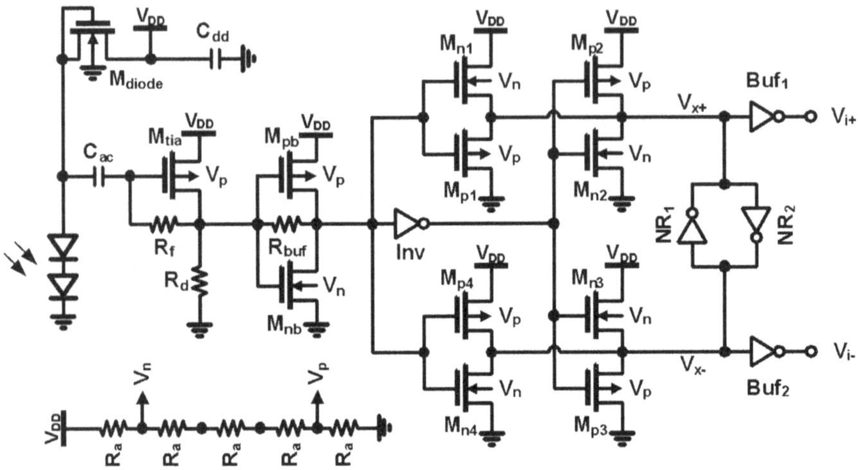

Fig. 33 TIA architecture

V_{TN} is increased. Thus the output of the amplifier becomes close to VDD/2 for a high-gain operation through the negative feedback biased scheme [36].

2.3.4 Operational Amplifier

The OPA in the class-D audio amplifier is composed of two-stage fully differential amplifiers in cascade. Figure 35a, c, respectively, illustrate the first and second stages of the operational amplifier. In order to comply with the DC level of the cascaded gain cells and also for a maximum output swing, their output common mode voltages are preset to 0.7 V_{dd} (0.42 V) and 0.5 V_{dd} (0.3 V) through the output common mode feedback amplifiers, as are shown in Fig. 35b, d. The common mode feedback is achieved by adjusting the back-gate bias of the transconductance gain stage (V_{n1}) and current source (V_{p1}). It reduces the threshold voltage for a maximum gain [37]. To boost the amplifier gain under a low supply voltage, cross coupled negative resistance (M_{nr1}-M_{nr2}) and (M_{nr3}-M_{nr4}) are also added in parallel with the output loads at the first and second stages, respectively.

2.3.5 Hysteresis Comparator

The hysteresis comparator is shown in Fig. 36. Its operation is similar to the gain cell of the OPA except that the input signal is coupled to the back gate for a wider dynamic range under ULV operation. Also, a stronger cross coupled negative impedance converter is adopted to adjust the hysteresis window. The gain cells and hysteresis comparator shares the same common mode feedback amplifier

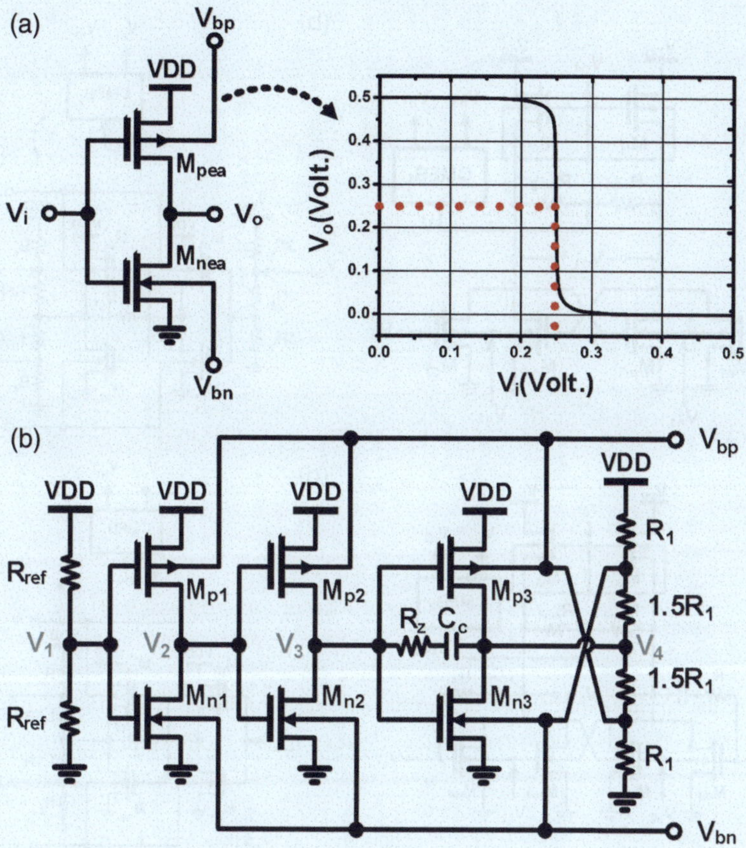

Fig. 34 (**a**) Inverter-based amplifier. (**b**) Bias generator

architecture. By adjusting the back-gate biased voltage, the threshold voltage of the inverter-based amplifier is preset to its output common mode voltage (V_{cm3}) for a high-gain operation.

2.3.6 Output Stage

The output stage behaves as a current switch that delivers pulse width modulated current to the transducer. To avoid short circuit current at the output stage, the input signal V_i passes through a non-overlapped clock generator before being fed into the power MOSFET M_p and M_n, as is shown in Fig. 37.

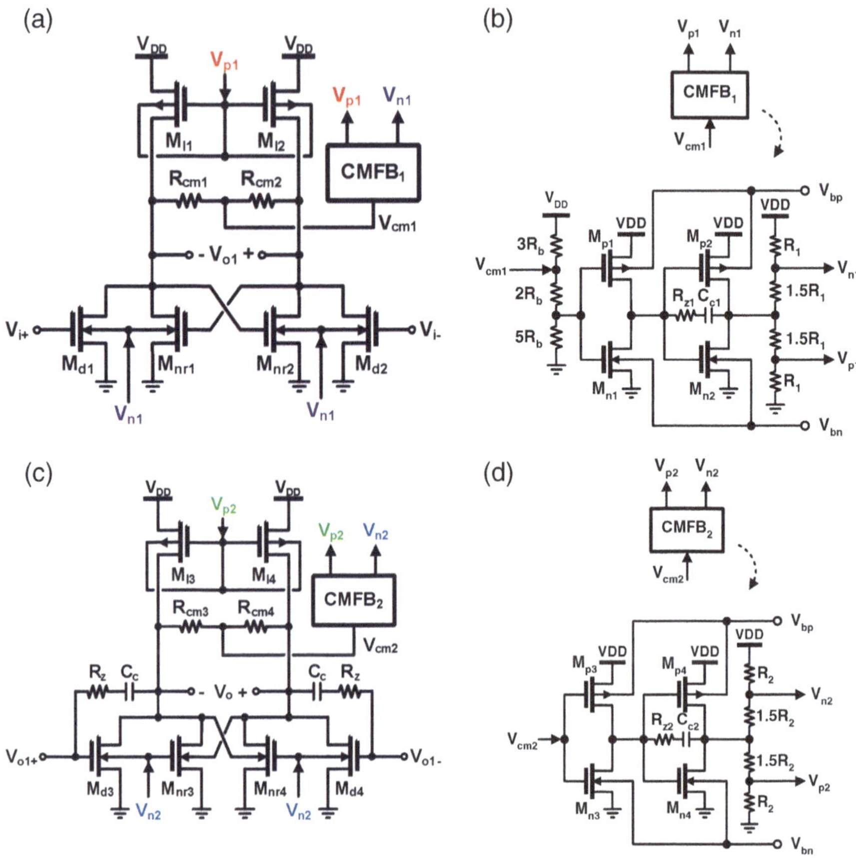

Fig. 35 (**a**) The first stage. (**b**) The first stage CMFB. (**c**) The second stage. (**d**) The second stage CMFB of the operational amplifier

3 Experimental Results

To demonstrate the design concept, PWM audio signal is utilized to modulate VCSEL TTR-1F45-427 at the sound processor. Meanwhile, the optical power is picked up by monitor PDs, whose responsivity is about 1.5 A/W. The optical power harvester regenerates a 0.6-V supply to power the class-D audio amplifier. The photocurrent generated from the PD array is also AC coupled to the audio driver. The overall audio performance is measured by Audio Precision AP2700.The audio driver for the transducer consumes 1.66 mW. The driver output is also connected to a transducer attached to an animal membrane for performance test.

Figure 38a shows the cross-sectional view of the mechanical actuator, which consists of a permanent magnet, an aluminum ring, two opposite direction winding coils, and a Provil novo membrane [33]. By alternatively driving the electrical wires

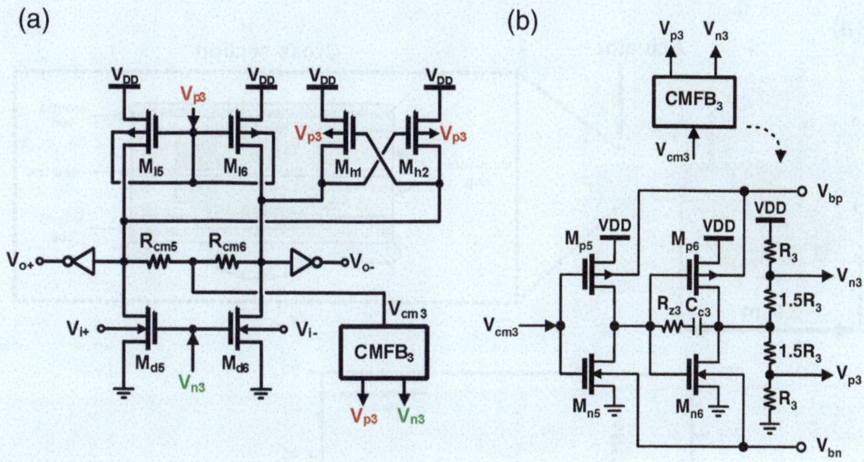

Fig. 36 (**a**) Hysteresis comparator. (**b**) Hysteresis comparator CMFB

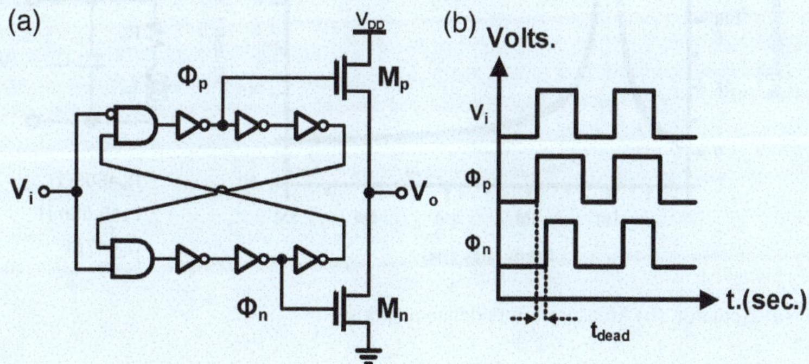

Fig. 37 (**a**) Gate driver. (**b**) Waveform

surrounding the magnets, the actuator is capable of generating stable vibration with minimum current consumption to simulate tympanic membrane, thus resulting in a significant improvement in energy efficiency. Figure 38b shows the measured impedance response of the actuator. The mechanical behavior of the membrane transducer is verified by using Polytec OFV508/OFV2802 laser Doppler vibrometer. Figure 39 shows the overall THD+N frequency response. It shows that the best-case THD+N is 0.4%. Figure 40 shows the power conversion efficiency, which is about 24.7% at the best case. The maximum output power delivered to the transducer is 0.408 mW. Figure 41 shows the chip micrograph. Fabricated in TSMC 90 nm CMOS process, the chip size is $0.88 \times 0.84\ \text{mm}^2$.

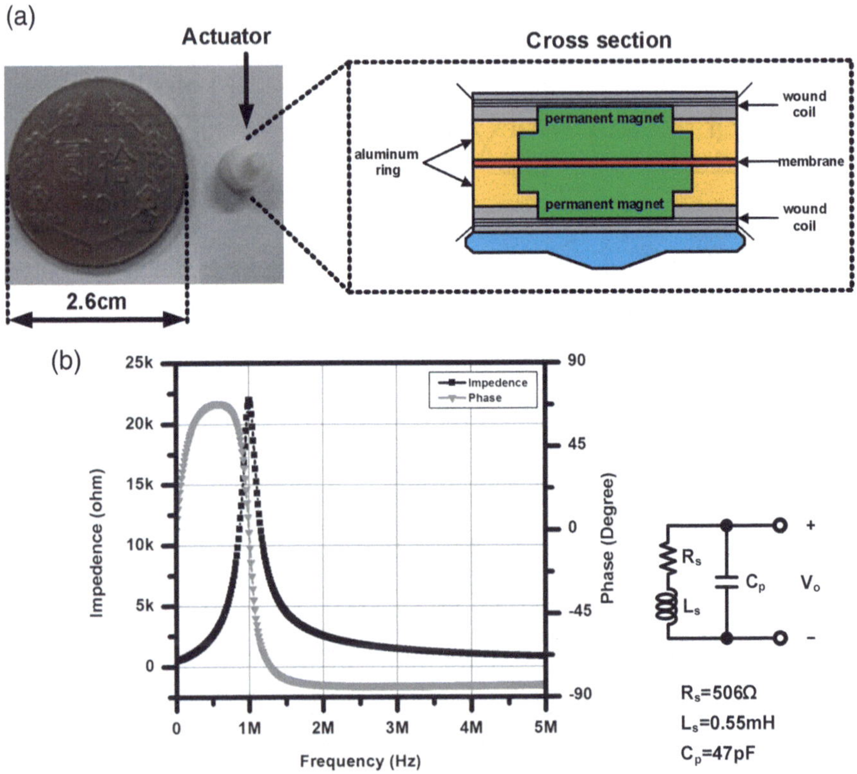

Fig. 38 (**a**) Actuator. (**b**) Measured impedance response

4 Conclusions

This chapter describes two application scenarios of OEICs, covering from high-speed interconnects to tympanic membrane transducer in a hearing aid device. Incorporating the proposed spatially modulated photodetectors, the fully integrated CMOS OEICs are capable of operating at tens of Gb/s and providing multichannel links with cross talk less than 0.1 dB. Additionally, nested-feedback transimpedance amplifier and comparator-based receiver are presented for low noise and high sensitivity operation. On the other hand, a light-driven tympanic membrane (TM) transducer of hearing aid device with signal and power transfer is presented. The energy harvester incorporates with ultralow-voltage audio driver to mechanically stimulate tympanic membrane (TM) transducer. It improves sound quality while avoiding occlusion effects. The class-D audio amplifier is based on a self-oscillation architecture, thus no extra clock source is required. The measured THD+N is 0.4% over the audio bandwidth with modulation index of 0.4, and the maximum power conversion efficiency is 24.7%. Compared to the prior art, the

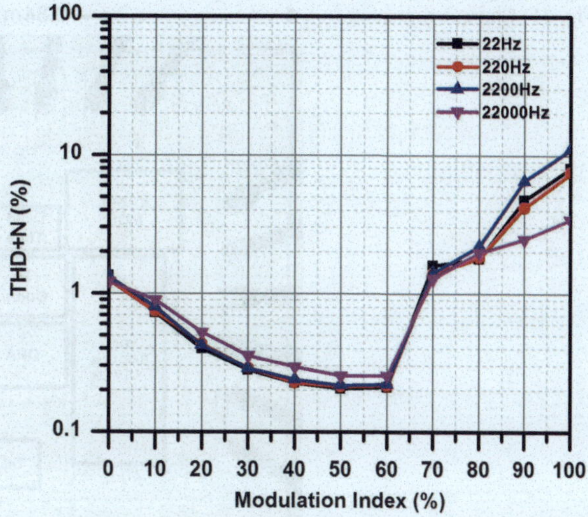

Fig. 39 THD + N performance

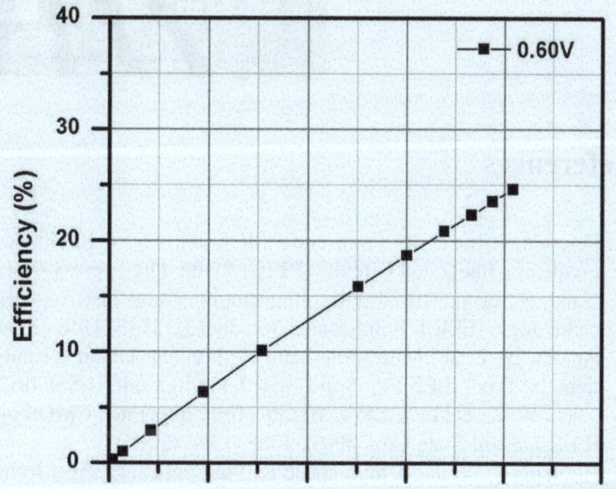

Fig. 40 Power conversion efficiency

proposed architecture requires only a single wavelength LD and PD. Thus no color filters are required to facilitate TM transducer design. By heterogeneous system integration, they successfully demonstrate the potentials of OEICs in the versatile of applications.

Fig. 41 Chip micrograph

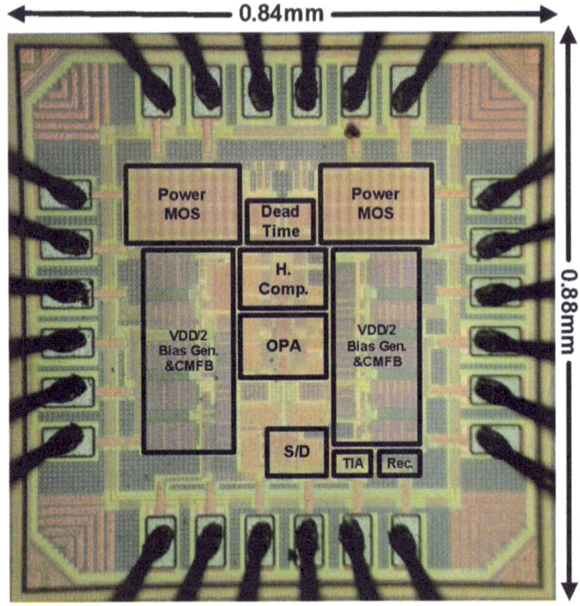

References

1. Petersen, A.K., et al.: Front-end CMOS chipset for 10 Gb/s communication. In: IEEE Radio Frequency Integrated Circuits (RFIC) Symp. Dig., pp. 93–96 (2002)
2. Galal, S., et al.: 10-Gb/s limiting amplifier and laser/modulator driver in 0.18-μm CMOS technology. IEEE J. Solid-State Circ. **38**(12), 2138–2146 (2004)
3. Analui, B., et al.: Multi-pole bandwidth enhancement technique for trans-impedance amplifiers. In: Proc. IEEE Eur. Solid-State Circuits Conf. (ESSCIRC), pp. 303–306 (2002)
4. Chen, W.-Z., et al.: A 1.8-V, 10-Gb/s fully integrated CMOS optical receiver analog front-end. IEEE J. Solid-State Circ. **40**(6), 1388–1398 (2005)
5. Hermans, C., et al.: A high-speed 850-nm optical receiver front-end in 0.18-μm CMOS. IEEE J. Solid-State Circ. **41**(7), 1606–1614 (2006)
6. Csutak, S.M., et al.: High-speed monolithically integrated silicon optical receiver fabricated in 130-nm CMOS technology. IEEE Photon. Technol. Lett. **14**(4), 516–518 (2002)
7. Swoboda, R., et al.: 11 Gb/s monolithically integrated silicon optical receiver for 850 nm wavelength. In: IEEE Int. Solid-State Circuits Conf. (ISSCC) Dig. Tech. Papers, pp. 904–911 (2006)
8. Chen, W.-Z., et al.: A 2.5 Gbps CMOS fully integrated optical receiver with lateral PIN detector. In: Proc. IEEE Custom Integrated Circuits Conference (CICC), pp. 293–296 (2007)
9. Woodward, T.K., et al.: 1-Gb/s integrated optical detectors and receivers in commercial CMOS technologies. IEEE J. Sel. Top. Quantum. Electron. **5**(2), 146–156 (1999)
10. Rooman, C., et al.: Asynchronous 250 Mb/s optical receivers with integrated detector in standard CMOS technology for optocoupler applications. IEEE J. Solid-State Circ. **35**(7), 953–958 (2000)
11. Jutzi, M., et al.: 2-Gb/s CMOS optical integrated receiver with a spatially modulated photodetector. IEEE Photon. Technol. Lett. **17**(6), 1268–1270 (2005)
12. Chen, W.-Z., et al.: A 3.125 Gbps CMOS fully integrated optical receiver with adaptive analog equalizer. In: Proc. IEEE Asian Solid-State Circuits Conference (A-SSCC), pp. 396–399 (2007)

13. Tavernier, F., et al.: Power efficient 4.5 Gbit/s optical receiver in 130 nm CMOS with integrated photodiode. In: Proc. IEEE Eur. Solid-State Circuits Conf. (ESSCIRC), pp. 162–165 (2008)
14. Radovanović, S., et al.: A 3-Gb/s optical detector in standard CMOS for 850-nm optical communication. IEEE J. Solid-State Circ. **40**(8), 1706–1717 (2005)
15. Sze, S.M.: Physics of Semiconductor Devices. Wiley. John Wiley & Sons, Inc., Hoboken, NJ, USA, Canada (2007)
16. Huang, S.-H., et al.: A A 2 × 20-Gb/s, 1.2-pJ/bit, time-interleaved optical receiver in 40-nm CMOS. In: IEEE Asian Solid-State Circuits Conference (A-SSCC), pp. 97–100 (2014)
17. Huang, S.-H., et al.: A 10-Gb/s OEIC with meshed spatially-modulated photo detector in 0.18-m CMOS technology. IEEE J. Solid State Circ. **46**(5), 1158–1169 (2011)
18. Säckinger, E.: Boardband Circuits for Optical Fiber Communication. John Wiley & Sons, Inc., Hoboken, NJ, USA (2005)
19. Palermo, S., et al.: A 90 nm CMOS 16 Gb/s transceiver for optical interconnects. In: IEEE Int. Solid-State Circuits Conf. (ISSCC) Dig. Tech. Papers, pp. 44–45 (2007)
20. Honarvar, M., et al.: An 18.6Gb/s double-sampling receiver in 65nm CMOS for ultra-low-power optical communication. In: IEEE Int. Solid-State Circuits Conf. (ISSCC) Dig. Tech. Papers, pp. 130–132 (2012)
21. Georgas, M., et al.: A monolithically-integrated optical receiver in standard 45-nm SOI. IEEE J. Solid State Circ. **47**(7), 1693–1702 (2012)
22. Liu, F., et al.: 10 Gbps, 530 fJ/b optical transceiver circuits in 40 nm CMOS. In: IEEE Symp. on VLSI Circuits Dig. Tech. Papers, pp. 290–291 (2011)
23. Raj, M., et al.: A 4-to-11GHz injection-locked quarter-rate clocking for an adaptive 153 fJ/b optical receiver in 28 nm FDSOI CMOS. In: IEEE Int. Solid-State Circuits Conf. (ISSCC) Dig. Tech. Papers, pp. 404–405 (2015)
24. Huang, S.-H., Chen, W.-Z.: A 25-Gb/s, −10.8-dBm Input Sensitivity, PD-Bandwidth Tolerant CMOS Optical Receiver. IEEE Symposium on VLSI Circuits, pp.120–121 (2015)
25. Proesel, J., et al.: 25Gb/s 3.6pJ/b and 15Gb/s 1.37pJ/b VCSEL-based optical links in 90nm CMOS. In: IEEE Int. Solid-State Circuits Conf. (ISSCC) Dig. Tech. Papers, pp. 418–419 (2012)
26. Huang, T.-C., et al.: A 28Gb/s 1pJ/b shared-inductor optical receiver with 56% chip-area reduction in 28nm CMOS. In: IEEE Int. Solid-State Circuits Conf. (ISSCC) Dig. Tech. Papers, pp. 144–145 (2014)
27. Nazari, M.H., Emami-Neyestanak, A.: A 24-Gb/s double-sampling receiver for ultra-low-power optical communication. IEEE J. Solid-State Circuits. **48**(2), 344–357 (2013)
28. Takemoto, T., et al.: A 4 × 25-to-28 Gb/s 4.9-mW/Gb/s −9.7 dBm high-sensitivity 65-nm CMOS optical receiver for board-to-board interconnects. IEEE Int. Solid-State Circuits Conf. (ISSCC) Dig. Tech. Papers, pp. 118–119 (2013)
29. Ricketts, T.A., Hornsby, B.W.: Sound quality measures for speech in noise through a commercial hearing aid implementing digital noise reduction. J. Am. Acad. Audiol. **16**, 270–277 (2005)
30. Perkins, R.: Earlens tympanic contact transducer: a new method of sound transduction to the human ear. Otolaryngol. Head Neck Surg. **114**, 720–728 (1996)
31. Lee, C.-F., Shih, C.-H., Yu, J.-F., Chen, J.-H., Chou, Y.-F., Liu, T.-C.: A novel opto-electromagnetic actuator coupled to the tympanic membrane. J. Biomech. **41**, 3515–3518 (2008)
32. Puria, S., et al (n.d.): Optical electro-mechanical hearing devices with separate power and signal components. US patent NO. 0048982
33. Jian, J.-T., Song, Y.-L., Lee, C.-F., Chou, Y.-F., Chen, W.-Z.: A 0.6 V, 1.66mW energy harvester and audio driver for tympanic membrane transducer with wirelessly optical signal and power transfer. IEEE International Symposium on Circuits and Systems, pp. 874–877 (2014)
34. Lu, J., Gharpurey, R.: Design and analysis of a self-oscillating class D audio amplifier employing a hysteretic comparator. IEEE J. Solid State Circ. **46**(10), 2336–2349 (2011)
35. Berkout, M., Dooper, L.: Class-D audio amplifiers in mobile applications. IEEE Trans. Circ. Syst. I: Regular Papers. **57**(5), 992–1002 (2010)

36. Chatterjee, S., Tsividis, Y., Kinget, P.: 0.5V analog circuit technique and their application in OTA and filter design. IEEE J. Solid State Circuits. **40**(12), 2373–2387 (2005)
37. Park, Y.-S., Lee, S.W., Kong, B.S., Park, K.I., Ihm, J.D., Choi, J.S., Jun, Y.H.: PVT invariant single-to-differential data converter with minimum skew and duty-ratio distortion. In: Proceedings of the IEEE International Symposium on Circuits and Systems (ISCAS), pp.1902–1905 (2008)

Depth Estimation Using Single Camera with Dual Apertures

Hyun Sang Park, Young-Gyu Kim, Yeongmin Lee, Woojin Yun, Jinyeon Lim, Dong Hun Kang, Muhammad Umar Karim Khan, Asim Khan, Jang-Seon Park, Won-Seok Choi, Youngbae Hwang, and Chong-Min Kyung

1 Introduction

There is a huge demand for depth sensing from many computer vision applications. The most popular depth-sensing technology is a two-camera-based stereo vision system, which resembles the human binocular vision system. In typical stereo vision systems [1], two cameras are displaced horizontally from each other to obtain two different views of the same scene. The depth of the scene can be obtained by observing the disparity of the two images, as the disparity is directly related to depth. Due to the low computational complexity and relatively simple hardware, numerous stereo vision cameras have been commercially available [2, 3]. The availability of these cameras has led to the popularity of depth-based applications, which include hand gesture recognition [4], face detection [5], foreground segmentation [6], touchless fingerprint recognition [7], etc.

Since the stereo camera typically uses two cameras, its form factor has the structural limitation. To overcome this limitation, many other depth-sensing technologies have been proposed. Among these, structured-light-based [8] or time-of-flight (TOF)-based [9] methods have recently gained popularity. However, TOF sensors need at least two hardware modules: the IR sensor and the IR emitter; therefore, it

H.S. Park (✉)
Kongju National University, Cheonan-si, Chungcheongnam-do, South Korea
e-mail: vandammm@kongju.ac.kr

Y.-G. Kim • Y. Lee • W. Yun • J. Lim • D.H. Kang • M.U.K. Khan • A. Khan • J.-S. Park
W.-S. Choi • C.-M. Kyung
Department of Electrical Engineering, Korea Advanced Institute of Science and Technology (KAIST), Daejeon, South Korea

Y. Hwang
Korea Electronics Technology Institute, Seongnam-si, Gyeonggi-do, South Korea

H. Yasuura et al. (eds.), *Smart Sensors at the IoT Frontier*,
DOI 10.1007/978-3-319-55345-0_7

consumes high power due to the active IR emission. Furthermore, the TOF sensors cannot be used in sunlight.

To avoid an additional module, single camera-based solutions also have been proposed. In optics, depth of field (DOF) is the range of distance from the camera (lens or sensor) within which objects in a scene that appear acceptably sharp in an image. Objects beyond the DOF will appear blurry as they are de-focused. Thus, the sharpness or blurriness of an object in an image is dependent on its distance from the camera. The aperture size and the focal length of the camera also affect the level of sharpness (or blurriness). Observing the same scene with different optical parameters can allow depth estimation of the scene. The practical way to capture such two images is to take two images sequentially with different optical parameters such as aperture size [10], aperture shape [11], or focal length [12]. Although such approaches show decent depth estimation performance, such approaches require objects to be static in the scene to maintain the correspondence between two temporally adjacent images.

Monocular depth extraction based on defocus or blur has also been proposed. In this approach, the level of blurriness along the edges is estimated and then translated to appropriate depth values [13]. The primary assumption of this approach is that all edges must be sharp when focused. In other words, the blur induced in the image is purely due to depth. In practice, however, there are lots of objects which do not follow this assumption as these are inherently blurry. For reliable depth estimation based on blur, the relative blur has to be considered instead of measuring the absolute levels of blur.

Regardless of practical limitations, depth estimation from blur is a cost-effective technique as it requires single lens optics with only one camera, there is no correspondence problem, and no active light source is used. These characteristics allow realization of a compact and portable depth camera. Recently, [14] proposed a novel single camera system with dual apertures, thereby called as dual-aperture (DA) camera. It is equipped with a RGB-IR image sensor with a larger aperture for visible light and a smaller aperture for IR light. Due to the different aperture sizes, an object will show different levels of blur in the RGB and IR channels which can be used to estimate depth. It can capture sharp and blurry grayscale images simultaneously with a single shot.

In this paper, we propose a depth estimation pipeline based on the DA camera. Reconstructing RGB and IR images from the RGB-IR sensor is not considered. This paper is composed as follows. In Sect. 2, we briefly describe the DA camera system and the principle of depth estimation. In Sect. 3, the proposed depth estimation pipeline is described. Section 4 shows the experimental results of the performance of the proposed method. The paper is concluded in Sect. 5.

2 Dual-Aperture Camera

The dual-aperture (DA) camera differs from conventional cameras in two ways: (1) it enables the sensor to respond to IR spectrum in addition to visible light spectrum, and (2) it uses two separate apertures through the optical path, one for visible

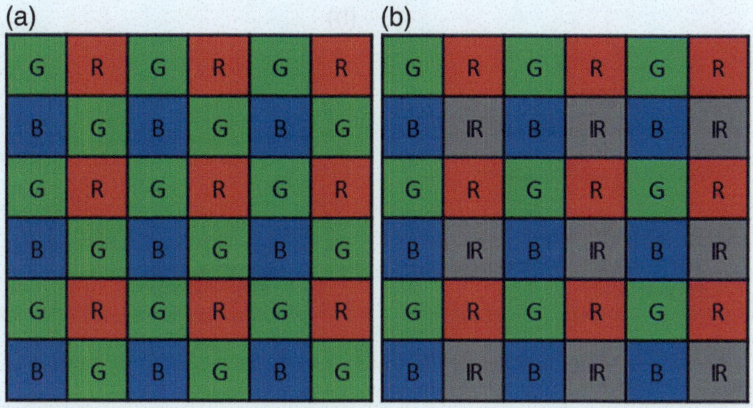

Fig. 1 CMOS image sensors where (**a**) shows typical 2 × 2 Bayer pattern and (**b**) shows the pattern used in DA

spectrum and the other for IR spectrum. Light in the visible spectrum is allowed through the larger aperture, while all incident lights (including visible and near-IR) are allowed through the smaller aperture. This results in the sensor being able to capture an image where the IR channel has a larger DOF than the other three-color channels. By comparing two grayscale images from any of the three-color channels with the IR image, depth can be extracted.

2.1 Color Filter Array

A conventional digital camera uses a sensor that is made up of light-sensitive photodiodes. Normally these pixels are coated with pigments composed of three primary colors such as red, green, and blue as shown in Fig. 1a. Each pixel captures light corresponding to one of the three colors. Currently one of the most common color filter arrays in image sensors is the 2 × 2 Bayer pattern. In a DA camera, the sensor is modified such that one of the two green pixels is replaced by an IR pixel as shown in Fig. 1b. Ideally the infrared wavelengths should only be sensitive to the IR pixels.

2.2 Camera Module Architecture

The aperture comprises of two parts. One is a lens aperture with a large hole for obtaining blurry RGB channels and the other smaller aperture for obtaining a sharp IR channel as shown in Fig. 2b. With these two apertures, a DA camera can simultaneously take two grayscale images with different levels of blur.

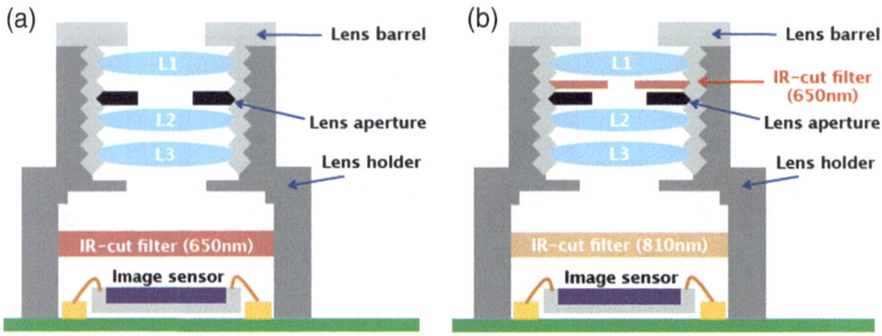

Fig. 2 DA camera module structure. (**a**) Conventional signal aperture camera. (**b**) Dual-aperture camera

2.3 Spectral Characteristic

The spectral characteristic of a DA camera is shown in Fig. 3b. In conventional cameras, a cutoff filter of 650 nm has to be coated before the image sensor such that IR signal is blocked as depicted in Fig. 3a. Thus the R, G, or B pigment is designed to pass the corresponding primary color as well as IR spectrum as shown in Fig. 3b. Typically, the IR sensitivity of R, G, and B pixels is not an issue as in a conventional camera the IR cutoff filter is already employed mandatory.

In a DA camera, the near-IR spectrum between 650 and 810 nm is used to form a sharp image channel. This is why an IR cutoff filter of 810 nm is used instead of 650 nm. However, the IR signal between 650 and 810 nm also is projected onto R, G, and B pixels, which can saturate them quickly. This IR crosstalk can be effectively reduced by using a small IR aperture. By reducing the relative IR aperture area to the RGB aperture area, in practice, the effective amount of IR crosstalk onto R, G, and B pixels is reduced. Thus, the small aperture size for IR spectrum is crucial for providing the sharper IR channel as well as reducing the IR crosstalk onto the blurry R, G, or B channel.

2.4 Depth Estimation Principle

Figure 4 provides the geometric representation of a camera system to formulate the relation between the blur size and the object distance. The amount of blur, B_{size}, of an object is given by

$$B_{\text{size}} = A_{\text{rgb}} \left| 1 - \frac{v}{v_0} \right|, \text{where } v_0 = \frac{\mathrm{F}x_z}{x_z - F} \tag{1}$$

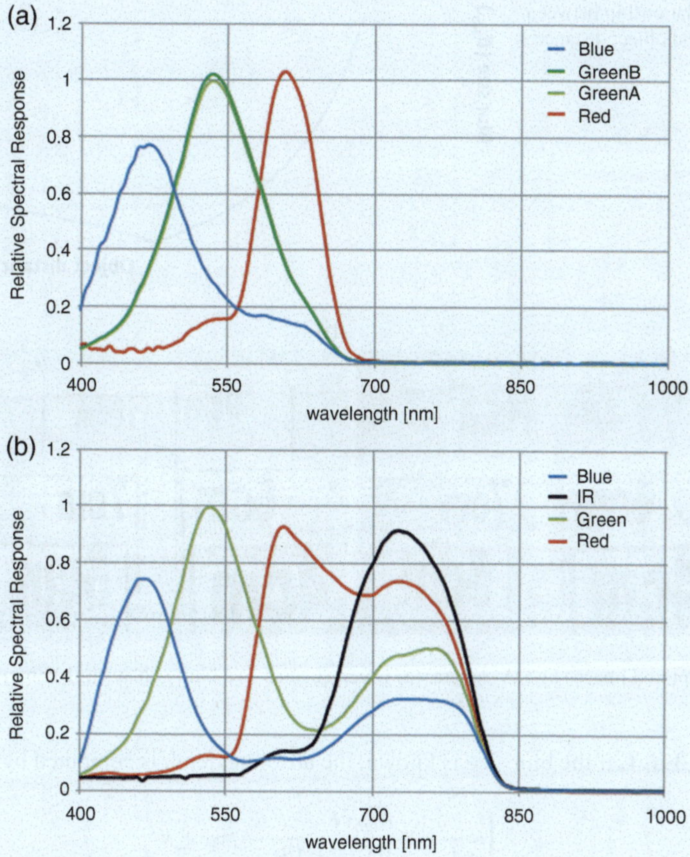

Fig. 3 Spectral characteristic (**a**) conventional signal aperture camera coated with an IR-cut filter and (**b**) DA camera

Fig. 4 Relationship between depth and blur size

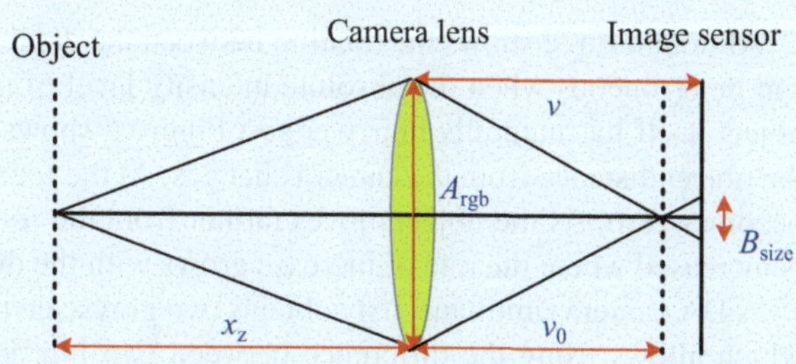

where F is the focal length (fixed for each lens module), A_{rgb} is the diameter of the RGB aperture, x_z is the absolute distance of the object from the camera, v_0 is the distance where the object would be when in-focus, and v is the distance between the lens and the image sensor.

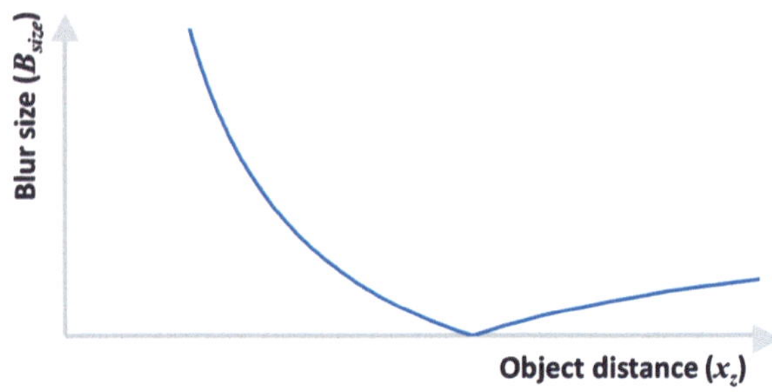

Fig. 5 Relationship between blur size and object distance

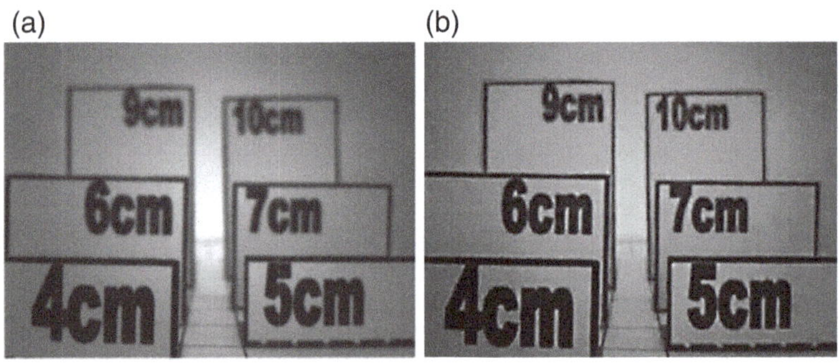

Fig. 6 Captured images by DA camera. (**a**) Green channel as a blurry channel. (**b**) Sharp channel

Using Eq. 1, if the blur size is known, the absolute depth is computed by

$$x_z = \begin{cases} \dfrac{A_{\mathrm{rgb}} v F}{v A_{\mathrm{rgb}} - \mathrm{F}A_{\mathrm{rgb}} + \mathrm{F}B_{\mathrm{size}}}, v_0 \geq v \\ \dfrac{A_{\mathrm{rgb}} v \mathrm{F}}{v A_{\mathrm{rgb}} - \mathrm{F}A_{\mathrm{rgb}} - \mathrm{F}B_{\mathrm{size}}}, v_0 < v \end{cases}. \tag{2}$$

Although Eq. 2 shows the relation between the depth and the blur size, the result can be erroneous when the absolute intensity level of a blurry object is low or the object itself has naturally blurry edges. Figure 5 shows how the blur size varies as the object distance from the camera changes. At the focal point, ideally the blur size becomes zero. As the object moves farther from the focused distance, the blur size is increased where the rate of increase grows with the distance.

A DA camera simultaneously obtains two grayscale images with different DOFs, which allows using the difference between two blur sizes for the same object in estimating its depth. The depth information is extracted by comparing the difference of blur between the sharp (IR) and blurred (RGB) channels. One of the visible color channels can be selected as a blurry channel and the IR channel as the sharp channel. Figure 6 shows an example pair of blurry and sharp channels.

Fig. 7 PSF-based image blurring

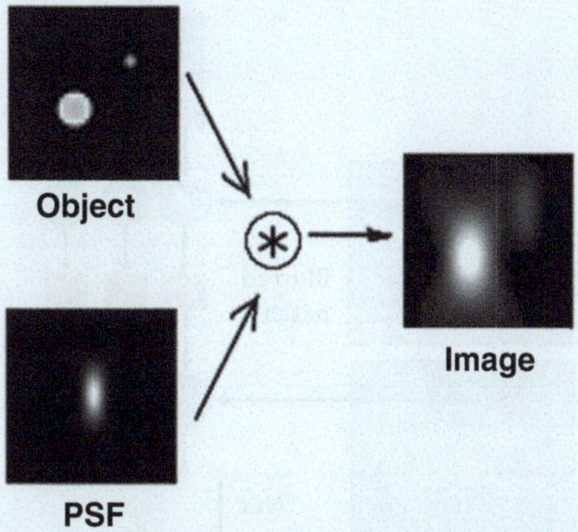

2.5 *Depth Estimation Algorithm*

In image-based depth estimation techniques, reliable depths can be obtained only along edges where the variation in intensity is observed. In stereo vision systems, the disparities can be properly estimated around edges that have a variation in intensity in directions other than parallel to the epipolar line. In DFD-based systems, the blur sizes are estimated around all edges. However, measuring the absolute level of blur is not considered to be reliable enough to estimate depth because some edges are naturally blurry. Accordingly, two images of the scene are observed through different apertures. The difference of two blur sizes for the same edge will be consistent regardless of the edge's inherent blurriness. This naturally follows that estimation of depth requires a technique for estimation of blur along the edges.

The point spread function (PSF) defines the response of an imaging system to a point source. In other words, the PSF is considered as the impulse response of the imaging system. The blurring function by the PSF can be modeled by a Gaussian distribution. The only parameter for a Gaussian distribution is σ, which can be regarded as a representative number for the level of blurriness. Figure 7 shows an example image which is blurred by a given PSF.

In a single shot, a DA camera captures an IR image with a small aperture and a RGB image with a larger aperture, respectively. The objects in the IR image will be all-in-focus and have focused sharp edges, while those in the RGB image will have blurry edges. According to Eq. 2, the degree of blurriness is dependent on the object distance, the aperture size, and the focal length. However, it can be simplified to be a function of the object distance since all other parameters typically remain constant for a given camera system.

The core of the depth estimation algorithm is to find the blur difference between the sharp and blurry images. The blur difference σ_B is chosen among possible N values of $\sigma_k \in \{\sigma_1, \sigma_2, \ldots, \sigma_N\}$, which shows the best similarity between the observed blurry channel and the Gaussian-blurred sharp channel with σ_B. Mostly,

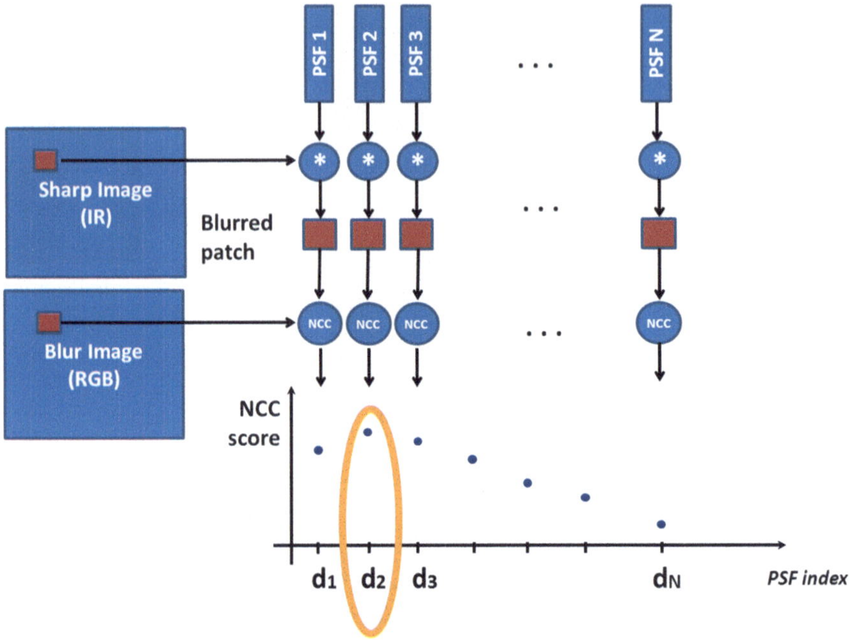

Fig. 8 PSF-based blur size estimation framework

the green channel is selected as a blurry channel due to its high SNR. According to Eq. 3, choosing the best blur size σ_B is same as deciding the PSF index p^* for the corresponding PSF function.

$$\begin{aligned}\sigma_B &= \sigma_{p^*} \\ p^* &= \underset{\sigma_k \in \{\sigma_1, \sigma_2, \cdots, \sigma_{\text{MAX}}\}}{\arg\max} \text{NCC}\left(I_{\text{IR}} {}^* G(\sigma_k), I_G\right) \\ \text{NCC}\left(I_{\text{IR}}, I_G\right) &= \frac{\text{COV}\left(I_{\text{IR}}, I_G\right)}{\sigma_{I_{\text{IR}}}\sigma_{I_G}} = \frac{\overline{I_{\text{IR}} \cdot I_G} - \overline{I_{\text{IR}}} \cdot \overline{I_G}}{\sigma_{I_{\text{IR}}}\sigma_{I_G}}\end{aligned} \tag{3}$$

The PSF estimation is performed in two steps, as depicted in Fig. 8. At each pixel along edges, a small image patch is taken from the sharp and blurry channels. The patch size must be large enough to cover the largest Gaussian PSF under consideration. In the first step, a matching score is calculated with each Gaussian PSF by performing NCC (normalized cross correlation) between the given blurry and PSF-blurred IR patches. The second step is to find the best matching PSF by finding the index whose corresponding NCC value is the largest. The distance of the object is then derived from the PSF index.

3 Proposed Depth Estimation Procedure

The proposed depth estimation pipeline is divided into edge extraction and sparse depth extraction stages. The edge extraction is a preprocessing step to prepare data for blur size estimation. It is composed of two functions, namely, demosaicking for inter-color edge alignment (DEA) and multi-scale space edge extraction (MEE). When the PSF-based blur size estimation is applied to flat regions where no edges exist it is extremely difficult to find the appropriate PSF index since all NCC score with different PSFs are very similar. The edge extraction step is to find appropriate pixels where reliable depth estimation is possible and to avoid redundant computational overhead at flat regions. As a result of the preprocessing step, an edge map is obtained.

The sparse depth extraction stage is composed of several functions which include adaptive blur channel selection (ACS), two-dimensional jittered matching (TJM), compensation for specular reflection (CSR), hierarchical selective blurred image interpolation (HIS), and depth noise reduction (DNR). Brief details are given in the following.

There are three possible candidates for a blurry image: R, G, or B channel. Due to illuminant's spectral distribution, some objects in the IR image may not appear in one of blurry channels. ACS is a process to choose the best blurry image which provides a more distinctive NCC score.

Due to chromatic aberration and/or aperture misalignment, the R, G, and B images are slightly misaligned. This effect is exaggerated especially at the periphery of the sensor. It is not reliable to determine the correlation between two misaligned patches. TJM is to compensate for the shift between color channels such that the appropriate correlation can be calculated.

Strong light reflection may happen at the edges of reflective objects, which adds impulsive noise near the center of the edges. Such specular reflection makes both edges from the sharp and blurred images to be seen highly dissimilar. For blur size estimation, the preservation of the edge slope at the right location is more important than their exact shapes. CSR is to shift two edges under comparison such that only the informative edge slopes can be compared through NCC. Although CSR cannot correct depth at specular highlights, it helps to remove resultant depth errors.

Basically PSF blurring is a computationally heavy process. To reduce such heavy computational overhead, PSF blurring can be performed only at the critical PSF indices, while interpolation between PSF-blurred images is conducted for the other PSF indices. Using HIS allows negligible quality degradation at the drastic reduction of computational cost.

The final step is noise reduction around edges with the assumption that the depth values along the connected edge can be modeled by a constant or linear model. The extracted depth values (or PSF indices) on a connected edge are gathered together to find the linear depth which minimizes the mean squared errors. The detailed description of each function will be given in the following subsections.

3.1 Demosaicking for Inter-color Edge Alignment

Image interpolation is the problem of generating a high-resolution image from low-resolution image with few artifacts. Actually, the interpolation model which describes the relationship between high-resolution image and low-resolution image plays a critical role in the performance of an interpolation algorithm. There are various kinds of interpolation models such as linear model (e.g., bicubic, bilinear, LADI [15]) and covariance-based model (e.g., NEDI [16]). Usually, the linear model interpolation is preferred for computational simplicity. However, when the linear model interpolation does not consider the right edge direction around an edge, results of the linear model interpolation show blurred images and various artifacts. Though many linear model interpolation algorithms which consider edge directions have been proposed, these algorithms are based on the three-color filter array as shown in Fig. 1a.

The goal of the proposed interpolation method is to have fewer artifacts in complex regions where irregular edges and texture details exist. In order to produce a better interpolated image, the green channel is reconstructed first because it has more edge information than the red or blue channels. Because the green channel possesses most of the spatial information of the image to be interpolated, it has great influence on the perceptual quality of the image. Once the reference green channel is fully produced, the green channel can be used to guide the red and blue channels by providing edge information.

For interpolation of multiple color channels, one of color channels needs to be interpolated first and then to be referenced to interpolate the other color channels. In most interpolation algorithms, the green channel is chosen as the reference channel that has to be interpolated first. In order to make reference green channel, we first utilize the color difference between channels to make the green channel in advance and then calculate the approximation of the second-order derivative and gradient values along interpolation directions. The approximation of the second-order derivative value is used for determining interpolation direction at first. After determining the interpolation direction, gradient values are used as weight factor for providing more accurate edge information. By using these calculated values, we can carry out the interpolation. Finally, we apply a post-processing to remove artifacts.

Then red and blue channels are interpolated referring to the green channel. This results in edge-aligned R, G, and B channels because when red and blue are interpolated, green channel's intensity is considered. As a result, red, green, and blue channel images are edge aligned with each other. Resultantly, the RGB image has less false color and has visually good quality. Figure 9 shows the simulation results by the proposed interpolation algorithm.

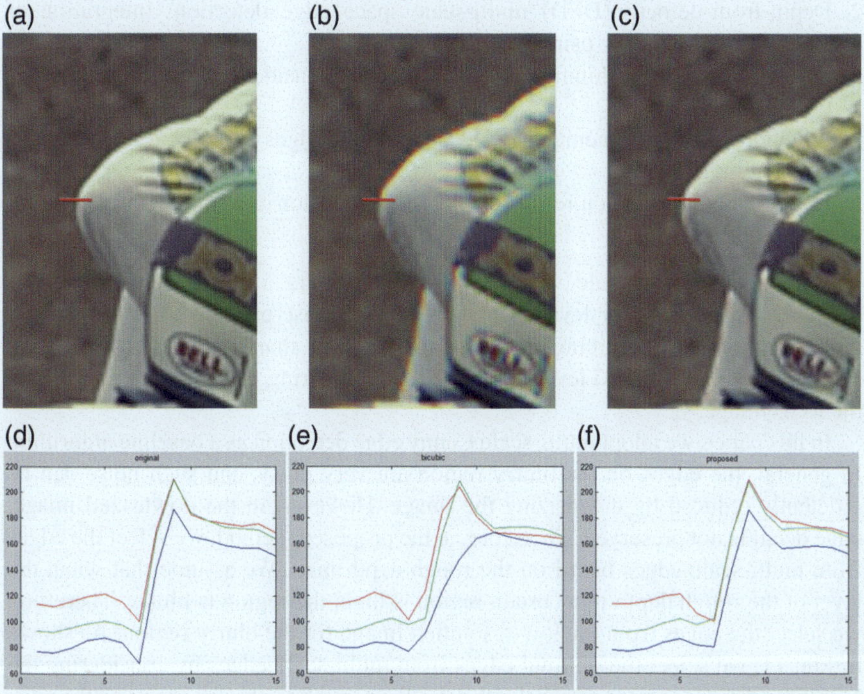

Fig. 9 RGB images with Kodak #5 with (**a**) original image, (**b**) result with bicubic interpolation, (**c**) result with proposed method, (**d**) edge profile for (**a**), (**e**) edge profile for (**b**), (**f**) edge profile for (**c**)

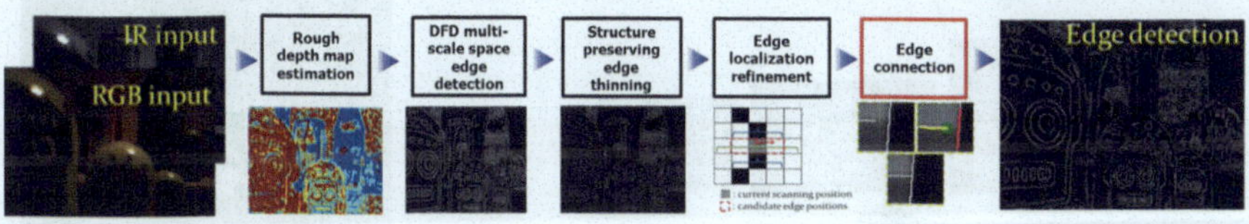

Fig. 10 Scale-space edge detection with rough depth map

3.2 *Multi-scale Space Edge Extraction*

The proposed edge detection algorithm robustly finds semantically meaningful edges in a blurry image. In addition, we incorporate the illumination insensitive normalized gradient (IING) approach [17] into the proposed algorithm to efficiently deal with illumination changes. The overall procedure of the proposed algorithm is described in Fig. 10. Functional descriptions are summarized as follows.

1. Rough depth map estimation: Estimation of a pixel-wise depth using NCC between IR and green channel images

2. Depth-from-defocus (DFD) multi-scale space edge detection: Integration of multi-scale space edge using rough depth map
3. Structure-preserving thinning: Edge thinning by considering local edge distribution
4. Edge localization refinement: Re-localization of edges at its local maximum of gradient magnitude
5. Edge connection: Structure-preserving edge extension and connection of disconnected edges

With user-defined scale space levels we generate a rough depth map that implicitly represents blur level of images in pixel-wise manner. The rough depth map is computed by enforcing the Gaussian blur to the sharp IR image, and then the best matched blur (depth) level is selected by comparing the blurry IR image and the RGB image.

In this paper, we adopt multi-scale Canny edge detection as a baseline algorithm. In general, the edges on the blurry region are very noisy, and such noise can be efficiently reduced by downsizing the image. However, in the downsized image, edge detail is not preserved. Therefore, in the proposed method, we select the edges from multi-scale edges based on the rough depth map. We assume that when the level of the rough depth in a certain region is high, the region is blurry. Therefore, we select the edges from the low-resolution image for the blurry region. As shown in Fig. 11, we select the optimal edges from multi-scale edges by considering the blur information from the rough depth map. By doing this, we can maintain the edge detail and reduce noisy edges.

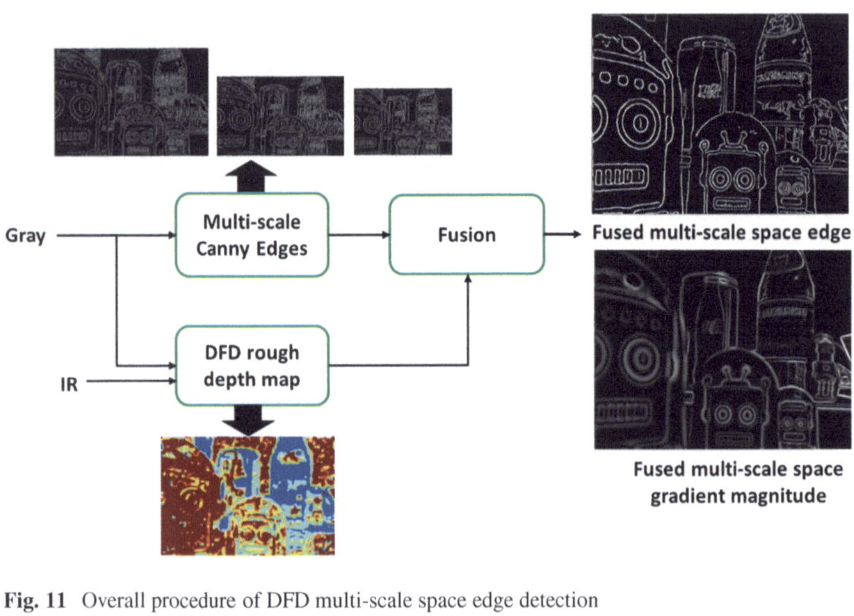

Fig. 11 Overall procedure of DFD multi-scale space edge detection

Robustness to illumination changes is an essential requirement of edge detector. In order to satisfy illumination insensitiveness, we adopt IING [17] instead of normalized gradient (NG) which has been widely used in edge detection. An input image can be divided into intrinsic and extrinsic parts. The intrinsic part is insensitive to illumination and sharp spatial variation and corresponds to textures. The extrinsic part is sensitive to illumination and low spatial variation and implies shading and illumination. IING suppresses extrinsic parts by element-wise division using $\widehat{W}$ and is described as follows:

$$\begin{aligned} \widehat{W} = I_{\mathrm{G}}{}^{*}G(\sigma_w), N_x = \frac{I_{\mathrm{G}}{}^{*}G_x(\sigma_s)}{\widehat{W}}, N_y = \frac{I_{\mathrm{G}}{}^{*}G_y(\sigma_s)}{\widehat{W}} \\ \nabla N = \sqrt{N_x^2 + N_y^2}, \nabla I_{\mathrm{IING}} = \frac{\nabla N}{\max(\nabla N)} \end{aligned} \tag{4}$$

where $G_x(\sigma_s)$ and $G_y(\sigma_s)$ are the one-dimensional horizontal and vertical Gaussian functions whose standard deviation is σ_s and $G(\sigma_w)$ is the two-dimensional Gaussian function whose standard deviation is $\sigma_w(>>\sigma_s)$.

Edge integration using DFD rough depth map contains different thickness of edges by pixel-wise depth level. In order to achieve the requirement of thin edge and preserve its original structure, we use the fast structure preserving thinning method [18]. Canny edge map often shows disconnected edges on junctions since a local magnitude maximum is not in the gradient direction. In order to connect them, we preemptively find sparse endpoints with detached edges having different orientation and extend them from endpoint until it reaches other edges using symmetry. Structure-preserving thinning violates the assumption that edge should reside on local maximum of gradient magnitude. In order to re-locate edges to its local maxima of gradient magnitude, we perform edge localization refinement by scanning in the horizontal and vertical direction and moving edges to their local maxima. We repeat scanning until the average IING on edge converges. Figure 12 shows the final edge detection results using the proposed edge detection scheme. "*Robots_G32_E330K*" and "*Robots_G32_E600K*" are from our test image set.

3.3 Adaptive Blur Channel Selection

To show the color dependency more clearly, three-color image sensors are to be considered. When the blue channel is chosen as the sharp channel, either red or green channel can be selected the blurry channel. In this case, the adaptive blur channel selection can use either red or green channel as the blurry channel adaptively pixel by pixel. This algorithm selects a blurry channel adaptively according to correlation values of all blurry channels for each patch. Then after

Fig. 12 DFD-MSS edge detection using IING

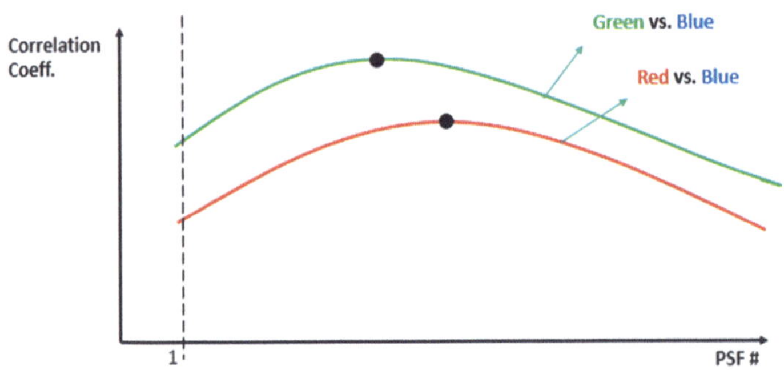

Fig. 13 Exemplary correlation coefficient curve where *blue* is used as sharp channel and either *red* or *green* is used as blur channel

the comparison of the correlation values of all blurry channels, we can choose the depth value which has the higher correlation in all channels as Eq. 5.

$$p^* = \underset{\sigma_k \in \{\sigma_1, \sigma_2, \cdots, \sigma_{\mathrm{MAX}}\}}{\arg\max} \max \left\{ \mathrm{NCC}\left(I_B{}^* G(\sigma_k), I_R\right), \mathrm{NCC}\left(I_B{}^* G(\sigma_k), I_G\right) \right\} \tag{5}$$

Direct application of Eq. 5 needs too much computational cost. From lots of experimental results, it is observed that desirable correlation curve tends to be higher than the other at all PSF indices. Figure 13 shows the case where the correlation curve of green vs. blue is always higher than the one of red vs. blue at all PSF indices. So we can select the blurry channel to use by calculating and comparing one correlation value.

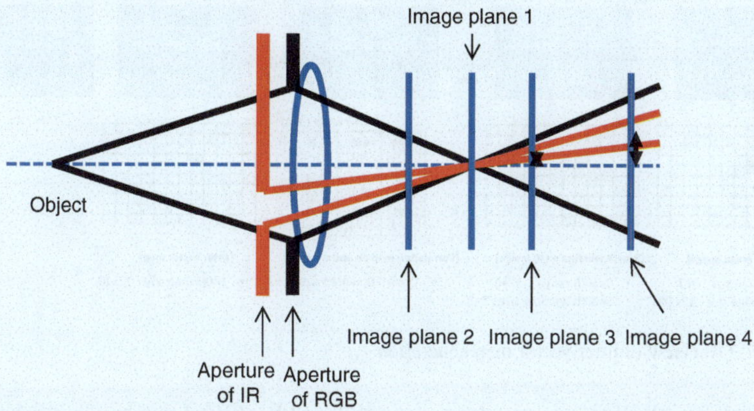

Fig. 14 Off-centered dual apertures

As mentioned above, the blurry channel to use can be selected only by calculating correlation for all blurry channels at the smallest value of PSF index. So, the blur channel selection equation is changed to Eq. 6. The revised equation takes almost same runtime compared to Eq. 5 without loss of quality.

$$\begin{aligned} i &= \underset{i \in \{R,G\}}{\arg\max} \{\mathrm{NCC}(I_i, I_B)\} \\ p^* &= \underset{\sigma_k \in \{\sigma_1, \sigma_2, \cdots, \sigma_{\mathrm{MAX}}\}}{\arg\max} \mathrm{NCC}\left(I_B{}^* G(\sigma_k), I_i\right) \end{aligned} \tag{6}$$

3.4 *Two-Dimensional Jittered Matching*

When a camera is focused at a near distance, vignetting problems occur at far objects. Major effect of vignetting problem is the edge misalignment between the sharp and blurry channels. Due to chromatic aberration and/or misalignment of two apertures, object boundaries in sharp and blurry channels become misaligned as the pixel position is far from the center of the lens. In Fig. 14, the effect of aperture misalignments is shown, where directions of edge misalignment on the image plane 2 and 3 are opposite to each other.

Depth extraction with jittered comparison is proposed to reduce such depth errors due to off-centeredness and is composed of two parts. The first is the generation of jitter vector map, and the second is to estimate depths with the generated jitter vector. The process of jitter vector map generation is shown in Fig. 15.

By utilizing the given edge map, a small patch is identified at an edge pixel in the green channel. The search window is defined as a region surrounding the co-located edge pixel in the IR channel. Then, one finds an IR patch which has the maximum

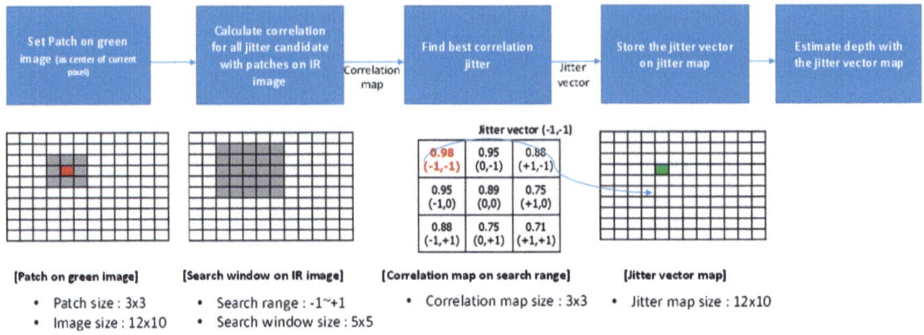

Fig. 15 Overview of jitter vector map generation

correlation with the green patch among all possible shifted patches in the search window. The jitter vector of the pixel is decided as a vector heading to the center of IR patch from the center of green patch. The jitter vector is found at all edge pixels. Then a jitter vector map is constructed. Considering the jittered matching and the adaptive blur selection, the best PSF index p^* is decided by Eq. 7.

$$\begin{aligned} i &= \underset{i\in\{R,G,B\}}{\arg\max}\left\{\mathrm{NCC}\left(I_i(x,y), I_{\mathrm{IR}}(x,y)\right)\right\} \\ \left(j_x, j_y\right) &= \underset{-1\le v_x, v_y\le 1}{\arg\max}\ \mathrm{NCC}\left(I_{\mathrm{IR}}(x,y), I_i\left(x+v_x, y+v_y\right)\right) \\ p^* &= \underset{\sigma_k\in\{\sigma_1,\sigma_2,\cdots,\sigma_{\mathrm{MAX}}\}}{\arg\max}\ \mathrm{NCC}\left(I_{\mathrm{IR}}(x,y)^* G(\sigma_k), I_i\left(x+j_x, y+j_y\right)\right) \end{aligned} \tag{7}$$

3.5 Compensation for Specular Reflection

Different spectral responses of blurry and sharp channels cause serious depth errors at those regions with high specular refection. In practice, not all specular reflection on edges introduces depth errors. The critical depth error happens with the specular reflection where the edge slopes on the sharp and blur channels have the opposite direction. Figure 16 shows an example of depth error which is caused by misaligned edges between sharp and blurry channels.

One remedy to fix those depth errors due to specular reflection is to align the edges at the blurry and sharp channels. Thus, this process includes the jittered matching described in the previous section. In our framework, the depth is decided by the PSF index which enables the highest NCC. The sign of NCC indicates whether the edge slopes at both channels have the same direction or not. Then it needs to be checked if there is specular reflection or not.

If there is a strong overshoot along the edge, as shown in Fig. 16, it is expected that there is specular reflection. If the maximum intensity around an edge pixel is bigger than that out of the edge, it is decided that the corresponding edge pixel is in specular reflection region. Before the detection process, thickening the edges

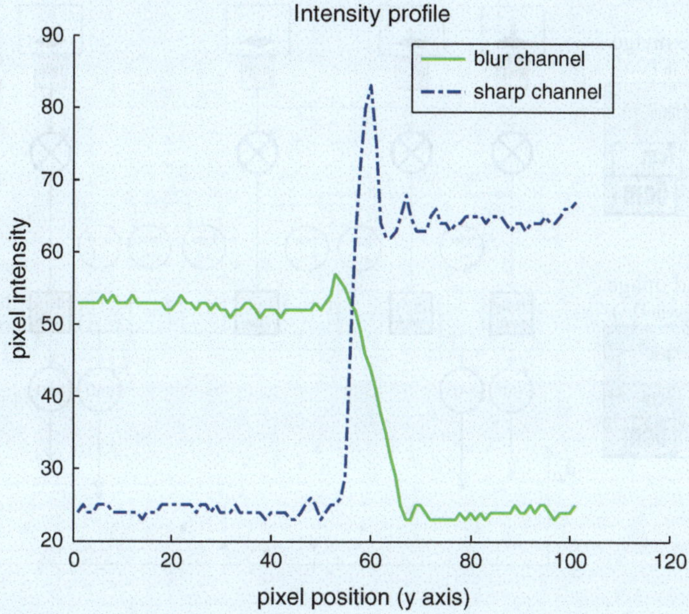

Fig. 16 Exemplary intensity profiles of blur and sharp channels with depth estimation failure

is necessary because the edges are misaligned from the center. To align the edges between IR and blurred patches, one patch has to be shifted within the search range. Therefore, if erroneous specular reflection is detected, the remaining process is same as jittered matching. However, the amount of search range for CSR is much bigger than that of TJM.

3.6 Hierarchical Selective Blurred Image Interpolation

The main idea of reducing computational complexity of depth extraction is to replace several repetitive convolutions with low-complexity interpolations. Figure 17 shows the proposed scheme of depth extraction. In the conventional structure [14], convolutions are used to generate blurred IR patches for all PSFs. However, our approach is to generate some blurred patches – called basis patches – by convolutions of reference patch and corresponding PSF or basis PSF. Then we generate other blurred patches by low-complexity interpolation utilizing basis patches.

To define the interpolation method, we employ one of characteristics of Gaussian function; a Gaussian PSF can be well approximated by weighted sum of two other Gaussian distributions. Let the basis index set be a set of PSF indices whose σ is sufficiently distinctive to the others. It is a good practice to collect the PSF indices

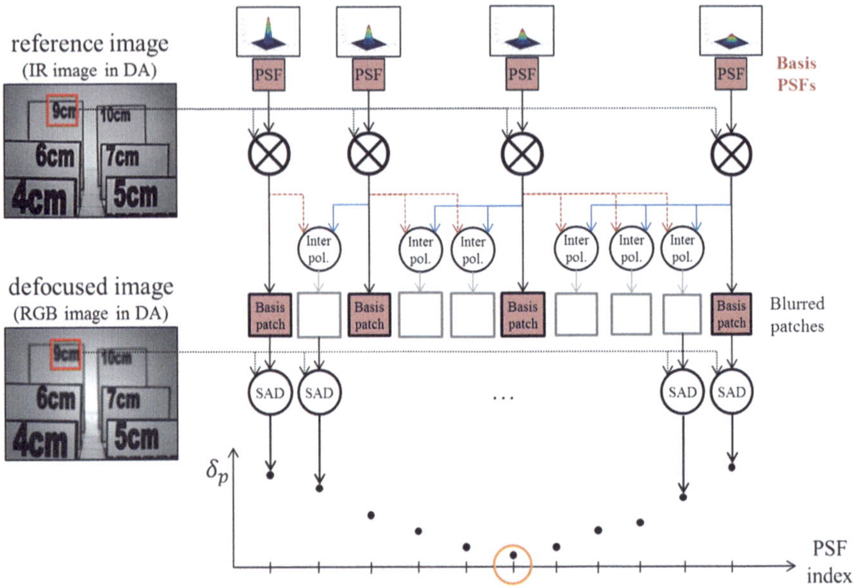

Fig. 17 Structure of hierarchical selective blurred image interpolation

whose σ is an integer. The set of all other PSF indices is regarded as the ground truth index set. Let $s_k = \{k|\, 0 \le k \le \text{MAX} \text{ and } k \in Z\}$ is the set of all PSF indices, where MAX is the largest index and **Z** is the set of integers. Then a basis set can be defined as $s_{k^*} = \{k^* | \sigma_{k^*} \in Z \text{ and } k^* \in Z\}$. Thus $G(\sigma_k)$ in Eq. 7 can be approximated by Eq. 8 when $k \notin s_{k^*}$.

$$G(\sigma_k) \cong \alpha_k G(\sigma_j) + \beta_k G(\sigma_{j+1}), \text{ where } \sigma_j \le \sigma_k < \sigma_{j+1} \text{ and } \{j, j+1\} \in s_{k^*} \quad (8)$$

3.7 Depth Noise Removal

DNR considers two perception-based cues to improve the depth map. First, the depth at a pixel is similar to the neighboring pixels. Second, the depth across a straight edge segment is typically continuous. DNR-0 improves the depth map using the first cue, and DNR-1 improves the depth map using the second. Note that in practice the result after DNR is a single pixel-wide depth map; however, for better visibility, we have dilated the result of DNR. Figure 18 shows the depth map before and after DNR.

It is observed that typically the depth is similar in local neighborhoods in natural objects, i.e., the depths of pixels which are located close to each other are similar. DNR-0 uses this property of natural objects to improve the depth map. To improve the depth map using DNR-0, we use a Markov random field (MRF)-based

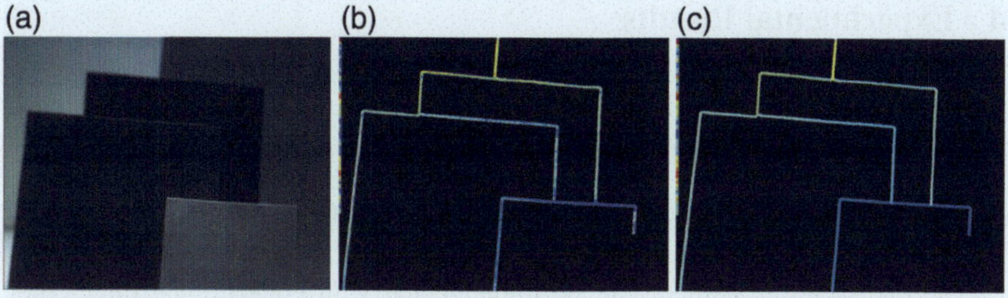

Fig. 18 (**a**) Original image, (**b**) depth map before DNR, (**c**) depth map after DNR

framework. With N denoting the total number of pixels and n denoting the pixel index, we find the value of x_n for which the following energy function is minimized using iterated convolutional modes (ICM) [19]

$$E = \sum_{n=1}^{N} \left(\frac{(x_n - y_n)^2}{2\sigma^2} + \gamma (x_n - \mu_n)^2 \right) \tag{9}$$

where y_n is the observed depth at the pixel n, σ^2 is the covariance of noise, μ_n is the average depth in neighborhood of n, and γ is a weighting factor. Larger value of γ ensures that E is strongly influenced by the neighborhood depths; therefore, optimizing E with a larger value of γ is equivalent to strongly enforcing the depth at a pixel to be similar to the neighborhood. The value of μ_n is obtained as

$$\mu_n = \frac{\sum\limits_{m=1}^{M} (y_m | y_m > 0)}{\sum\limits_{m=1}^{M} (y_m > 0)} \tag{10}$$

where M is the neighborhood of pixel n. In words, μ_n is the average of nonzero depths in the neighborhood of n in the depth map. This result can be achieved by applying two box filters. First, a box filter of size $M \times M$ is applied to the depth map which results in a depth map denoted by M_{sum}. Next, a box filter of size $M \times M$ is applied to 1{depth map >0} which results in M_{count}, where 1{.} denotes the indicator function. In words, a pixel in M_{count} is assigned a value 1 if its depth is greater than 0. μ_n is obtained by pixel-by-pixel division of M_{sum} by M_{count}.

An edge segment is defined a set of connected nonzero pixels which terminate when it butts another edge. An edge segment should belong to a single surface. It is known that depth across a surface does not change abruptly in natural scenes. In DNR-1, we exploit this property of surfaces by estimating depth estimates of edge segments using regression.

4 Experimental Results

Figure 19a was taken by a Hi-342 image sensor while focused at the nearest object. Hi-342 is a four-color RGB-IR image sensor manufactured by SK hynix. The selected ADC gain is 32 (minimum) for low noise level. The resolution of the camera is 1024 × 768@8bpp, and 30 PSFs are defined for depth estimation. The resultant sparse depth map is shown in Fig. 19.

The dual-aperture camera can be implemented with conventional three-color image sensors. In this case, a small aperture is placed for red channel and the larger aperture for green and blur channels. The principle for depth estimation has no difference between three- and four-color image sensors. Figure 20a was taken by a Nikon D60 with focusing at the nearest object. The selected ISO level is ISO100 (minimum) for low noise level. The resolution of the camera is 3872 × 2592 and is down-sampled to 968 × 648, and 20 PSFs are chosen for fast simulation. The resultant depth map is shown in Fig. 20.

5 Concluding Remarks and Future Work

In this paper we have presented the depth estimation pipeline. The input to the pipeline is a CFA image based on either three-color or four-color image sensor. The major modification to the conventional camera module is the introduction of another small aperture which enables some color channel to have a longer depth of field (DOF). The color channel with a longer DOF can be IR or red according to the spectral characteristic of the small aperture. The use of IR is preferable since IR data can be a benefit for many other applications in the field of computer vision.

The CFA image is converted to a full-color image through edge-preserving interpolation. Besides, the edge map for the entire image is also generated since the depth values are estimated only at object boundaries. The sharp channel is the color channel with a small aperture, and the other channels with the larger aperture are regarded as blurry channels. The blur difference between the sharp and blurry channels is used for depth estimation. To get robust depth values, the following functions have been proposed:

- Adaptive blur channel selection
- Two-dimensional jittered matching
- Compensation for specular reflection
- Depth noise reduction

Although the proposed depth pipeline shows a remarkable quality of depth, it still needs further improvements until to arrive at the same quality as stereo imaging depth. The color channel dependency is one of the crucial problems of our approach. The best performance is expected when the sharp and blurry channels are of the same color. We are developing a new sensor architecture where two different DOFs

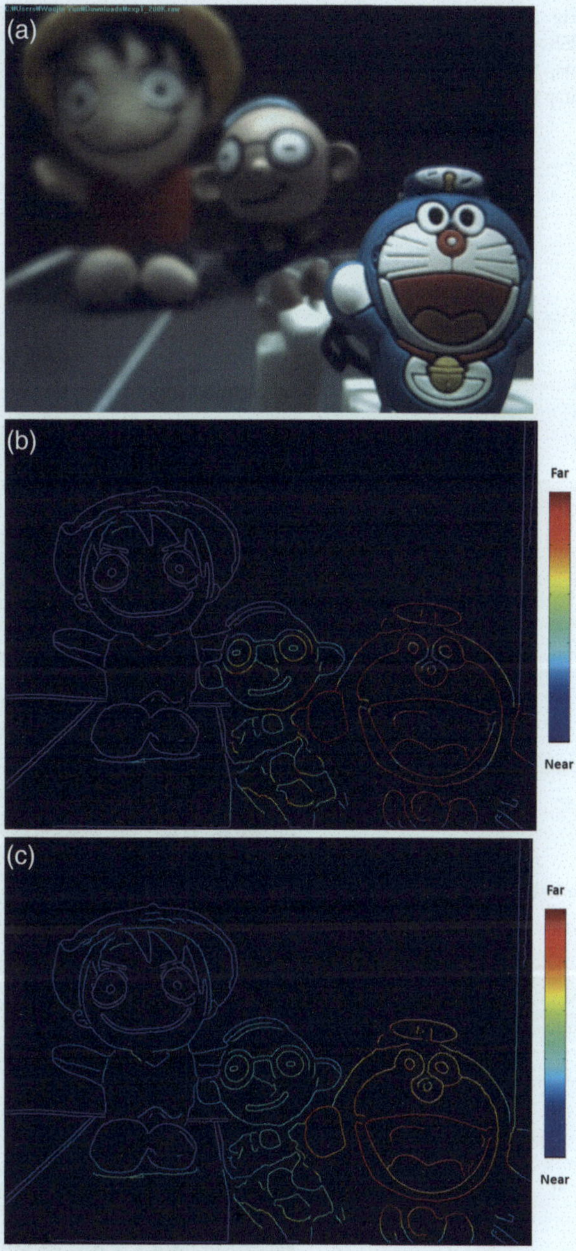

Fig. 19 (**a**) Test image: 1024 × 768@8bpp, (**b**) depth map before DNR, (**c**) depth map after DNR

are realized on the same color pixels. Besides, as in stereo imaging, disparity-based depth estimation is being investigated since it is much more robust to noise compared to blur-based one.

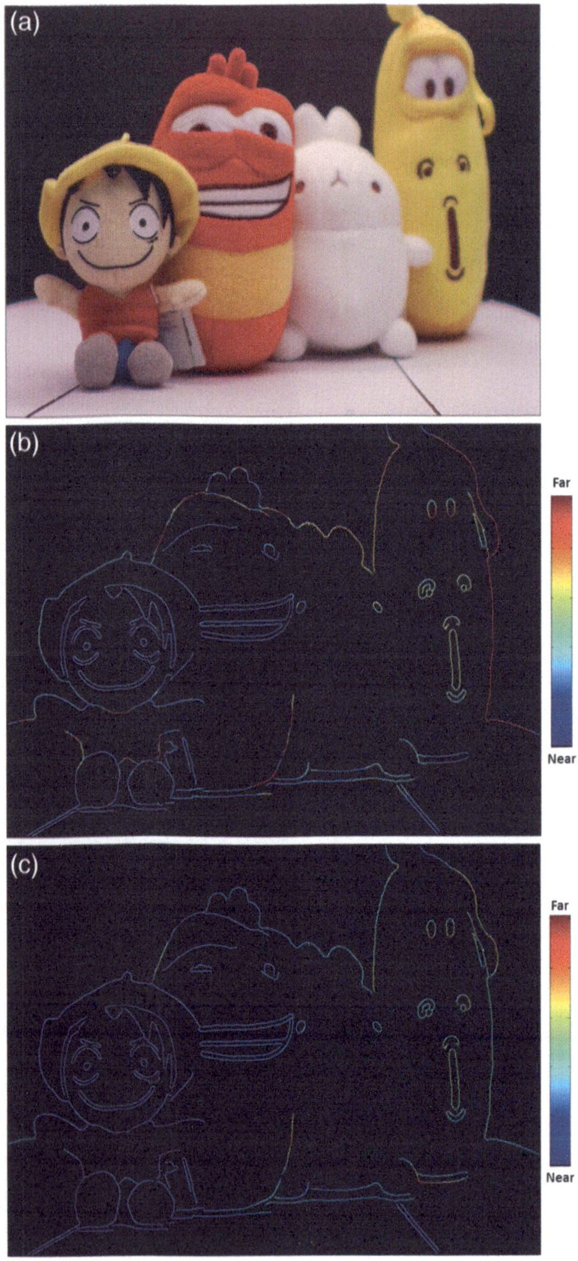

Fig. 20 (**a**) Test image: 968 × 648@8bpp, (**b**) depth map before DNR, (**c**) depth map after DNR

Acknowledgments This work was supported by the Center for Integrated Smart Sensors funded by the Ministry of Science, ICT and Future Planning as the Global Frontier Project.

References

1. Brown, M.Z., Burschka, D., Hager, G.D.: Advances in computational stereo. IEEE Trans. Pattern Anal. Mach. Intell. **25**(8), 993–1008 (2003)
2. https://www.ptgrey.com/stereo-vision-cameras-systems
3. https://www.stereolabs.com/
4. Ren, Z., Yuan, J., Zhang, Z.: Robust hand gesture recognition based on finger-earth mover's distance with a commodity depth camera. In: Proceedings of the 19th ACM International Conference on Multimedia, pp. 1093–1096 (2011)
5. Burgin, W., Pantofaru, C., Smart, W.D.: Using depth information to improve face detection. In: Proceedings of the 6th International Conference on Human-Robot Interaction, pp. 119–120 (2011)
6. Harville, M., Gordon, G., Woodfill, J.: Foreground segmentation using adaptive mixture models in color and depth. In: Proceedings of IEEE Workshop on Detection and Recognition of Events in Video, pp. 3–11 (2001)
7. Labati, R.D., Genovese, A., Piuri, V., Scotti, F.: Touchless fingerprint biometrics: a survey on 2D and 3D technologies. J. Internet Technol. **15**(3), 325–332 (2014)
8. Salvi, J., Pages, J., Batlle, J.: Pattern codification strategies in structured light systems. Pattern Recogn. **37**(4), 827–849 (2004)
9. Gokturk, S.B., Yalcin, H., Bamji, C.: CA time-of-flight depth sensor-system description; issues and solutions. In: Proceedings of IEEE Conference on Computer Vision and Pattern Recognition Workshop, pp. 35–35 (2004)
10. Green, P., Sun, W., Matusik, W., Durand, F.: Multi-aperture photography. ACM Trans. Graph. **26**(3), (2007)
11. Zhou, C., Lin, S., Nayar, S.: Coded aperture pairs for depth from defocus. In: Proceedings of IEEE International Conference on Computer Vision, pp. 325–332 (2009)
12. Hiura, S., Matsuyama, T.: Depth measurement by the multi-focus camera. In: Proceedings of IEEE Computer Society Conference on Computer Vision and Pattern Recognition, pp. 953–959 (1998)
13. Subbarao, M., Surya, S.: Depth from defocus: a spatial domain approach. Int. J. Comput. Vis. **13**(3), 271–294 (1994)
14. Martinello, M., Wajs, A., Quan, S., Lee, H., Lim, C., Woo, T., Lee, W., Kim, S.S., Lee, D.: Dual aperture photography: image and depth from a mobile camera. In: Proceedings of IEEE International Conference on Computational Photography, pp. 1–10 (2015)
15. Chen, X., He, L., Jeon, G., Jeong, J.: Local adaptive directional color filter array interpolation based on inter-channel correlation. Opt. Commun. **A324**, 269–276 (2014)
16. Li, X., Orchard, T.: New edge-directed interpolation. IEEE Trans. Image Process. **10**, 1521–1527 (2001)
17. Hwang, W., Wang, H., Kim, H., Kee, S., Kim, J.: Face recognition system using multiple face model of hybrid Fourier feature under uncontrolled illumination variation. IEEE Trans. Image Process. **20**(4), 1152–1165 (2011)
18. Zhang, T.Y., Suen, C.Y.: A fast parallel algorithm for thinning digital patterns. Commun. ACM. **27**(3), 236–239 (1984)
19. Besag, J.: On the statistical analysis of dirty pictures. J. R. Stat. Soc. Ser. B Methodol. **48**(8), 259–302 (1986)

Scintillator-Based Electronic Personal Dosimeter for Mobile Application

Gyuseong Cho, Hyunjun Yoo, Daehee Lee, Jonghwan Park, and Hyunduk Kim

1 Introduction

Radiation is energetic subatomic particles or electromagnetic waves such as gamma-ray or X-rays emitted from unstable nuclei or atoms. Among various radiation types such as alpha-ray, beta-ray, gamma-ray, X-ray, and neutron, gamma-ray is the most interesting because of its deep penetration power and abundance in natural and man-made environment. Since the radiation exposure may induce a biologically harmful effect such as cancer genesis if there is a possibility that the dose is over a certain level, the personal radiation dose of the people who work or reside in those environment must be monitored periodically. Any possibility of overexposure to human must be avoided according to the national regulation. For the radiation protection point of view, thermoluminescence dosimeters (TLDs) are generally most used as a legal dosimeter for many countries. In a personal TLD badge case, three or four types of TLD pieces are placed side by side in order to measure different types of radiation, for example, high- and low-energy gamma-rays, electrons, and neutrons simultaneously. Though the TLD is capable of measuring various radiation types with a change of materials, and its measurement accuracy is very high, it has a disadvantage that it can measure the accumulated dose only. TLD cannot measure instantaneous dose or dose rate in real time. Therefore in a certain situation such as a radiological accident, the instantaneous dose can be very high and may affect a serious biological damage to persons who are exposed. However TLD cannot warn or alarm such situation.

An electronic personal dosimeter (EPD) is a radiation detector which measures the real-time radiation dose absorbed by a person in a radiation-exposed environment. The nuclear companies or radiological clinics that treat nuclear and radiation

G. Cho (✉) • H. Yoo • D. Lee • J. Park • H. Kim
Korea Advanced Institute of Science and Technology, Daejeon, Republic of Korea
e-mail: gscho@kaist.ac.kr

H. Yasuura et al. (eds.), *Smart Sensors at the IoT Frontier*,
DOI 10.1007/978-3-319-55345-0_8

sources must provide EPDs to the workers in addition to TLDs to prevent accidental exposure to high radiation. After the Fukushima nuclear power plant accident in March 2011, even the general public's interest in EPD is continuously increased recently. Also the food contamination becomes a concern in neighboring countries, especially for educational institutes such as kindergarten and elementary schools. A general radiation detector type used in EPDs is an energy-compensated gamma counters such as a Geiger Muller tube (GM tube) or a metal-filtered photodetector. These EPDs are portable and convenient devices conventionally used ever since the beginning of radiation use by man, because they can measure the dose. However since they cannot measure the energy of gamma-ray, so they cannot identify the radioisotope sources that emit gamma-rays. The measurement of individual radiation energy is called the radiation spectroscopy, and it requires normally a complicated, stationary, and expensive system, such as an NaI(Tl) scintillation detector or a high-purity germanium detector for gamma-ray spectroscopy. Recently portable spectrometers using the room temperature semiconductors such as a CdZnTe (CZT) have been introduced for field workers, but their prices are quite high, so they are not affordable by the general public.

The main topic of this chapter is a description of a new smart device-based and inexpensive EPD with a function of gamma spectroscopy for both the experts and general public. The gamma energy range of interest to measure is from 20 keV to 1.5 MeV [1]. The proposed EPD is composed of a compact radiation sensor, an application-specific integrated circuit (ASIC), a microcontroller unit (MCU), and an Android phone. The compact radiation sensor is a combination of a sub-centimeter size CsI(Tl) scintillator and a silicon photodiode which convert the deposited gamma energy in the scintillator into charge packets through this combination. The detection efficiency varies depending on the incident angle of gamma-rays. The criteria for angular response were suggested by the International Electrotechnical Commission (IEC) to be used as a legitimate EPD [1]. The ASIC includes a preamplifier, a shaping amplifier, and a peak detector to pass a voltage signal to the MCU. The MCU converts the peak voltage signal of a single interacted gamma-ray into an energy channel number, called the energy bin; then if the detector senses many gamma-rays for a given time, counts for every channel make a histogram, called the energy spectrum. A new fast dose conversion algorithm embedded in MCU is proposed to calculate the $H_p(10)$ in real time periodically. In addition, another downloadable application program for a smart device identifies the gamma-emitting nuclide type and informs the users. Finally, we evaluate the performance of the proposed EPD by comparing difference ratio (DR) values depending on the gamma energy and gamma fluence. The angular response also was measured to check the satisfaction of IEC guidelines.

2 System Design

2.1 Design of a Compact Scintillation Detector

The compact radiation sensor is a combination of a sub-centimeter size CsI(Tl) scintillator and a silicon photodiode. A single gamma-ray interacts with the CsI(Tl) scintillator and emits a few hundred or thousand visible photons of 540 nm wavelength, and then the photodiode absorbs these photons and converts them into electronic charges. This combination has been used as a good radiation spectroscopic detector with good properties such as a relatively high light yield, a high effective Z-number of CsI(Tl), and a good optical matching of CsI scintillation light to the Si PIN diode.

The geometry of the compact radiation sensor should be optimized in order to have the maximum detection efficiency as well as the best energy resolution, while it is small enough to be integrated in personal mobile devices such as a smartphone or a tablet. The suggested basic structure is a cylinder with tapered part as a light guide to the smaller PIN diode. The geometry of the scintillator is determined by the total length and tapered length of the cylinder. The optimum geometry is determined by estimating two key performance parameters: the figure of merit (FOM) and the angular response. The FOM is defined as the absolute detection efficiency divided by the relative energy resolution. The absolute detection efficiency (ADE) is the total number of detected gamma-rays divided by the total number of gamma-rays emitted from a check source for a measurement time. The relative energy resolution (RER) is the full width at half maximum (FWHM) of the gamma-ray photo-peak (PP) divided by the peak channel. For a given structure of the sensor and the way that it attached to the system, the detection efficiency varies depending on the incident angle of gamma-rays. The optimum geometry of the compact radiation sensor is determined in order to have the highest value of FOM while satisfying the international criteria for angular response [1].

The area of the silicon photodiode in the compact radiation sensor is chosen to be 3×3 mm^2 because of generality. The maximum diameter of the CsI(Tl) scintillator is chosen to be 5 mm, considering the thickness of the smart devices and the diameter of the coupling surface of the scintillator to the 3 mm photodiode. So, the scintillator column is a tapered cylinder structure to smoothly connect the two different surfaces. The geometry and the value of each parameter are stated in Table 1.

To decide the optimum geometry of the CsI(Tl) scintillator, we vary two geometrical parameters as total length and tapered length. Firstly, the light output depending on the tapered length and total length was simulated using the light transport simulation program to compare the light collection efficiencies [2], because the higher light output makes the energy resolution better [3]. The simulation condition is that 1 Watt of light is generated in the center of the scintillator, and the light output at light collection surface was calculated and compared. The simulation result for the light output is shown in Fig. 1. The light output had the maximum value at 1 mm

Table 1 The geometry and specifications of the suggested scintillator are presented in this table

Geometry	Parameter	Value (mm)
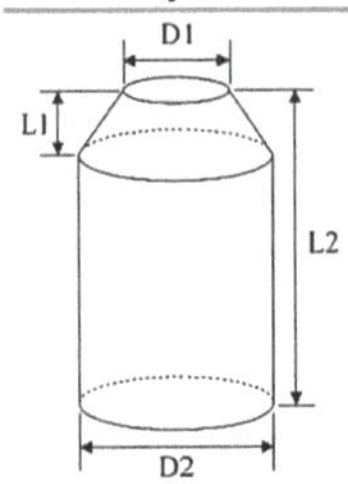	Diameter of the light collection surface, D1	3 (fixed)
	Diameter of the cylindrical body, D2	5 (fixed)
	Tapered length, L1	0 L2 for each total length
	Total length, L2	3, 5, 10, 15, 20, 30

It is a cylinder with a tapered head structure. All surfaces of the crystal were covered with a reflector (Teflon tape) except for the coupling surface to photodiode as denoted for D1

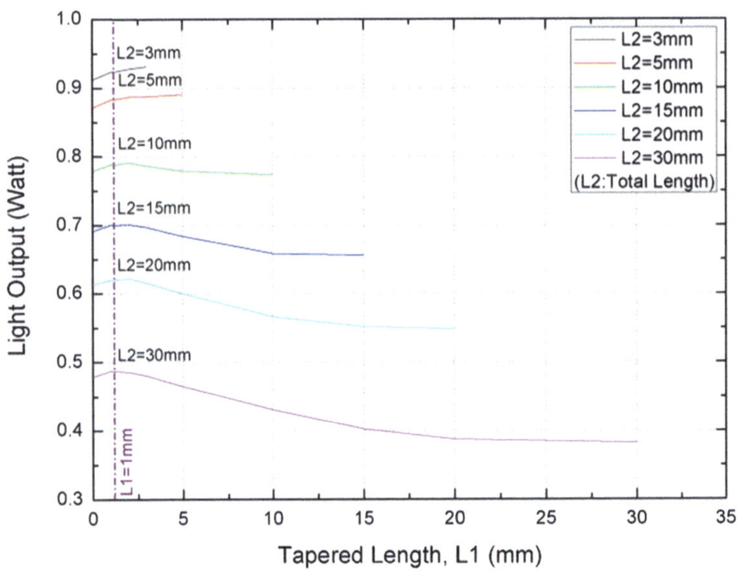

Fig. 1 The light output depending on the total length, L2, and the tapered length, L1

of the tapered length for 10, 15, 20, and 30 mm total length. And the light output for 3 and 5 mm total lengths proportionally increased with tapered length, but there was little changes over 1 mm tapered length. So, the tapered length was decided as 1 mm to have the maximum light output for all total lengths.

Secondly, the total length which satisfies the criteria for angular response was selected after comparing the rate difference of angular responses (RDARs). The criteria for this angular response were suggested by the International Electrotechnical Commission (IEC) to be used as a legitimate EPD [1]. The check sources to measure the gamma energy spectra were Am-241, Cs-137, and Co-60 to cover the energy range. The RDAR of each total length is shown in Fig. 2. Among the seven total lengths, 3 and 5 mm total lengths satisfied the criteria of angular response from 0 to 120°. The RDAR of each total length was significantly decreased at 150° and 180°, because the parts of low-energy radiation were absorbed by the printed circuit board on which the sensor is attached.

Finally, the FOMs of 3 and 5 mm total lengths were estimated to decide the optimum total length. The comparison result of the FOM for each total length is shown in Fig. 3. The absolute detection efficiency and relative energy resolution for FOM were calculated based on the measured energy spectra. The 5 mm total length cylinder had higher FOM than 3 mm one, because the geometrical detection efficiency of 5 mm total length is almost two times larger than 3 mm one. The energy resolution of the two total lengths showed little difference due to the similar light outputs as shown in Fig. 1. So, the optimum geometry of the CsI(Tl) scintillator in the compact radiation sensor is finally decided to be a cylinder with 1 mm tapered length and 3–5 mm total length. This scintillator optimization process was published elsewhere [4].

2.2 *Design of Front-End ASIC*

To measure the charge signal generated at the compact radiation sensor, three components are required in the front-end ASIC of the proposed EPD: a charge-sensitive amplifier (CSA), a shaping amplifier, and a peak and hold circuit. The final voltage output is processed in the digital domain by the following MCU to produce a spectrum through an analog-to-digital conversion.

The CSA is the first stage to convert the signal charge from the compact radiation sensor to a voltage pulse. Since CSA is the dominant noise source among the components of the front-end ASIC [5], so the optimized low noise design of the CSA is required to measure the charge correctly. The current pulse which is generated at a PIN diode is amplified at the CSA. We designed the CSA with a cascode amplifier geometry to increase the gain further among a number of possible topologies [6]. The designed amplifier for CSA is shown in Fig. 4. To minimize the power consumption, the amplifier was biased with 1 μA bias current. The right side of Fig. 4 shows the CSA connection with a feedback capacitor and reset switch block.

The amplifier has the gain of 55 dB, the phase margin of 70°, and the bandwidth of 25 MHz. Figure 5 shows the simulation results. In the case of the reset block, it can be implemented typically through a resistor, an active resistor, or a reset witch. In this design, we used a leakage compensation circuit not only acting as a reset component but also for leakage compensation which was developed by Krummenacher [7]. The leakage compensation circuit is shown in Fig. 6 with the CSA. This configuration provides a constant current fast return to zero through the reset path which is controlled by the IKrum current. Both negative leakage currents

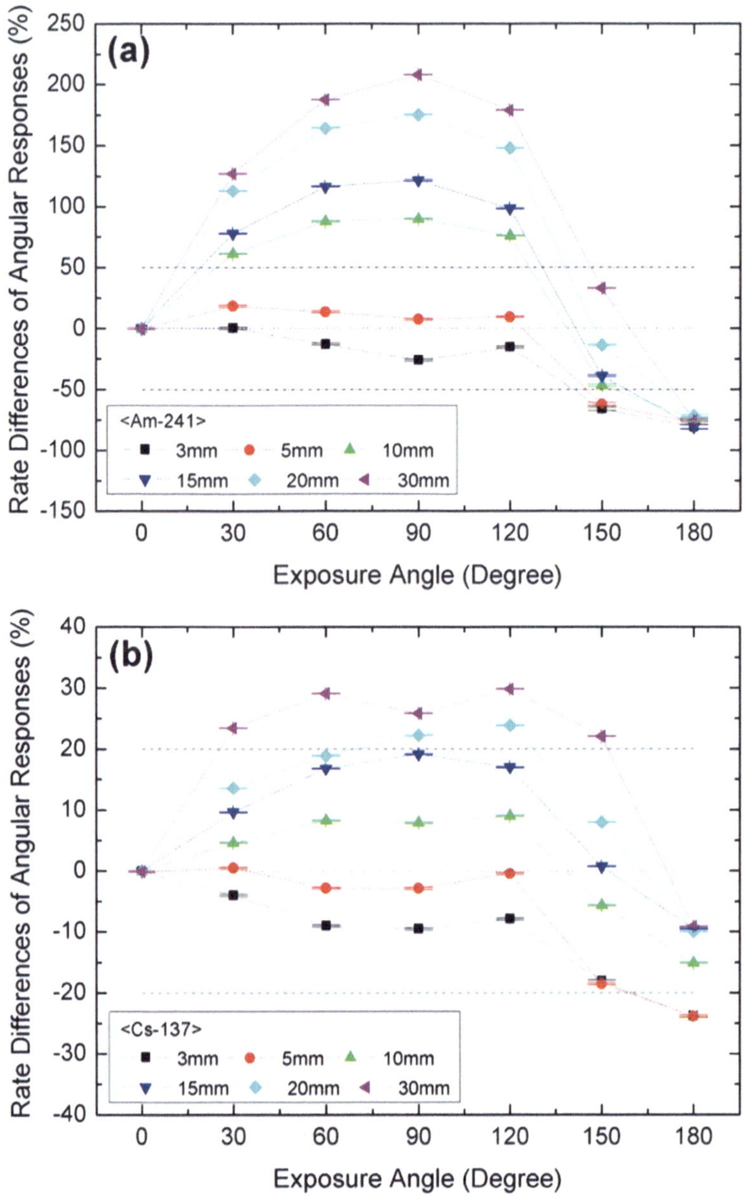

Fig. 2 RDAR depending on the exposure angle and gamma energy

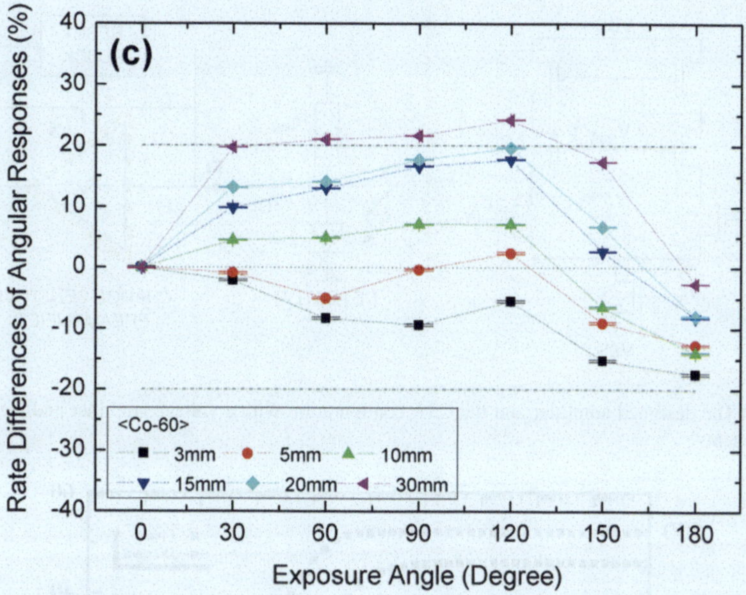

Fig. 2 (continued)

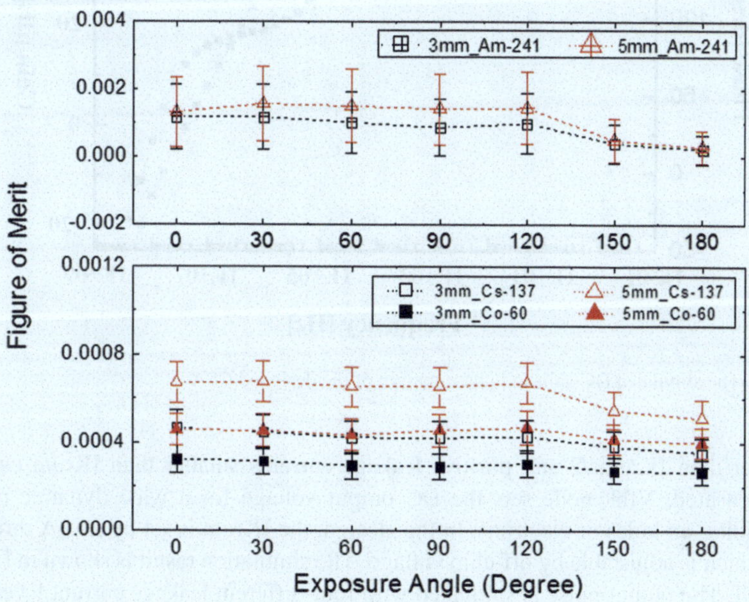

Fig. 3 The FOM depending on the exposure angle, gamma energy, and total length

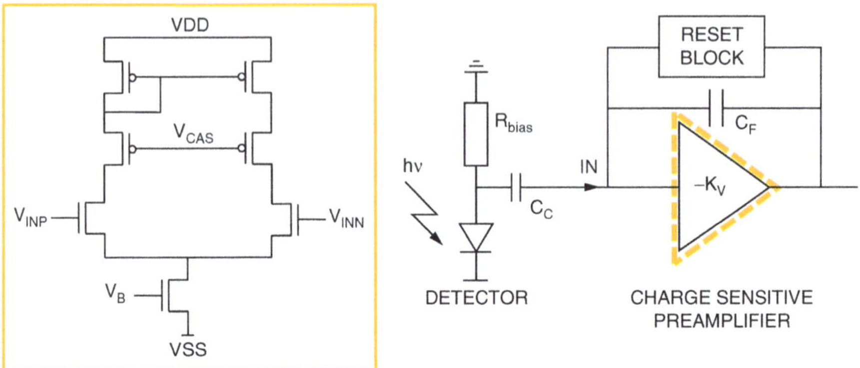

Fig. 4 The designed amplifier and the CSA configuration with a voltage amplifier and feedback components

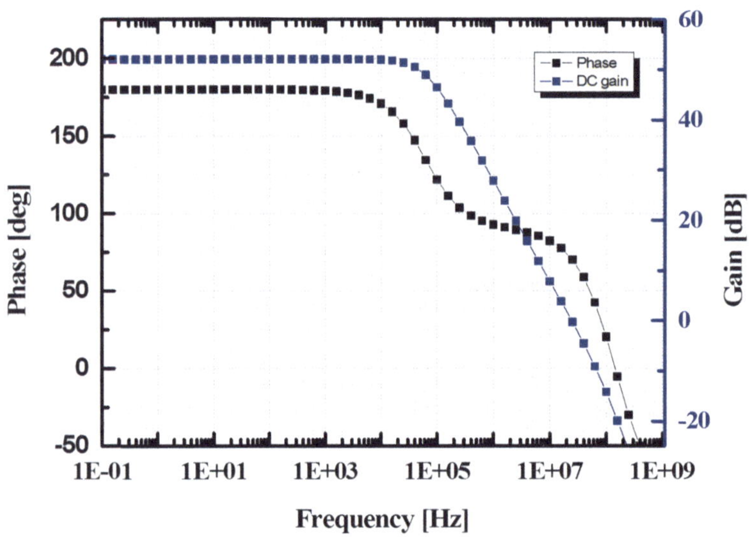

Fig. 5 The simulated DC gain and phase margin of the designed CSA

smaller than IKrums/2 and positive leakage currents smaller than IKrum can be compensated. Vfbk node sets the DC output voltage for a wide dynamic range depending on holes or electrons. In this design, the IKrum is set by 20 nA through VB which is adjustable by off-chip voltage. The simulation result is shown in Fig. 6 as well. The output pulse is simulated with four different leakage current levels: 0, 2, 4, and 6 nA. The output pulse is independent of the leakage current as shown in Fig. 6.

The two 50 fF capacitors are used as feedback in the CSA in parallel. Each capacitor can be selected through a gain control signal from off-chip signal. The

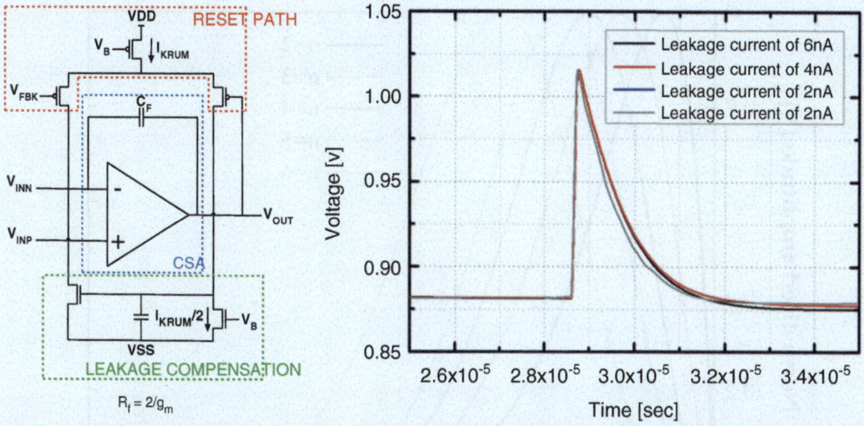

Fig. 6 A leakage compensation circuit for the CSA and the simulated CSA output signal for various leakage current levels

charge-to-voltage conversion gain of 3.2 mV/ke – or 6.4 mV/ke – can be achieved. The gamma energy range, 50 keV to 3 MeV, is converted to 600 mV difference, which is the output dynamic range of the CSA.

The pulse shaping stage after the CSA is added for the following tasks: to improve the signal-to-noise ratio in the system by filtering the CSA output, to add another gain stage in the chains, and to reduce the possibility of the pulse pileup by shortening the pulse duration. The selection of the order of filtering, type, and shaping time strongly depends on the target resolution of the dosimeter [8–10]. The types of shaping amplifiers often use are a unipolar and a bipolar shaper. The unipolar shaper uses a single stage of differentiator and a multiple stage (order) of integrators. As shown in Fig. 7, as the order of integrating stages increases, the output pulse shape becomes a true Gaussian pulse for the impulse input. As the order becomes high, the signal-to-noise becomes better, so at least fourth or fifth order of integrator is required. The integrator stage can be implemented by a simple passive R-C stage, but as the order increases, its area becomes large, so it is not commonly used as an ASIC. Here, we used fifth-order true Gaussian pulse shaping with an active integrator for smaller area as shown in Fig. 8. The first stage is a differentiator, and the last two identical stages are synthesized active integrators with a multiple feedback structure [11] and produce two integrating poles.

In the case of the bipolar Gaussian shaper, two CR differentiators and multiple integrators are used. This type of shaper produces a negative undershoot at the output pulse shape. Since the crossing time of this pulse is independent of the pulse amplitude, it can used in the circuit which requires a good timing resolution [12].

Figure 9 shows a simulation result with the CSA and the fifth-order active filter shaping amplifier. The red line represents the output of the CSA, and the purple line represents the output of the shaping amplifier. The output from shaping amplifier has a shorter pulse width, which reduces the probability of the pulse pileup when the

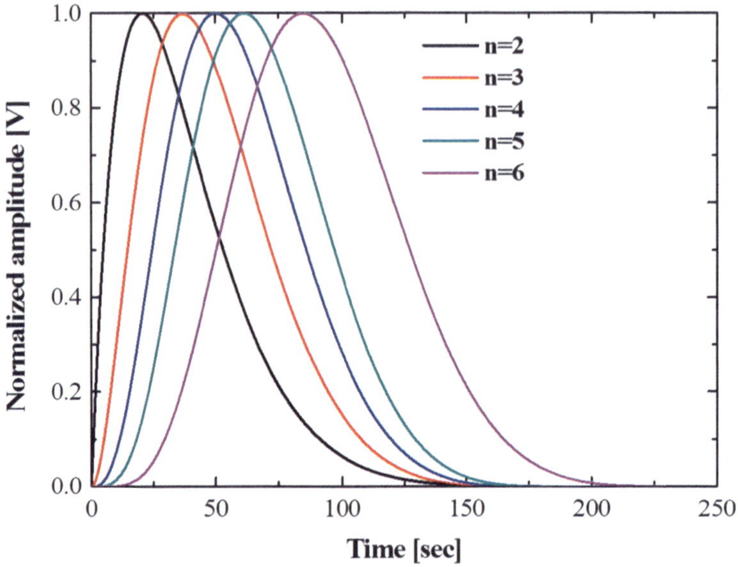

Fig. 7 The normalized outputs of the Gaussian shapers for different order of integrators

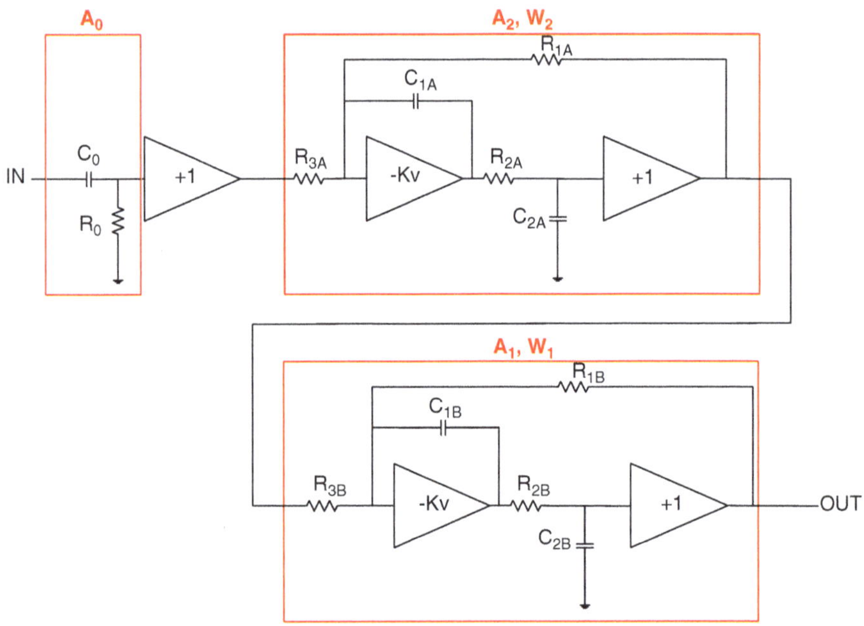

Fig. 8 The fifth-order true Gaussian shaper with a differentiator and two active filters with a multiple feedback structure

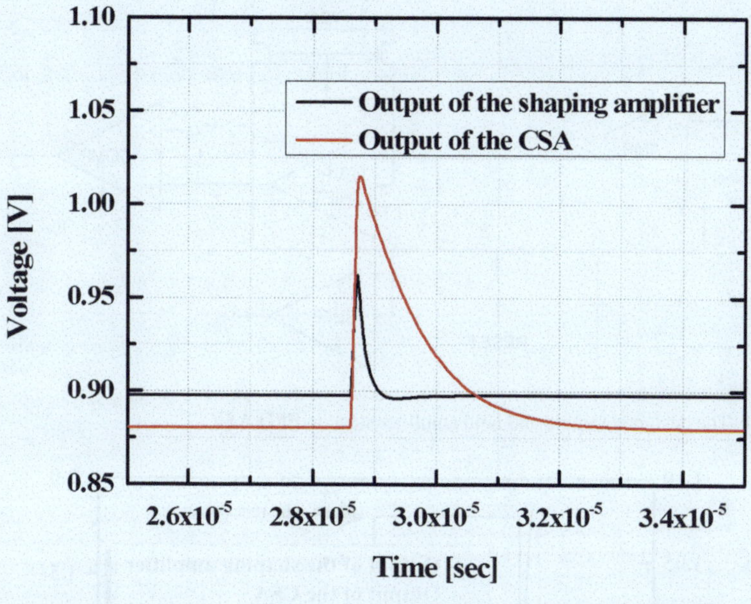

Fig. 9 The output pulse shapes of the CSA and the shaping amplifier used for EPD ASIC

incident gamma-ray flux is high. The overall gain of the shaping amplifier stage is about unity. The dynamic range of the shaping amplifier and CSA has about 700 mV voltage difference from 500 to 1200 mV. This dynamic range covers the incident gamma-ray energy range from 50 keV to 3 MeV. The power consumption for fifth-order shaping amplifier is 5 μW in total.

Finally, a sample and hold circuit must be incorporated after the shaping amplifier. The sample and hold circuit detect the peak voltage of the shaping amplifier output pulse for the analog-to-digital conversion (ADC) in order to measure the energy deposited to the scintillator by the interaction of a single incident gamma-ray at a time. The ADC is embedded in the MCU; however, the MCU ADC is not fast enough to detect the peak from shaping amplifier. Thus, the sample and hold circuit maintain the peak analog value for that ADC. The sample and hold circuit are shown in Fig. 10. A trigger signal for the MCU triggering is produced by a comparator [13–15]. The sampled signal from the sample and hold circuit maintains the peak value until the reset signal is enabled. The period for the hold time can be adjusted in the MCU program. A reset for the next signal can be generated at the MCU. If a reset signal is supplied to the sample and hold circuit, the output level of the sample and hold returns to the baseline to be ready to detect the following signal, as shown in Fig. 11.

The front-end ASIC for the proposed EPD, composed of three components, was designed using the 0.18 μm standard CMOS process with six metals and one ploy. Figure 12 shows the layout of the designed chip.

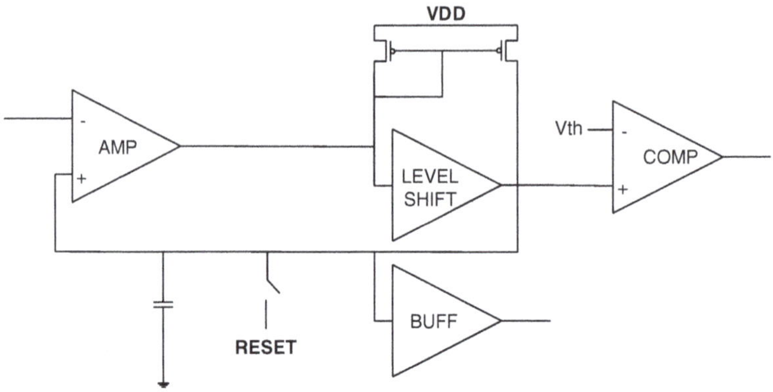

Fig. 10 The designed sample and hold amplifier stage in EPD ASIC

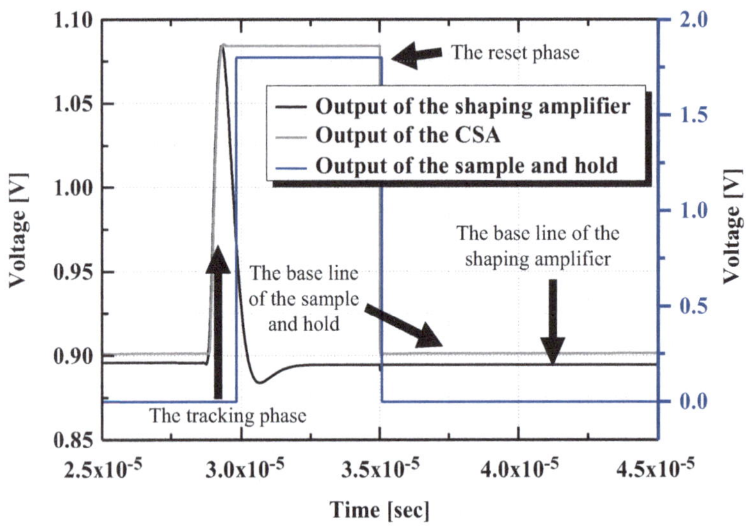

Fig. 11 The outputs of the shaping amplifier and sample and hold circuits

2.3 *System Design*

2.3.1 Design of EPD System for Mobile Phones

Since mobile phones became the most pervasive form of personal communication, many engineers have tried to directly connect with various peripheral devices that have been requested by customers. However, the technology must basically solve several engineering issues such as energy harvesting and data transfer [16–19].

In EPD development, the main components of the device are a compact radiation sensor, a front-end ASIC chip for signal processing, and a system board with an

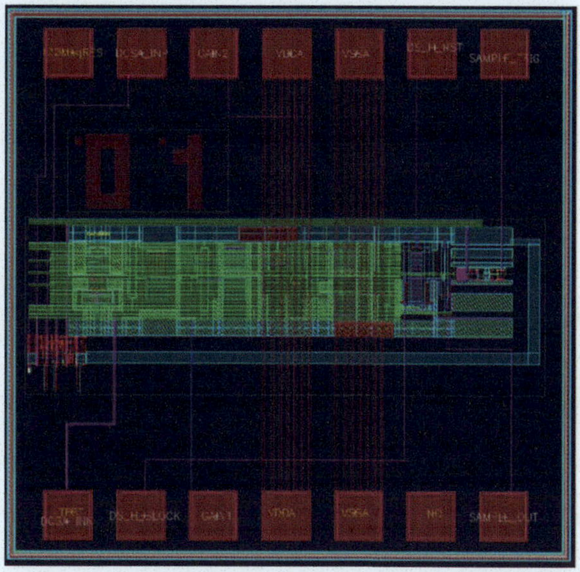

Fig. 12 The layout of the front-end ASIC for EPD

MCU for data processing and communication. Figure 13 shows the system concept of a dosimeter device and the design of EPD used for a mobile phone. In order to supply power from a phone and to control bidirectional data transfer, we need to choose a four-conductor 3.5 mm audio-jack interface (TRRS type, CTIA) for stereo sound and microphone input because this has been standardized and widely accessible among the various analog and digital interfaces. In this project, the microphone on the sleeve of the pin is used to transfer data from peripheral device. Furthermore, the left audio on the tip and right audio on the first ring of the stereo sound are assigned for energy harvesting and for command signals from the mobile phone, respectively.

2.3.2 Power Harvesting Through Audio Jack

The power harvesting using audio jack of the mobile phone is the most interesting issue and challenge for the engineer who seeks to enable additional devices. It is not only impractical but also hard to implement this technology as it is limited by the fact that the power delivered from the phone is not substantial. Hence, many developers have tried hard to find a proper technique to develop more friendly and useful devices [18, 19].

The technology using the audio jack interface converts AC waveform such as sine or square wave sent out from the audio output port into multiplied DC voltage signal, which is commonly based on the rectification method. The radiation dosimeter works in a similar manner.

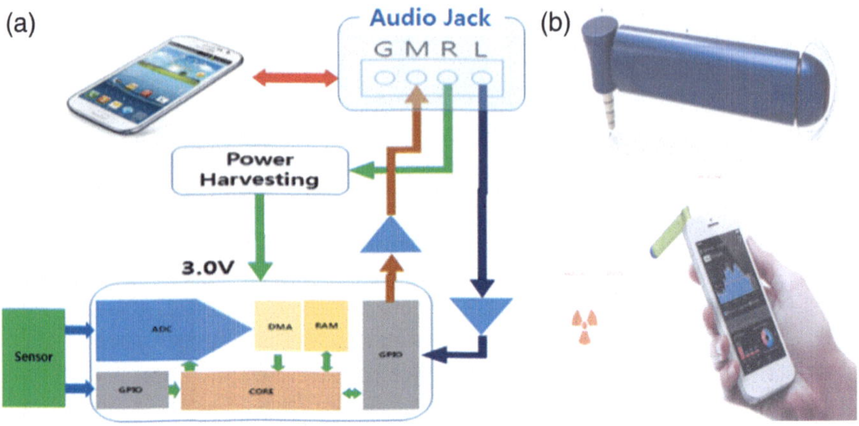

Fig. 13 (**a**) Conceptual diagram of the electronic personal dosimeter for mobile phone. A 3.5 mm audio jack interface is adopted for power harvesting and data communication. (**b**) The outfit of the proposed EPD

In this project, two power harvesting systems are selected for performing the evaluation of how much power can be harvested using an audio jack from a mobile phone. Figure 14a shows a microtransformer type which boosts the input AC voltage conveyed from the left audio on the tip to the high voltage. After the microtransformer, the rectifier is used in order to convert the transformed AC voltage to the DC voltage without a voltage drop; a regulator is placed at the end of the position on the power harvesting circuit since this device continuously needs 3 V for operating systems such as the photodiode and the MCU [18]. In the case of the diode voltage multiplier method as shown in Fig. 14b [1], the diode that boosts the input AC voltage in proportion to the number of diodes is used instead of the microtransformer. This circuit is designed for converting low voltage signal from the left audio on the tip to six times higher voltage.

A mobile phone can generate various waveforms which differ slightly between an iPhone and Android phones. iPhone supports a higher performance compared with the Android phone from the point of power harvesting from the mobile. Hence, the Android phone was selected to test a prototype development. In Fig. 15, the lower yellow line shows the AC 44.1 kHz waveform input signal through the left audio on the tip from an Android, and the upper green line is about 5 V DC output after AC to DC conversion implemented by the diode voltage multiplier.

2.3.3 Data Communication Through Audio Jack

Most peripheral devices commonly use the Bluetooth wireless technique to communicate with host mobile phone. In this project, a 3.5 mm audio jack, which is another widely used interface technique with mobile phones, is adopted to transfer the data

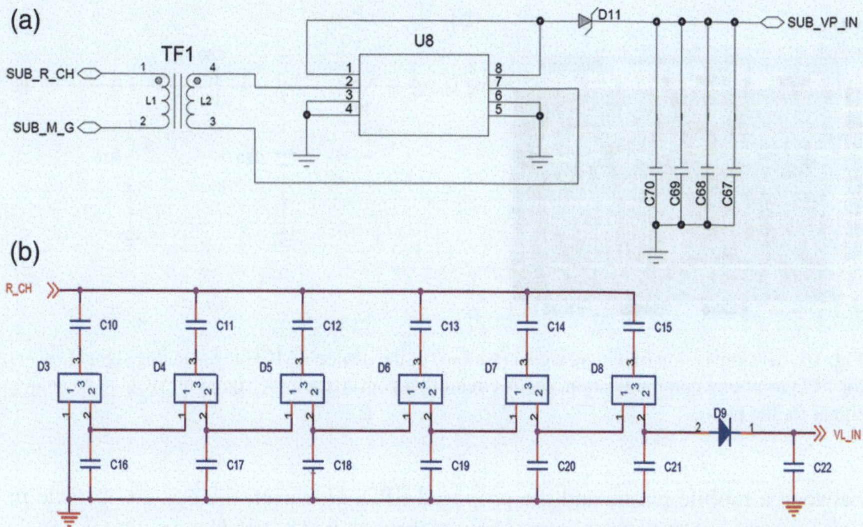

Fig. 14 The circuits for power harvesting from the mobile phone to peripheral device. The multiplication from the low-input voltage to high-output voltage is boosted by the two methods. (**a**) The first one is used to the microtransformer and (**b**) the other is the diode voltage multiplier

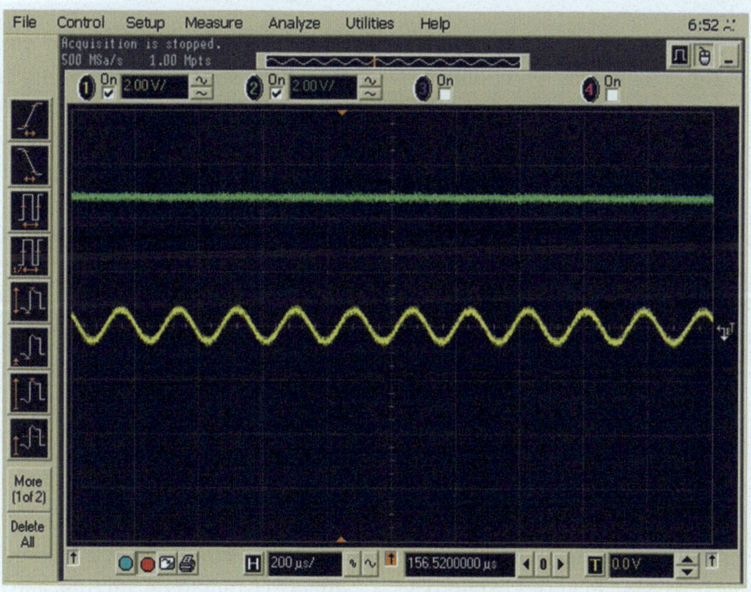

Fig. 15 The upper DC output signal voltage boosted by power harvesting circuit when the lower input AC signal voltage comes through the left audio tip from an Android phone

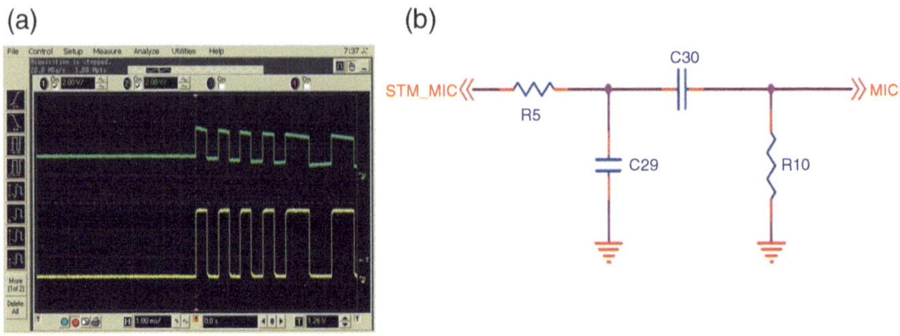

Fig. 16 (**a**) Conversion of a logic signal (*bottom*) of the device's MCU to an analog signal (*upper*) for EPD-to-phone communication. (**b**) A circuit that converts a logic signal of MCU to an analog signal for the phone

between a mobile phone and the proposed EPD. However, there is an obstacle in that an analog voltage signal can only pass through it [18, 19]. Hence, digital signals generated by the processor of the phone or the EPD as shown in the bottom spectrum of Fig. 16a must be first converted to analog signals with a sample rate of 44.1 kHz (upper spectrum of Fig. 16a). In the case of the MCUs digital signal of the device, the simple circuit is adapted to allow communication with the phone as shown in Fig. 16b.

In order to encode the signal by the MCU, a signal containing the information of a measurement result is modulated with the phase shift keying (PSK) method. Next the transfer of this encoded data is quickly implemented using the audio jack interface to the phone, and it is translated to digital signals (0 or 1) by the Manchester encoding method. In the case of an Android phone, there are many manufacturers such as Samsung, LG, and Huawei that produce various mobile phones. This creates a problem in that all these manufacturers cannot adopt this audio jack communication method since the performance and specification slightly differ from phone to phone. Especially, an MIC signal impedance would unexpectedly cause a signal transfer error since each mobile phone has its native property of the impedance as reported in a previous study [19].

2.4 Dose Conversion Algorithm

Since gamma-radiation interaction mechanism with a scintillator or a human body has three possible categories (a photoelectric effect, a Compton scattering, and a pair production), the deposited energy to the scintillator is smaller than or at least equal to the gamma-ray energy. The photoelectric effect is the only process that all gamma energy is deposited to the target material and the other two processes lose some of the gamma energy through escaping process of scattered gamma-ray or annihilated gamma-ray. So the radiation absorbed dose or dose equivalent is not the same as the total gamma energy but is related to the deposited energy.

The International Commission on Radiation Units and Measurements (ICRU) has defined and reported three operational dose equivalents, and $H_p(10)$, the personal dose equivalent estimated at depth of 10 mm below the skin, is one of them [20–22].

For many events of mono-energy gamma-ray interactions for a given measurement time, a spectroscopic sensor system can obtain a histogram of counts as a function of absorbed energy bins called a gamma energy spectrum. A dose conversion algorithm is required to calculate the $H_p(10)$ using this measured energy spectrum for a given time and repeats it periodically. This algorithm will be programmed in the MCU of the system board. Traditionally the incident gamma energy spectrum is calculated by a deconvolution of the measured spectrum with a system response inverse matrix, and this method requires a high calculation time. Another general method to confirm the gamma energy of the measured energy spectra is a pattern recognition method such as the principal component analysis (PCA) [23, 24], but it also requires a substantial amount of calculation time to analyze the gamma spectra so it is not appropriate for the real-time spectrum-based EPD.

A new fast dose conversion algorithm is proposed to convert the measured energy spectrum into $H_p(10)$ in real time. This $H_p(10)$ is called a measured-spectrum-based $H_p(10)$ (MSBH). To calculate dose rate $\dot{H}_p(10)\ (\mu\mathrm{Sv/h})$, $H_p(10)$ (μSv) as the sum of all products of the spectral bin counts in the measured energy spectra and the direct dose conversion factor (DDCF) for each spectral bin must first be calculated and then, $H_p(10)$ is divided by a measurement time, T (sec). The DDCF for each spectral bin energy is determined based on an innovative assumption to calculate the $H_p(10)$ without the gamma energy information. So the median bin energy assumption (MBEA) is suggested to directly calculate the $H_p(10)$ by multiplying the DDCF values and the spectral bin counts. This MBEA presumes that the spectral bin energy is the same with the median energy in the energy spectrum of a new corresponding imaginary gamma energy, E'_g.

In the case of a low spectral bin energy such as 50 keV, the median bin energy of the energy spectrum is almost the same with the photo-peak energy due to the high probability of photoelectric absorption. It makes the $H_p(10)$ calculation for low gamma energies accurate. In the case of a high spectral bin energy such as 1.3 MeV, the imaginary gamma energy for that median energy is significantly higher due to the high probability of Compton scattering. So the underestimation of the spectral bin count at a low spectral bin energy in the energy spectrum can be supplemented by the overestimation of the spectral bin count in a high spectral bin energy. The DDCF is defined as the HCF (fluence to $H_p(10)$ conversion factor) suggested by ICRU divided by the detection efficiency of the corresponding imaginary gamma energy [20, 21] as follows:

$$\mathrm{DDCF}\left(E_{\mathrm{bin}}\right) = \frac{\mathrm{HCF}\left(E'_g\right)}{1 - e^{-\mu\left(E'_g\right)x}}$$

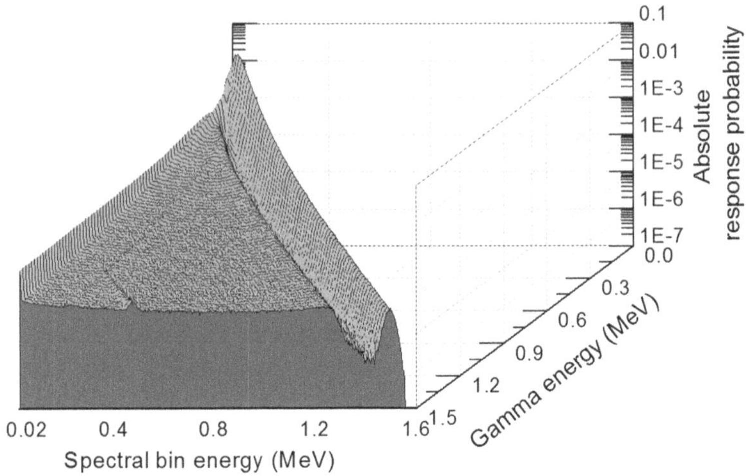

Fig. 17 The energy spectra to estimate the median energies of gamma-rays with the energy between 20 keV and 1.5 MeV

where HCF (μSv/γ-fluence) is $H_p(10)$ conversion factor suggested by ICRU [21, 25], μ is the attenuation coefficient (/cm), x is the sensor thickness (cm), and the denominator is the detection efficiency of a gamma with energy E'_g. To calculate the imaginary gamma energy of each spectral bin energy, the all-energy spectra of the gamma energy from 20 keV to 13 MeV with the interval of 10 keV were simulated by MCNPX [26]. The simulated energy spectra are shown in Fig. 17. These results were obtained through the Gaussian broadening of the simulated spectrum for the compact radiation sensor in order to incorporate the realistic fluctuation of measured data. The fitting parameters of MCNP Gaussian broadening models were obtained by measurement of the energy spectra of three reference isotope sources. The energy resolution, gamma energy, and isotope type of the three sources are 37.6% for 0.059 MeV from Am-241, 5.1% for 0.662 MeV from Cs-137, and 1.6% for 1.33 MeV from Co-60. The median bin energy of the gamma in interested energy range is shown in Fig. 18. There are two discontinuous points at 0.3 MeV and 1.7 MeV, because the attenuation coefficient of photoelectric absorption and pair production significantly changes at these energies. The DDCF depending on the imaginary gamma energy is shown in Fig. 19.

The performance of the dose conversion algorithm was evaluated by difference ratio (DR) depending on the gamma energy. DR is the relative difference defined as a ratio of the difference between the measured spectrum-based $H_p(10)$ (MSBH) and the source spectrum-based $H_p(10)$ (SSBH) divided by SSBH. SSBH is a theoretical value using the source activity, distance, and $H_p(10)$ conversion factor (HCF). MSBH was calculated by the algorithm on the simulated energy spectra of interested gamma energy range as shown in Fig. 20. There were three discontinuous points at 50, 300, 1500 keV. Firstly, the negatively high DR at the energy below 50 keV is caused by the fact that a part of radiation energy is absorbed at the reflector

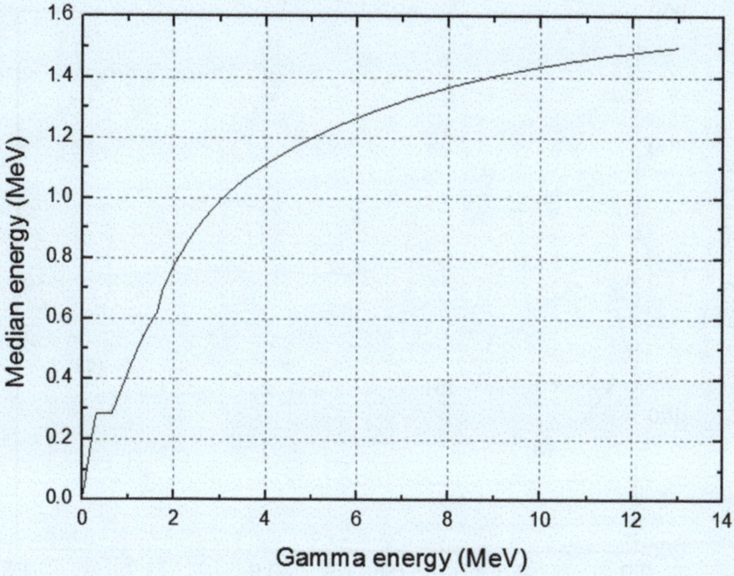

Fig. 18 The median energy depending on the gamma energy

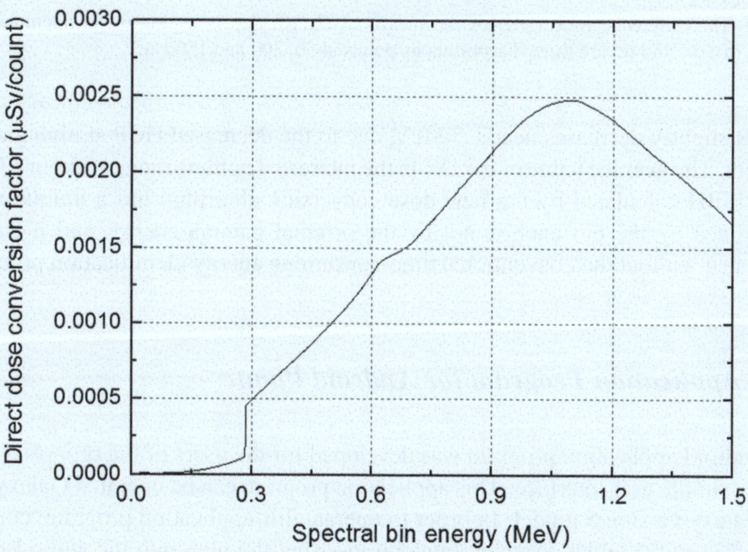

Fig. 19 The direct dose conversion factor (DDCF) depending on the measured spectral bin energy

of the scintillator, because the general reflector was made of a high Z-number material, such as TiO_2. Secondly, the abrupt change from the underestimation to the overestimation of the DR value at around 300 keV is caused by the fact that the attenuation coefficient near the photoelectric absorption rapidly decreased. Thirdly,

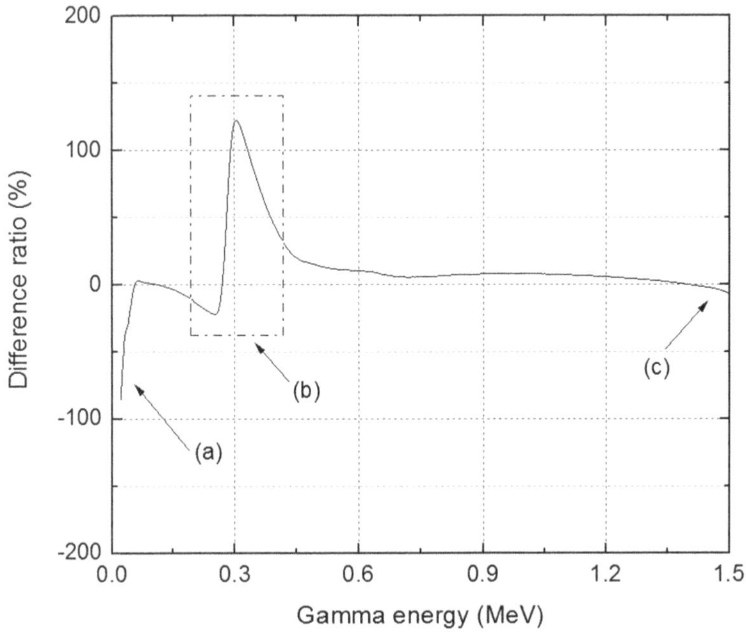

Fig. 20 The difference ratio (DR) of the measured $H_p(10)$ depending on the incident gamma energy. (**a**), (**b**) and (**c**) are three discontinuous points at 50, 300 and 1500 keV

the DR slightly decreases near 1.5 MeV, due to the decreased HCP at high gamma energies. The average value of the DR in the interested gamma range is about 17.3%. The MSBH calculated by the new dose conversion algorithm has a unique value determined by the bin energy, not by the original gamma energy, and it can be calculated without the conventional time-consuming energy identification process.

2.5 Application Program for Android Phone

An Android application program was developed for the users of the proposed EPD using a simple user interface. This application program can be operated Galaxy Tab and Galaxy version 5 model. In order to operate this application program, connect the device to the tablet or smartphone by inserting the plug into the audio jack of the user's device. Android Studio is used for the development environment instead of Eclipse. Android Studio is the official integrated development environment (IDE) for Android platform developments.

The user interface is composed with the measurement mode, the recode mode, and the analysis mode. The measurement mode is designed to display the dose rate $\dot{H}_p(10)$ (μSv/h) and counts in real time. The degree of radiation hazard will be

Fig. 21 User interface for the measurement mode window

displayed as three levels: a normal radiation dose, a threshold radiation dose, and an intolerable dose when the measurement is being performed. The user interface window for the measurement mode is shown in Fig. 21. Some characters in Figs. 21, 22, and 23 captions are written in Korean because the beta version of the proposed EPD will be tested in Korea first.

The second and third modes are the recode and analysis modes. The user can save the dose rate after the measurement is finished. Saved dose rate data can be displayed daily, monthly, and yearly according to the user's need. Also saved dose rate data can be displayed as a graph as shown in Fig. 22.

The analysis mode displays a graph with which the user can easily identify the radionuclide easily. The analysis can be also displayed as a list which is shown in Fig. 23.

3 Test Results and Discussion

3.1 Measurement of Gamma Energy Spectrum

To evaluate the performance of the EPD proposed in this study, we measured the energy spectra of seven kinds of the radioisotope check sources such as Am-241, Co-57, Ba-133, Na-22, Cs-137, Mn-54, and Co-60 to compare the energy

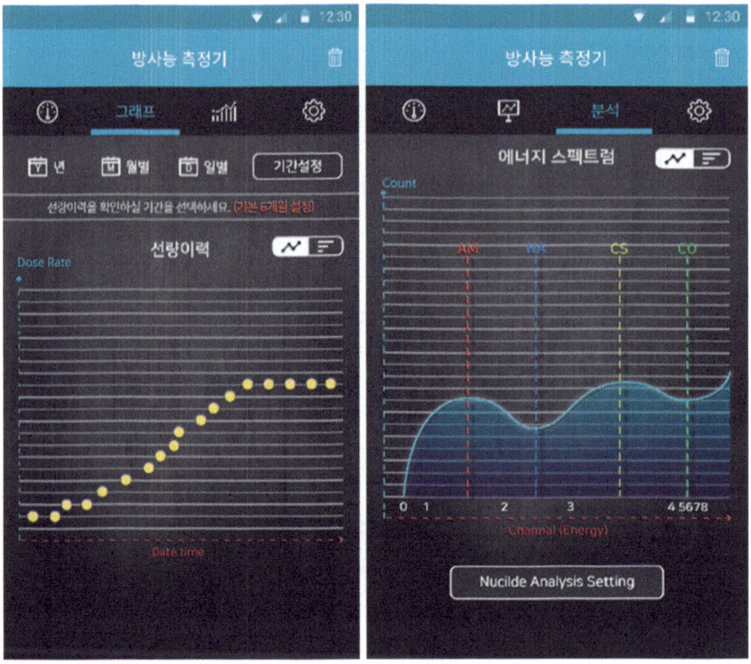

Fig. 22 Graph of record and analysis mode

resolution. The measurement distance between the sensor surface and the check source was 30 mm, and the measurement time was 3600 s for all check sources. The measurement result of the energy spectra is shown in Fig. 24. The measured energy resolution of 59.5 keV (Am-241), 662 keV (Cs-137), and 1330 keV (Co-60) were 37.6%, 5.1%, and 3.3%, respectively. These values of the relative energy resolutions are acceptable for the sub-centimeter size scintillator [27]. The energy resolution values of the compact radiation sensor are also used for the MCNP Gaussian broadening correction as shown in Fig. 17.

3.2 *Measurement of Personal Dose*

The accuracy of the dose conversion algorithm was evaluated by the DR defined in the above. The DRs depending on the gamma energy and the fluence level are shown in Fig. 25a–g. The DRs of each gamma energies fluctuate at fluence levels below $10^3 \sim 10^4$ γ-ray/0.09cm^2. The fluctuation above $10^3 \gamma - \text{ray}/0.09\ \text{cm}^2$ becomes stable for gamma energies from all check sources. The $H_P(10)$s at a fluence level of $10^3 \gamma - \text{ray}/0.09\ \text{cm}^2$ are listed on the Table 2.

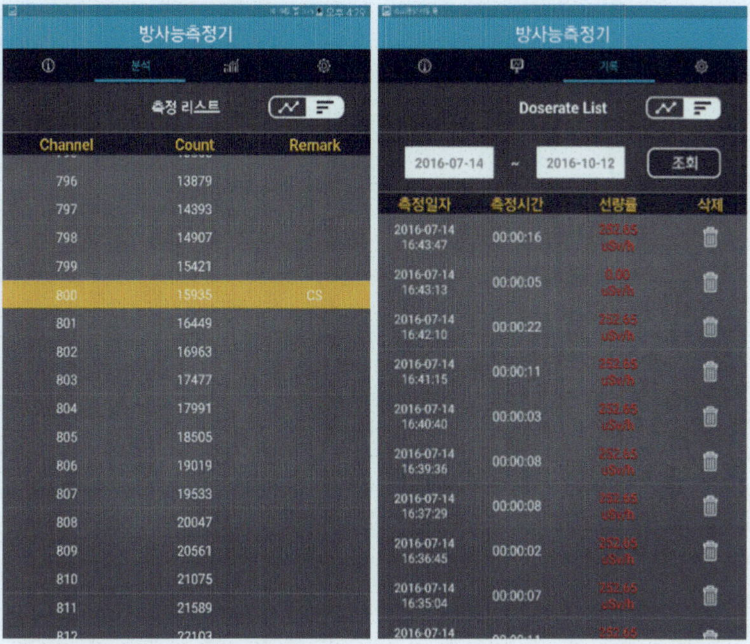

Fig. 23 List of record and analysis mode

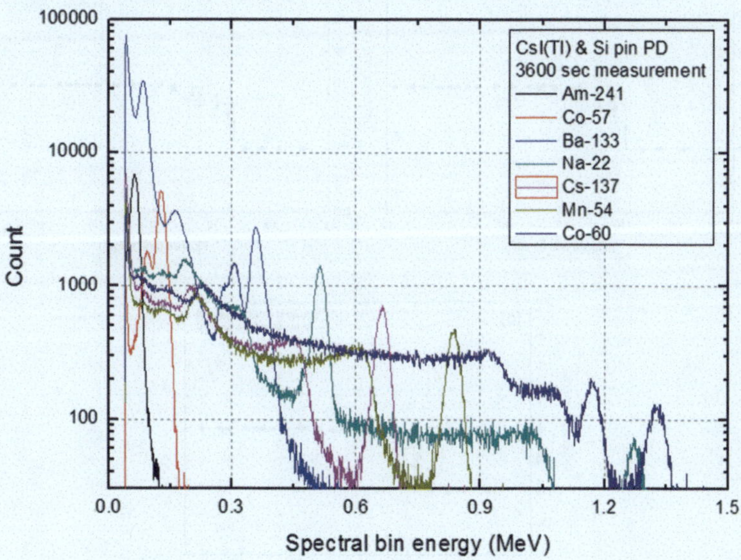

Fig. 24 The energy spectra of the seven radioisotope check sources with 3600 s measurement time. The measured energy resolution of 59.5 keV (Am-241), 662 keV (Cs-137), and 1330 keV (Co-60) were 37.6%, 5.1%, and 3.3%, respectively

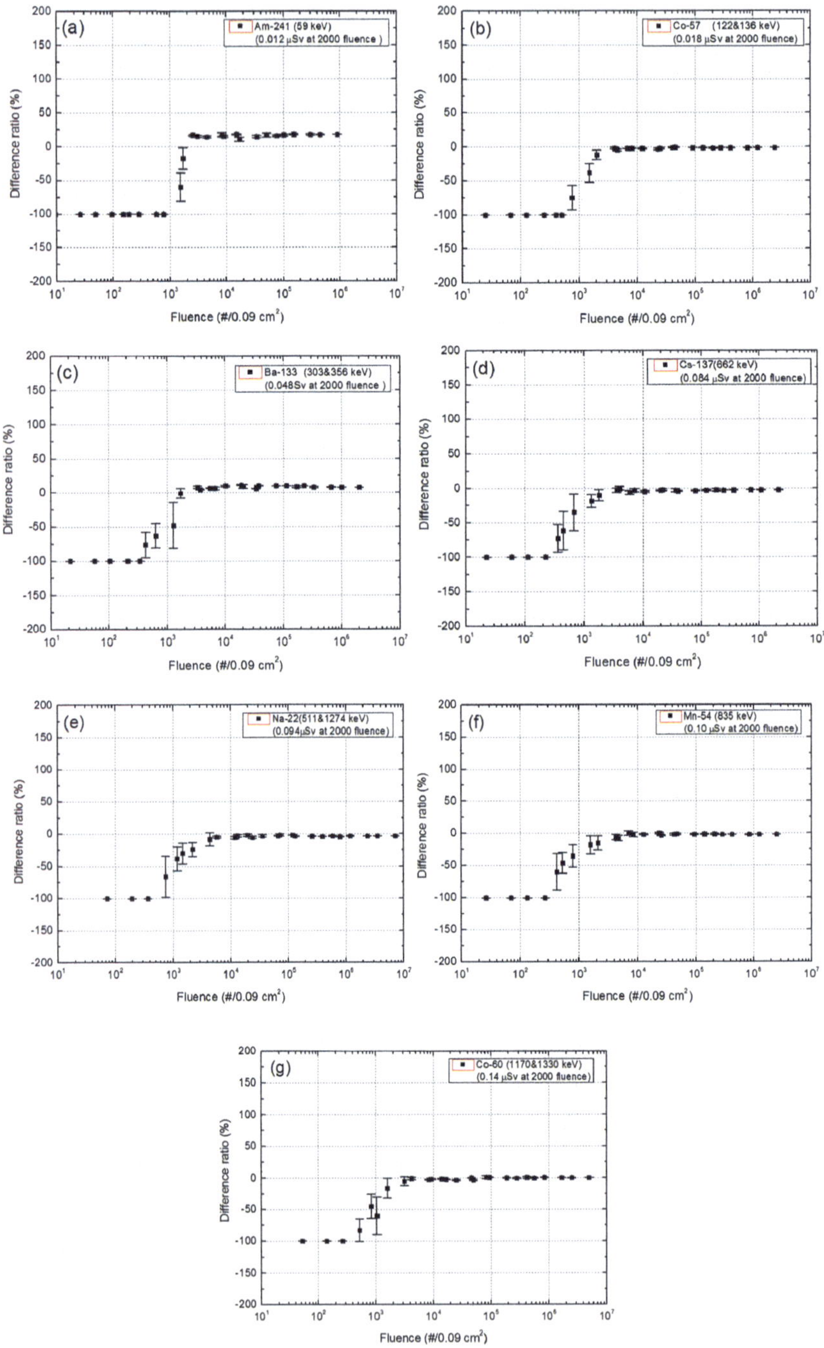

Fig. 25 DR depending on the gamma fluence level for seven radioisotope check sources: (**a**) Am-241, (**b**) Co-57, (**c**) Ba-133, (**d**) Na-22, (**e**) Cs-137, (**f**) Mn-54, and (**g**) Co-60

Table 2 The specification of seven check sources and their personal doses at $10^3\gamma - \text{ray}/0.09\ \text{cm}^2$ of the fluence

Radio isotopes	Am-241	Co-57	Ba-133	Na-22	Cs-137	Mn-54	Co-60
Gamma energy 1 [MeV] (Decay yield [%])	0.059 (36.0)	0.122 (85.5)	0.303 (18.3)	0.511 (180)	0.662 (85)	0.835 (100)	1.170 (100)
Gamma energy 2 [MeV] (Decay yield [%])	–	0.136 (10.7)	0.356 (61.9)	1.274 (100)	–	–	1.330 (100)
Half-life [year]	432.2	0.2	10.5	2.6	30.2	0.8	5.3
Activity [μCi]	23.7	23.7	23.7	23.7	23.7	23.7	23.7
Personal dose [μSv]	0.006	0.009	0.024	0.047	0.042	0.051	0.069

3.3 Measurement of Angular Response

The criteria for angular response is ±20% of the rate difference of angular response (RDAR) from 0° to 60° at 662 keV (Cs-137) and ±50% of RDAR from 0° to 60° at 59.5 keV (Am-241) [1]. The RDAR is defined as the relative difference between the total count in energy spectra at reference exposure angle at 0° and the rotated exposure angles at 0°, 30°, 60°, 90°, 120°, 150°, and 180°. To estimate the angular response of the developed EPD, we measured the $H_P(10)$ with three isotopes such as Am-241, Cs-137, and Co-60 at seven exposure angles from 0 to 180° with the 30° step.

The maximum RDAR of the developed EPD was 18.9% at 30°, and the criteria were satisfied in the exposure angle from 0° to 120°. The angular response was relatively uniform in this range of exposure angle, because the CsI(Tl) scintillator in the compact radiation sensor had the similar size of diameter (3 mm) and total length (3–5 mm). However, the RDAR for Am-241 rapidly decreases between 150° and 180° as shown in Fig. 26 because the partial amounts of low-energy gamma-ray were absorbed by the printed circuit board of EPD system. So, the angular response of the proposed system satisfies the criteria in the range of exposure angle from 0° to 120°.

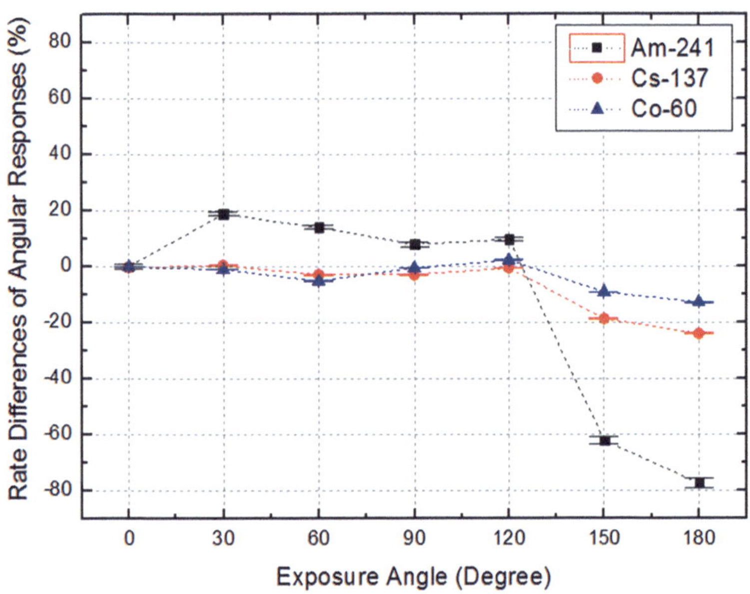

Fig. 26 The rate difference of angular response of three check sources and seven exposure angles

4 Conclusion

In this study, we recommend EPD to simultaneously measure the gamma energy spectra and personal dose, $H_P(10)$, and quickly identify the gamma energy source. To develop the EPD, firstly, the optimum geometry of the compact radiation sensor was decided by light and gamma transport simulations to have the maximum FOM while satisfying the international criteria of angular. Finally a 3×3 mm^2 active area of the silicon photo diode and the tapered structure of the CsI(Tl) scintillator with total thickness 3–5 mm were selected. The individual radiation pulse signal in the compact radiation sensor is converted to voltage signal by the designed front-end ASIC and digitized by MCU in a system board, which communicates to Android phone via audio jack for transferring the spectrum data and control signals. Then, the measured energy spectra by CsI(Tl) scintillator-Si PIN photodiode combination is converted to $H_P(10)$ by a new fast direct dose conversion algorithm. The accuracy of the algorithm depending on the gamma energy and gamma fluence was evaluated by the difference rate (DR) to a theoretically calculated dose. The average DR in the interested gamma energy ranging from 20 keV to 1.5 MeV is 17.3%, and the DR becomes stable at a fluence level above $10^3\gamma - \text{ray}/0.09\ \text{cm}^2$. All these results prove that the proposed spectroscopic EPD can be useful not only for the radiation workers but also for the general public because it can measure the real-time dose rate as well as the real-time radioisotope identification at very low cost.

Acknowledgment This work was supported by the Center for Integrated Smart Sensors funded by the Ministry of Science, ICT and Future Planning as Global Frontier Project.

References

1. International Electrotechnical Commission.: Radiation Protection Instrumentation. Measurement of Personal Dose Equivalent Hp(10) and Hp(0.07) for X, Gamma, Neutron and Beta Radiation: Direct Reading Personal Dose Equivalent and Monitors. International Standard IEC 61526 (2005)
2. Lighttools, http://optics.synopsys.com/lighttools/
3. Knoll, G.F.: Radiation detection and measurement. Wiley, Hoboken (2010)
4. Yoo, H., et al.: Optimal design of a CsI (Tl) crystal in a SiPM based compact radiation sensor. Radiat. Meas. **82**, 102–107 (2015)
5. Noulis, T., et al.: Noise analysis of radiation detector charge sensitive amplifier architectures. In: Topical Workshop on Electronics for Particle Physics, Naxos, Greece (2008)
6. Johns, D.A., Martin, K.: Analog Integrated Circuit Design. Wiley, Hoboken (2008)
7. Krummenacher, F.: Pixel detectors with local intelligence: an IC designer point of view. Nucl. Instrum. Methods Phys. Res. Sect. A. **305**(3), 527–532 (1991)
8. Gatti, E., Manfredi, P.F.: Processing the signals from solid-state detectors in elementary-particle physics. La Rivista del Nuovo Cimento (1978–1999). **9**(1), 1–146 (1986)
9. Colliding, F.: Signal Processing for Semiconductor Detectors. Lawrence Berkeley National Laboratory, Berkeley (2010)
10. Chong, Z.Y., Sansen, W.: Low-Noise Wide-Band Amplifiers in Bipolar and CMOS Technologies, vol. 117. Springer Science & Business Media, Berlin (2013)

11. Ohkawa, S., Yoshizawa, M., Husimi, K.: Direct synthesis of the Gaussian filter for nuclear pulse amplifiers. Nucl. Inst. Methods. **138**(1), 85–92 (1976)
12. Rossi, L., et al.: Pixel Detectors: From Fundamentals to Applications. Springer Science & Business Media, Berlin (2006)
13. De Geronimo, G., O'Connor, P., Kandasamy, A.: Analog CMOS peak detect and hold circuits. Part 1. Analysis of the classical configuration. Nucl. Instrum. Methods Phys. Res. Sect. A. **484**(1), 533–543 (2002)
14. O'Connor, P., De Geronimo, G., Kandasamy, A.: Amplitude and time measurement ASIC with analog derandomization: first results. IEEE Trans. Nucl. Sci. **50**(4), 892–897 (2003)
15. De Geronimo, G., Kandasamy, A., O'Connor, P.: Analog peak detector and derandomizer for high-rate spectroscopy. IEEE Trans. Nucl. Sci. **49**(4), 1769–1773 (2002)
16. Kuo Y.S., Schmid, T., Dutta, P.: Hijacking Power and Bandwidth from the Mobile Phone's Audio Interface. International Symposium on Low Power Electronics and Design (ISLPED'10) Design Contest. Austin, TX (2010)
17. Hall, J.C.: Sensor Data to iPhone Through the Headphone Jack(Using Ardunino). www.creativedistraction.com (2011)
18. SILICON LABS.: Connect the EFM32 with a Smart Phone through the Audio Jack. www.silabs.com (2013)
19. NXP AN11552.: OM13069 Smartphone Quick-Jack solution. www.nxp.com, Jun (2014)
20. International Commission on Radiation Units and Measurements (ICRU).: Determination of Dose Equivalents Resulting from External Radiation Sources. ICRU Publication 39, ICRU (1985)
21. International Commission on Radiation Units and Measurements (ICRU).: Determination of Dose Equivalents from External Radiation Sources- Part 2. ICRU Publication 43, ICRU (1988)
22. International Commission on Radiation Units and Measurements (ICRU).: Measurement of Dose Equivalents from External Photon and Electron Radiations. ICRU Publication 47, ICRU (1992)
23. Jolliffe, I.: Principal Component Analysis. Wiley, Hoboken (2002)
24. Stapels, C., et al.: Comparison of two solid-state photomultiplier -based scintillation gamma-ray detector configurations. Technologies for Homeland Security, 2009. HST'09. IEEE. Conference on. IEEE. Big Sky, MT (2009)
25. Veinot, K.G., Hertel, N.E.: Personal dose equivalent conversion coefficients for photons to 1 GeV. Radiat. Prot. Dosimetry. **145**(1), 28–35 (2011)
26. Pelowitz, D.B.: MCNPX user's manual version 2.5. 0. Los Alamos National Laboratory 76, Santa Fe (2005)
27. Sakai, E.: Recent measurements on scintillator-photodetector systems. Nuclear Science. IEEE Trans. Nucl. Sci. **34**(1), 418–422 (1987)

Part III
System and Application

LED Spectrophotometry and Its Performance Enhancement Based on Pseudo-BJT

Seongwook Choi and Young June Park

1 Spectroscopy for a Smart Sensor

Matters in our universe have their own "fingerprint" which is related to the optical spectrum from the material. The optical spectrum contains key information about their own molecular structure. For example, we can identify the elements consisting the sun – 91.2% of hydrogen, 8.7% of helium, etc. – by the emitting spectrum from its photosphere and chromosphere [1]. If the detection target does not emit the light by itself, we can excite the molecular state of the object by the light incident and then identify the scattered or excited light like the Raman spectroscopy [2] or the fluorescence spectroscopy [3], respectively. Also, if it is hard to get such effects, just obtaining the transmitted spectrum for each wavelengths can reveal the material property as an absorption spectroscopy [4] including the UV-VIS(ible) [5] and FTIR (Fourier transform infrared) spectroscopy [6, 7] that are widely used for the material analysis.

One of the advantage of the spectroscopy over other sensing methods is that it is not necessary to modify the target material in a chemical or physical way, as we do not destroy our finger when using the fingerprint authentication. For example, one can monitor the water quality in the pipeline at a house directly without using a chemical ligand as a dye [8]. Or, diabetic patients can check the blood glucose level via the blood spectroscopy at a fingertip or an earflap because a light can penetrate through the thin skin here [9]. (Even though the light incident does not deform the target material, some cases require a sample preparation for the use of spectroscopy.) On the contrary, other methods, such as a bio or chemical sensor, modify the target material via the chemical reaction. Therefore, an in situ or in vivo characteristic of the spectroscopy can make it a suitable solution for the "smart sensor" as the

S. Choi (✉) • Y.J. Park
Department of Electrical Engineering, Seoul National University, Gwakak-gu, Seoul, South Korea
e-mail: church7@snu.ac.kr

H. Yasuura et al. (eds.), *Smart Sensors at the IoT Frontier*,
DOI 10.1007/978-3-319-55345-0_9

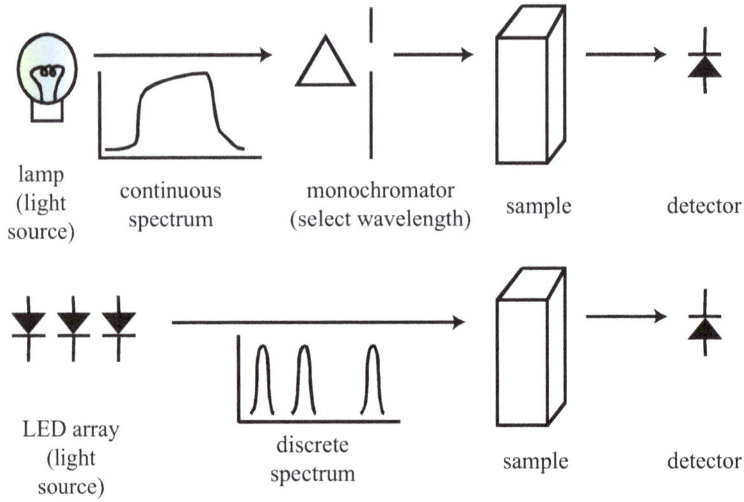

Fig. 1 (**a**) Typical system configuration of the conventional spectrometer. (**b**) System configuration of LED spectroscopy. It replaces the lamp and monochromator of the conventional system with a LED array and generates a discrete spectrum

Internet of Things (IoT) applications by seamlessly integrating the sensors in our living environment.

However, most conventional spectroscopies were hard to meet all the virtue of "smart," such as the low-cost and tiny platform size so have been utilized for the LAB devices. The reason for a high cost and a big size is mostly due to a light source and monochromator; for example, a lamp, a precise laser system, and optical components should be used for the FTIR [7]. Figure 1a shows the typical optical layout of a conventional absorption spectrophotometer [4]. The system includes a light source that emits the required wavelengths. The monochromator selects a specific wavelength, and the selected wavelength interacts with the sample and then detected at the receiver. This procedure is repeated until the monochromator scans all the range of wavelengths of interests.

The semiconductor optoelectronic device could fill up the gap between the spectrometer and smart sensor by adopting the limit emitting diode (LED) as a light source. This is called as a LED spectrophotometer [10–12] and already used for variety of applications. It replaces the light source (mostly lamp) and monochromator part with a LED array as shown in Fig. 1b. Since LED can be regarded as a point light source, the size of sensor can be extremely scaled down to a chip size as the case of spectroscopy on a chip [13]. Besides, the adoption of LED has many advantages over the lamp system; the advantages include a low power consumption, a low cost, a stable constant light power, a long lifetime, a lower heat generation, no warming time, and no heavy metal. Hence, the LED spectrometer can meet any form factor that smart sensor requires in a variety of situations with a viable cost.

In this chapter, a guide to set up the LED-PD system is presented for the LED spectrophotometry covering a device selection, driving circuit composition and applications. Especially, we will deeply focus on the technology that can enhance the sensitivity and the sensing range exceeding the capability of the selected devices and system. One method makes the silicon junction to detect the NIR photons utilizing the Franz-Keldysh effect [14], and the other enhances the limit of detection (LOD) based on the pseudo-bipolar junction transistor (BJT) [15].

2 Optoelectronic Devices for the LED Spectrophotometry

The LED spectroscopy consists of an array of LEDs and photodiodes. Both devices are optoelectronic devices which converts the electronic signal to the optical signal and vice versa. The basic structure of the LED and PD is a p-n junction. If it is forward biased, it operates as a LED. On the other hand, it operates as a PD if it is reverse biased. The range of the wavelength is determined by their material and structures. Hence, according to the wavelength range of interest, the selection of LED and PD should be different. Here, we review the available choices of the optoelectronic devices according to the wavelengths and their proper usage. Then, we discuss about how the material limitation (bandgap energy) of photodetector can be overcome with an aid of the Franz-Keldysh effect.

2.1 Technology of LED an Its Usage

A LED converts the electron to the photon. The wavelength of photons emitted by a LED is determined by its band structure. Hence, an adequate material should be chosen for the spectroscopy. Figure 2 summarize the currently available LED for each wavelength. Due to the recent success of UV-LED fabrication, LED spectroscopy can be available from UV to IR (250–2000 nm). In addition, the light-emitting p-n junction can be combined with a semiconductor cavity to form a laser diode (LD). This laser diode has a sharp wavelength peak and even enables a single mode beam emission. In order to identify a fingerprint – distinguish a single target material from a mixed sample – at least two or three LEDs near the absorption peak is needed as shown in Fig. 1b. A statistical method, such as a regression analysis [16], can be applied to estimate the concentration of target material from the absorption data using a multiple LED array.

The LED can be biased with a constant voltage or a constant current scheme. For the sensor application, the light intensity variation due to the thermal fluctuation should be suppressed or else the sensor readings on the same sample are different for every measurement. Since the light intensity is more proportional to the LED current than the voltage, it is recommended to use a constant current scheme for the LED spectrophotometer.

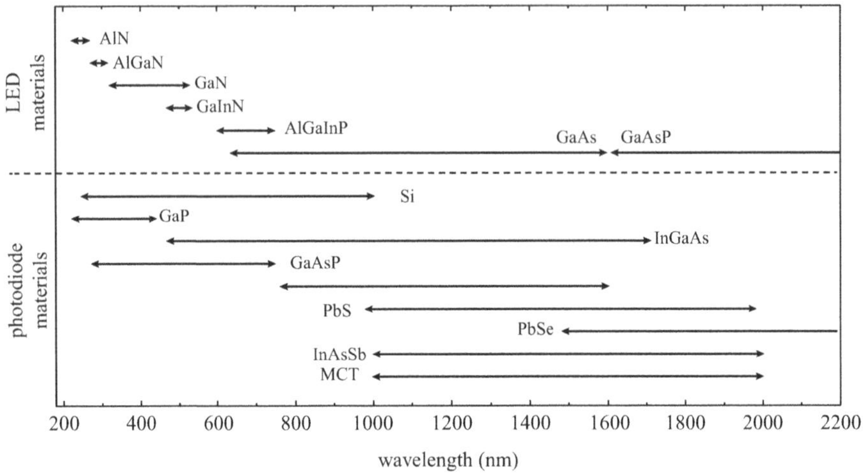

Fig. 2 Materials for LED and photodiode according to the wavelength of interest (from UV to near IR)

2.2 *Technology and Usage of Photodiode*

A photodiode converts the photon energy to the electrical signal. The photon excites the valence band electron to the conduction band (electron-hole pair generation). This excitation is not limited to the p-n junction, but only the excited carriers in the depletion region can affect to the terminal current by the electric field in the depletion region, otherwise, electrons will recombine with holes again.

The range of wavelength that PD can detect is also determined by the bandgap property because the photon energy should be larger than the bandgap energy. Available material choices for PD (along with LED) are summarized in Fig. 2 for various wavelengths. Generally, the range of wavelength of PD is much wider than that of LED, and PD should not be one-to-one matched to LED but can detect the multiple LEDs in sequence.

According to the operation mechanism, the biasing method of PD will change. Figure 3 depicts the operation range of photodiode including an avalanche breakdown point. When PD is biased at a zero voltage, the PD operates in a photovoltaic mode [17] and the short circuit current is measured in this case. An extension of the depletion width can enhance a responsivity of the PD by applying a voltage bias, which is done in a photoconductive model. When the bias voltage increases, the avalanche multiplication factor can be more than unity. In this case, called as an avalanche mode, the photocurrent is amplified. When the bias is larger than avalanche breakdown voltage is applied, the device operates in the Geiger mode. In this case, the gain has no practical meaning because of very high multiplication factor. Therefore, even a single photon can trigger the photocurrent, and it is often called as a single-photon avalanche diode (SPAD). The most usually recommended

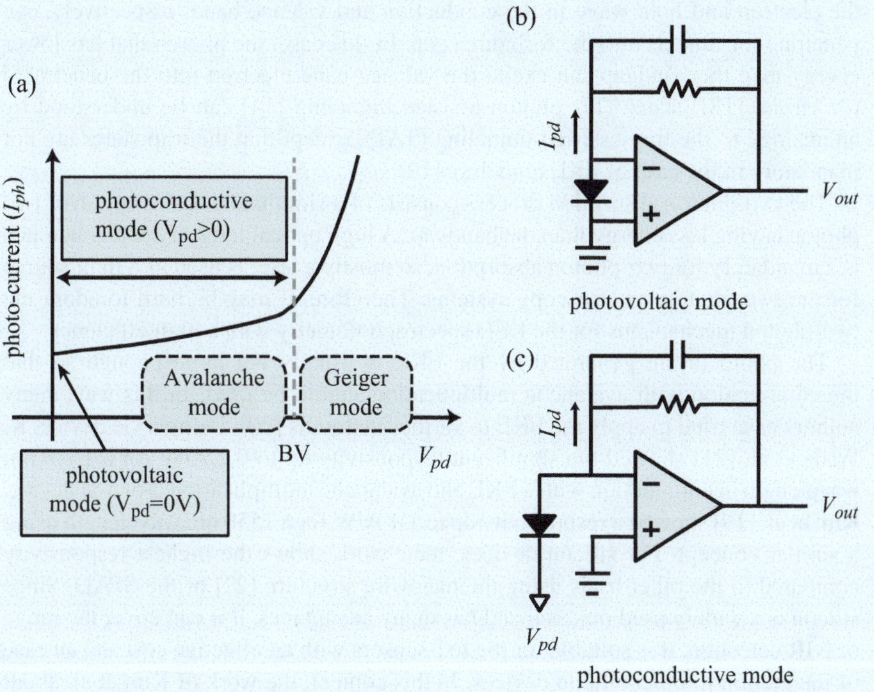

Fig. 3 (**a**) Operation mode of PD (reverse bias). (**b**) Typical circuit for a photovoltaic mode. (**c**) Typical circuit for a photoconductive mode

circuit for the photovoltaic and photoconductive mode is shown in Fig. 3b, c, respectively. These circuits convert the photocurrent to the voltage output that will be directly converted to the digital data through analogue-digital converter (ADC).

2.3 *Si Photodiode as a Near-Infrared Detector*

As stated earlier, the origin of the photocurrent in the semiconductor photodetector is an electron-hole pair generation after the photon absorption. The criterion for the absorption is that the energy of photon should be larger than that of bandgap energy of the material. In other words, the detection wavelength of the photodiode is limited by the bandgap energy of the material, e.g., the bandgap energy of Si is 1.12 eV (1107 nm), Ge is 0.67 eV (1850 nm), PbS is 0.37 eV (3350 nm), and InSb is 0.17 eV (7293 nm). Hence, for a near-infrared spectrophotometry, silicon is not a suitable material.

There are several methods to overcome the limitation by the bandgap; one is the Franz-Keldysh effect (FKE) and the other is the two-photon absorption. In case of the FKE, a high electric field introduces a stiff slope of band edge. Hence, the electron and hole wave in the conduction and valence band, respectively, can penetrate (or tunnel) into the forbidden gap. In this case, the photon that has lower energy than the bandgap can excite the valence band electron into the penetrated (or virtual [18]) state. This photon-assisted tunneling [14] can be understood by an analogy to the trap-assisted tunneling (TAT), except that the trap states are not mandatory in the case of FKE tunneling [19].

The two-photon absorption process consists of a simultaneous absorption of two photon having less energy than the bandgap. A high optical intensity and coherency is a mandatory for two-photon absorption, so mostly a laser is used as a light source for the two-photon spectroscopy systems. Therefore, it may be hard to adopt the two-photon mechanisms for the LED spectrophotometry with a high efficiency.

The photocurrent generated by the FKE is usually not large enough so that the combination with avalanche multiplication should be used. In this way, many authors have tried to apply the FKE to various materials [20]. Using a Ge device, K. Wada et al. [21] showed the significant responsivity up to 0.2 A/W for a 1640 nm wavelength in conjunction with a FKE and avalanche multiplication. For a Si device, Kim et al. [19] showed a responsivity up to 1.1 A/W for a 1550 nm wavelength using a similar concept. For silicon devices, their work shows the highest responsivity compared to the other trials using the nanowire structure [22] or the SPAD. Since silicon is a widely used material and has many advantages, if it can cover the range of NIR detection, it is suitable for the IoT sensors with an effective cost and an ease of integration to other silicon devices. In this context, the work of Kim et al. about the silicon IR photodiode (>1550 nm) [19] is reviewed in detail.

2.3.1 FKE in Zener Diode Structure

Considering the fact that the tunneling is most important part for the FKE, a Zener tunneling junction is a simple and proper structure because the band-to-band tunneling (BTBT) is a current mechanism of the Zener junction. It is formed when an abrupt p^{+}-n^{+} junction with high doping is made so that the applied voltage is focused on the narrow region, resulting a high electric field. In this case, the tunneling probability that is mandatory for the FKE increases as shown in the Fig. 4a. When a higher voltage is applied, an avalanche breakdown is followed by the Zener breakdown. (The order of Zener and avalanche breakdown is determined by the doping profile [23]). In this case, the generated electrons and holes are multiplied by the avalanche multiplication.

The responsivity of the Zener junction vs. applied reverse bias is shown in Fig. 4b under the illumination of 800 nm, 1310 nm, and 1550 nm of wavelengths [19]. The result can be divided into three regions according to the Zener and avalanche breakdown voltage (BV). Apart from the 808 nm wavelength (higher energy than the bandgap), 1310 and 1550 nm wavelengths (sub-bandgap energy) show the clear voltage dependence inferring the FKE. However, when only the FKE works (before the Zener BV), the responsivity of 1330 nm and 1550 nm by the FKE is somewhat small. Y. Zhou et al. [22] suggest using a nanowire structure to enlarge this small

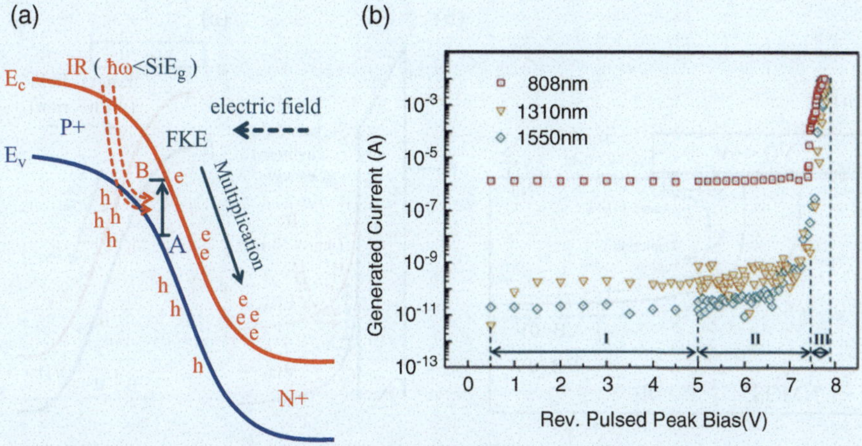

Fig. 4 (**a**) FKE and avalanche multiplication in the Zener junction. (**b**) Responsivity of Zener diode according to the applied bias voltage under the illumination of 808 nm, 1310 nm, and 1550 nm wavelengths. Reprinted from Kim et al. IEEE Trans. Electron Devices 2016;63:377–383, with permission [19]

responsivity but their 3D structure requires a high cost for the fabrication. H. Kim et al. suggested using the pulsed bias mode, enabling an avalanche multiplication as described in Fig. 4a. The pulse method is used to mitigate the reliability degradation caused by a high current. With an aid of the multiplication, as shown in Fig. 4b, they could obtain higher responsivity up to 1.1 A/W using a commercial Zener diode package with a cheap price (<s$0.01/1ea). The sub-bandgap photodiode based on the similar concept (FKE and avalanche effect) was also demonstrated using a planar p^+-n^+ device in a SOI wafer with a waveguide [24].

2.3.2 FKE in MOSFET GIDL Range

Another tunneling junction in the silicon device where the FKE can be applied is a MOSFET S/D junction under the GIDL condition as shown in Fig. 5a. Here, the high electric field is applied in the drain surface within the gate-drain overlap region. In this case, the direction of the BTBT field responsible for the FKE is normal to that of the avalanche multiplication, while they are the same for the case of Zener diode.

To make this condition, V_{gd} (for a BTBT field) and V_{db} (for an avalanche field) are biased by a negative voltage. Since the electron-hole pair generation range is only limited to the surface, the array of common drain MOSFETs is used to enlarge the active area. The benefit of this structure is that the conventional MOSFET fabrication process can be used without even using a n^+-p^+ junction (Zener diode), so the integration to the conventional logic device fabrication is much easier.

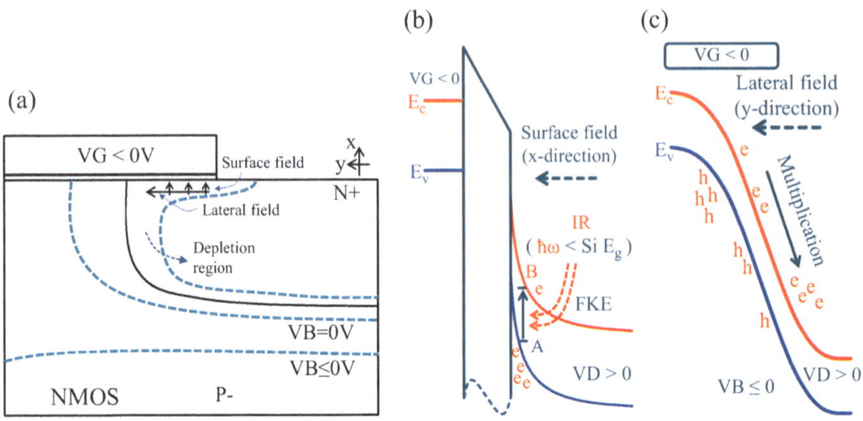

Fig. 5 (**a**) The structure of S/D in MOSFET and equipotential under the application of GIDL bias (**b**) The band diagram along A-A′ direction. The photon-assisted e-h pair generation is affected by the electric field in this direction. (**c**) The band diagram along B-B′ direction. Avalanche multiplication is affected by the electric field in this direction. Reprinted from Kim et al. IEEE Trans. Electron Devices 2016;63:377–383, with permission [19]

However, it shows small responsivity (0.1 A/W) than the Si Zener junction even though the avalanche multiplication is applied.

In Fig. 6, the responsivity of Si photodiode is compared with Ge and GeSn for 1550 nm PD. In the figure, both cases using a normal incidence (NI) and a waveguide (WG) are compared for Ge and GeSn [25–35]. Since the waveguide can deliver the light to the junction with a lower optical loss than the case of the normal incidence, the responsivity is usually higher. The comparison indicates that the Si PD from [19] shows comparable or even higher performance in terms of the responsivity than the more expensive materials.

3 Performance Enhancement Based on Pseudo-BJT Optical System

Including the LED spectrophotometer, a sensitivity, and limit of detection (LOD) are most important performance specifications of the sensor system. The most general way to enhance these performances of an optical sensor system can be summarized as follows: (1) enhance the performance of optical devices such as LED and PD, (2) increase the signal absorption by the sample, and (3) amplify the detection signal using an electrical circuit.

Regarding (1), in most cases, using a high-performance device requires an additional cost. Sometimes, a technological breakthrough is needed for a high-performance device without sacrificing the cost. About (2), increasing the light path length as long as possible helps according to the Beer-Lambert law. By using a

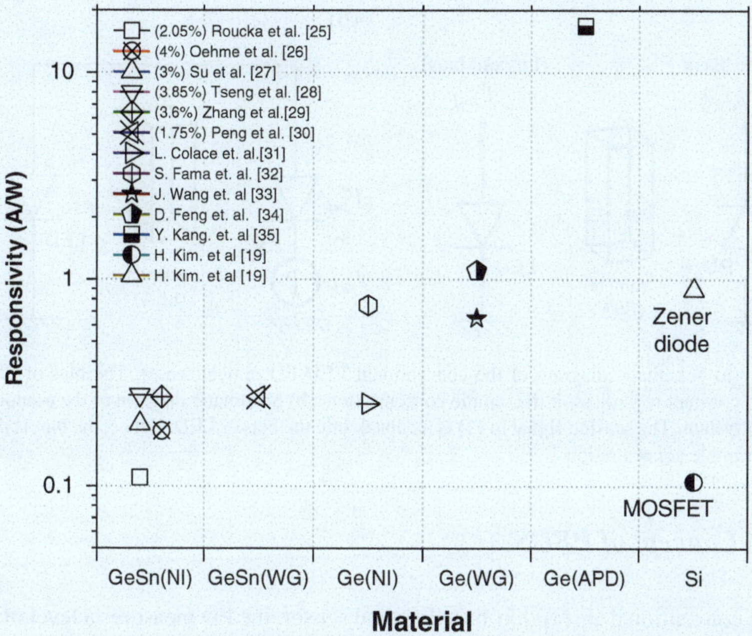

Fig. 6 The responsivity of Si photodiode [19] compared with GeSn, Ge photodiodes [25–35] under the illumination of 1550 nm wavelength. Modified from Kim et al. IEEE Trans. Electron Devices 2016;63:377–383, with permission [19]

mirror or a waveguide, the length of the light path can be increased maintaining the same sample volume or the system size. A multi-scattering enhanced absorption spectroscopy is another example that can enlarge the optical path length [36]. Regarding (3), the highest limit of amplification is defined by the signal-to-noise ratio (SNR) of photodetector, so the limitation of the optical device cannot be overcome.

Choi et al. [15] proposed the new concept of optical sensor system, by mimicking the operation mechanism of bipolar junction transistor (BJT). The benefit of this system, apart from the previously described methods, is that the only slight variation of the system can boost the performance, so an additional cost is negligible. Since its operation theory is analogous to the bipolar junction transistor, they call it as the pseudo-BJT optical system (PBOS).

In this section, we describe the mechanism, modeling, and practical usage of PBOS. In the new system, the most significant point is that the negative differential resistance (NDR) is found after the breakdown voltage (BVceo). As the NDR region is sensitive to the transistor α, the NDR characteristics of the PBOS is sensitive to the absorption of the light.

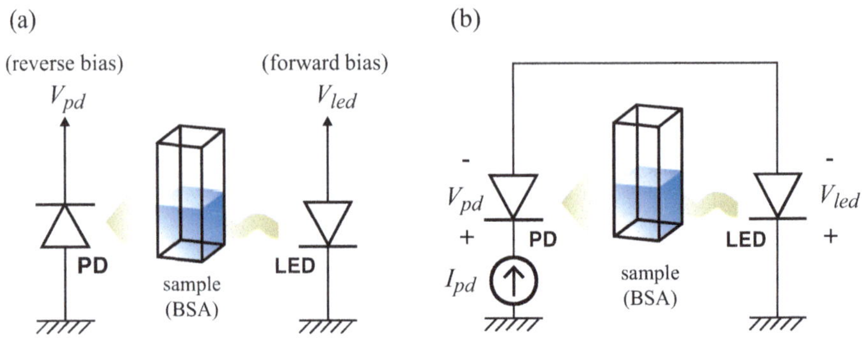

Fig. 7 (**a**) Schematic diagram of the conventional LED-PD optical sensor. The bias of PD and LED is constant regardless of the sample concentration. (**b**) Schematic diagram of the pseudo-BJT optical system. The sensing signal in PD is feedback into the bias of LED. Hence, the bias is related to the sample concentration

3.1 Concept of PBOS

In the conventional absorption-based optical sensor, the PD measures a level of light after interacting with the sample as shown in Fig. 7a. In this case, the bias of LED and PD are fixed. Usually, the light source is biased constantly to emit a constant light intensity. The photodiode is biased with a constant 0 V for the photovoltaic mode or a negative voltage for the photoconductive mode.

In the PBOS, a simple positive feedback is introduced with a series connection of the LED (forward bias junction) and PD (reverse bias junction) system as shown in Fig. 7b, where the feedback path is formed by illuminating the light generated from the LED on the PD junction. In this case, the bias voltage of LED is not fixed but determined by the level of the photocurrent from the PD. In this respect, the connection may be considered as a BJT with infinite base width. Then, the PD (like the base-collector junction) receives the light from the LED, and thus the PD current is increased due to generation current. This in turn increases LED light intensity followed by increase in the LED current which again triggers increase in the generation current forming a system with feedback loop. This positive action and the back-to-back connection of PD (n-p) and LED (p-n) junction are analogues to the n-p-n transistor. To understand the equivalence more clearly, junctions and the current components of BJT and PBOS are shown in Fig. 8a, b, respectively. In Fig. 7b, the p-region of both LED and PD is regarded as the base region of pseudo-BJT. When there is an avalanche multiplication (M) in base-collector or PD junction, M is more than unity ($M > 1$), otherwise, M equals to one ($M = 1$). The current component between BJT and pseudo-BJT is compared in Table 1.

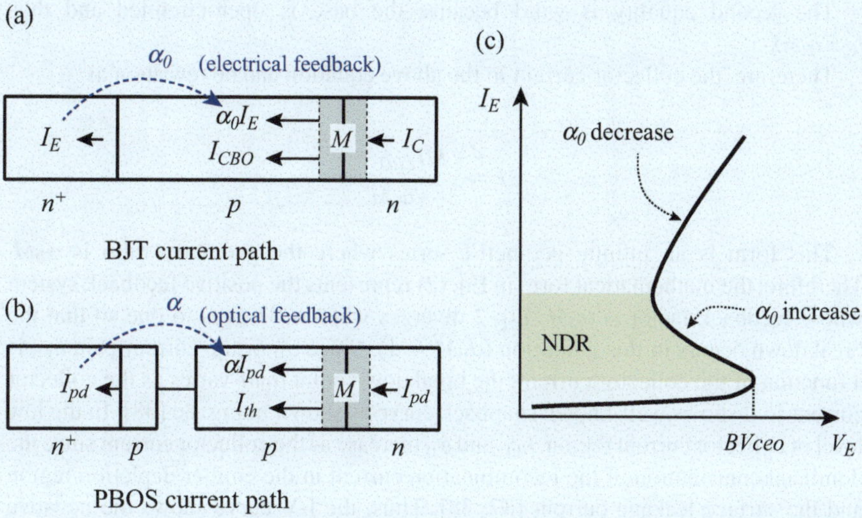

Fig. 8 (**a**) The current component in the BJT under the open-base operation (**b**) The current component in the pseudo-BJT t is analogues to that of BJT. (**c**) A typical I-V curve of BJT under the open-base mode. The negative differential resistance (NDR) is shown due to the change of the α_0

Table 1 Correlation between the parameters of BJT and pseudo-BJT

BJT		Pseudo-BJT	
Notation	Meaning	Notation	Meaning
I_E	Emitter current	I_l	LED current
I_C	Collector current	I_{pd}	PD current
I_{CBO}	Saturation current of C-B junction	I_{th}	Dark current
α_0	Common-base current gain	α	Optical current gain
M	Multiplication factor of C-B junction	M	Multiplication factor of PD

3.2 *Mathematical Model of PBOS*

3.2.1 A Basic Pseudo-BJT Model

The operation mechanism of the pseudo-BJT can be understood by the mathematical way as summarized in [15]. According to the BJT operation in Fig. 8a, the collector terminal current is composed of $\alpha_0 I_E$ and the reverse current of collector-base junction (I_{CBO}) multiplied by the multiplication factor M. So, I_C can be written as [37]

$$I_C = M\left(\alpha_0 I_E + I_{CBO}\right) = I_E \tag{1}$$

The second equality is valid because the base is open-circuited and thus $I_C = I_E = I$.

Therefore, the collector current in the above equation can be rewritten as

$$I = \frac{MI_{CBO}}{1 - a_0 M} \tag{2}$$

This form is an infinite geometric series where the geometry ratio is $\alpha_0 M$. Therefore, the mathematical form in Eq. (2) represents the positive feedback system whose feedback factor is $\alpha_0 M$. Eq. 2 diverges when $\alpha_0 M$ goes to one so that the breakdown occurs in this condition ($\alpha_0 M = 1$). Since α_0 or the current gain h_{FE} is a function of the collector current, the breakdown point $\alpha_0 M$ varies as the collector current so it shows well-known snapback curve as shown in Fig. 8c [38]. In the low level of collector current region, h_{FE} and α_0 increase as the collector current since the dominant contribution of the recombination current in the emitter depletion region and the surface leakage current [37, 38]. Thus, the I-V curve shows the negative differential resistance (NDR) as shown in Fig. 9c. In the high-level current region, α_0 decreases as the collector current due to the effective base doping increase (Webster effect) [37, 38]. Hence, the differential resistance becomes positive again in the high-level injection condition, resulting in the snapback curve in Fig. 7c.

In pseudo-BJT, the current component (in Fig. 8b) is very similar to that of BJT in the open-base operation as shown in Fig. 8a. Thus, the PD current can be described as

$$I_{pd} = M\left(I_{th} + I_{ph}\right) = M\left(I_{th} + \alpha I_{pd}\right) \tag{3}$$

where I_{th} is a thermal or dark current of the PD and I_{ph} is a photo current of the PD. In the above equation, α is an optical current gain (the ratio between the electron generation in a PD to the electron flow in a LED) and is defined as

$$\alpha = \eta_{pd} T_f \eta_{led} \tag{4}$$

where T_f is the transmittance of a sample and η_{pd} and η_{led} are responsivities of the PD (A/W) and the LED (W/A), respectively. In most of the operation conditions, η_{pd} and η_{led} are approximately constant. Since the responsivity and the transmittance are less than one, so the α is ($\alpha < 1$). The information of the sample (T_f) is contained in the parameter α so the sensing signal is desired to be sensitive to α. In a same manner with Eq. (2), the PD current of pseudo-BJT in Eq. (3) can be rewritten as

$$I_{pd} = \frac{MI_{th}}{1 - a_0 M} \tag{5}$$

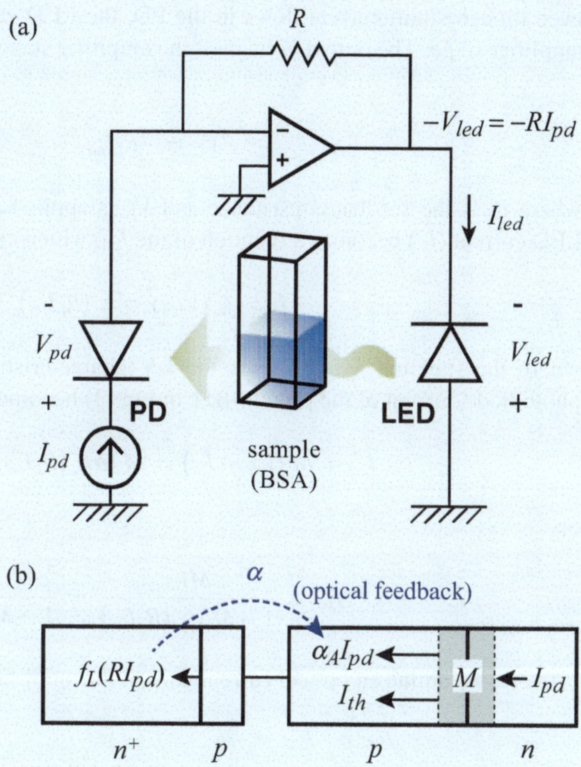

Fig. 9 (**a**) A pseudo-BJT with an amplifier stage. The small optical gain of PBOS is amplified by the transimpedance amplifier. (**b**) The current component of the amplified PBOS

It has completely the same form with that of BJT in Eq. (2) and diverges when $\alpha_0 M = 1$. Therefore, like the case of BJT, the current behavior after the breakdown point is determined by α, i.e., the NDR appears when α increase as the PD current.

In a practical case, the operation of the pseudo-BJT circuit in Fig. 7b may not work. This problem can be understood as follows: αI_{pd} in Eq. (3) is so small because α is less than the order of 10^{-2} and I_{pd} cannot exceed the saturation current of the PD, $I_{\mathrm{s,p}}$ (usually the order of nA). Hence, the photocurrent term in Eq. (3) is negligible and the I-V characteristic of the PD is the same as the dark current of PD. The reason can be understood that the current flowing in the pseudo-BJT is so small to turn on the LED. In this case, there is negligible optical feedback and thus no pseudo-BJT operation occurs.

3.2.2 An Amplified Pseudo-BJT

Use of the amplifier as shown in Fig. 9a can solve the current limitation problem in LED, preserving the pseudo-BJT operation. Here, the transimpedance amplifier is added between the PD and LED to amplify the I_{pd} that is fed into the LED. Then, even though small current flows in the PD, the LED can be turned on due to the amplifier stage. The output voltage of the amplifier stage (V_L) is

$$V_L = -R_f I_{pd} \tag{6}$$

where R_f is the feedback resistance and V_L is applied across the LED. Then, the LED current (I_L) becomes a function of the I_{pd}, which can be written as

$$I_L = f_L(V_{led}) = f_L\left(R_f I_{pd}\right) \tag{7}$$

where the function f_L represents the I-V characteristic of LED. Therefore, the photodiode current of the pseudo-BJT in Eq. (3) becomes

$$I_{pd} = M\left(I_{ph} + I_{th}\right) = M\left(\alpha f_L\left(R_f I_{pd}\right) + I_{th}\right) \tag{8}$$

or

$$I_{pd} = \frac{M I_{th}}{1 - M\alpha f_L\left(R_f I_{pd}\right)} = \frac{M I_{th}}{1 - M\alpha_A} \tag{9}$$

Here, we define an optical current gain of the amplified pseudo-BJT, α_A as

$$\alpha_A = \alpha f_L\left(R_f I_{pd}\right) \tag{10}$$

which means that the original optical current gain α is enlarged by the transimpedance amplifier. With this increased optical gain, the LED can be turned on and the optical feedback pathway can work. The current component of the amplified PBOS is described in Fig. 9c that shows the similar operation with that of non-amplified PBOS in Fig. 8b.

The NDR region of the amplified pseudo-BJT can be described with simple analytic forms as follows. When we use the ideal diode relation with $I_{th} = I_{s,p}(1 - \exp(-qV_{pd}/kT))$ that is the reverse current equation of the p-n junction and $I_L = I_{s,l}(\exp(-qV_L/kT) - 1)$ that is the forward current equation, the PBOS equation in Eq. (8) can be written as

$$\begin{aligned} I_{pd} &= \frac{1}{1-\left(\frac{V_{pd}}{V_b}\right)^m}\left[I_{s,p}\left(1-\exp\left(-qV_{pd}/kT\right)\right) + \alpha I_{s,l}\left(\exp\left(-qV_{pd}/kT\right) - 1\right)\right] \\ &\approx \frac{1}{1-\left(\frac{V_{pd}}{V_b}\right)^m}\left[I_{s,p} + \alpha I_{s,l}\left(\exp\left(-qV_{pd}/kT\right) - 1\right)\right] \end{aligned} \tag{11}$$

where V_b is the breakdown voltage of p-n junction and $I_{s,l}$ and $I_{s,p}$ are saturation currents of LED and PD, respectively. The approximation stands when the reverse

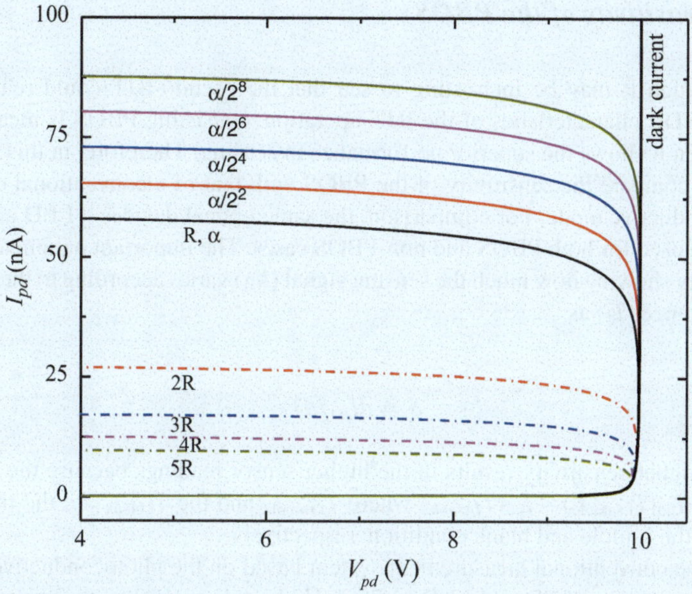

Fig. 10 Calculated I-V characteristics of PBOS with an ideal diode model using Eq. (11). The changes can be seen due to the various values of the feedback resistance (R_f) and α

current is almost saturated since V_{pd}>>0 near the breakdown point. Therefore, the V_{pd} can be readily expressed as

$$V_{pd} = V_b \left[1 - \frac{I_{s,p} + \alpha I_{s,l} \exp\left(qR_f I_{pd}/kT\right)}{I_{pd}} \right]^{1/m} \quad (12)$$

Note that the denominator in the bracket term is proportional to I_{pd} while the nominator is proportional to the exponential of I_{pd}. Therefore, when I_{pd} is small, the denominator dominates the exponential term in the nominator and V_{pd} increases as I_{pd} increases. However, when I_{pd} becomes large, the exponent in the nominator dominates so V_{pd} now decreases as I_{pd} increases (NDR region). In the Fig. 10, the I-V characteristics of pseudo-BJT is plotted using Eq. (12). It clearly shows that the NDR region appears in a pseudo-BJT system. When R_f increases, the current at a snapback point (I_{SB}) decreases and the snapback will start earlier. This can be understood in terms of the LED turn-on voltage since the larger R_f value turns on the LED with a smaller I_{pd}. In addition, I_{SB} decreases as the amplified optical gain α_A increases. Hence, the concentration of sample is reflected to the snapback point and the NDR slope.

3.3 Sensitivity of the PBOS

Even though it may be interesting to see that the pseudo-BJT could realize the similar NDR characteristics of the BJT operation, consisting PBOS is meaningful only when it shows the superior performance in sensing. Therefore, in this subsection, we compare the sensitivity of the PBOS with that of a conventional one in a photoconductive mode. For comparison, the same optical devices – LED and p-i-n PD – are used for both PBOS and non-PBOS cases. The important parameter is the sensitivity showing how much the sensing signal (I_{pd}) varies according to the sample transmittance (T_{f}) as

$$S = dI_{pd}/dT_f \tag{13}$$

The higher sensitivity results in the higher sensor readings because the sensing signal is $I_{\text{pd}}(T_{\text{f,sample}}) - I_{\text{pd}}(T_{\text{f,blank}})$ where $T_{\text{f,sample}}$ and the $T_{\text{f,blank}}$ are the transmittance of the sample and blank condition, respectively.

For the conventional measurement system based on the photoconductive mode, the bias condition of PD and LED is fixed. Under this condition, the sensitivity of the system can be obtained by differentiating Eq. (3) by the transmittance as

$$\left.\frac{dI_{pd}}{dT_f}\right|_{V_{pd}} = M\eta_{pd}\cdot\eta_{ld}\cdot I_{ld} \equiv I_{ph0} \tag{14}$$

In case of the PBOS under the fixed PD voltage bias, the optical output power from the LD is now a function of PD current so the sensitivity can be written as

$$\left.\frac{dI_{pd}}{dT_f}\right|_{V_{pd}} = \frac{I_{ph0}}{1 - T_f\frac{dI_{ph0}}{dI_{pd}}} = \frac{I_{ph0}}{1 - T_f\frac{dI_{ld}}{dI_{pd}}\frac{dI_{ph0}}{dI_{ld}}} = I_{ph0}\left(\frac{1}{1 - T_f\cdot G_E\cdot G_O}\right) \tag{15}$$

where G_{E} is the rate of change of the LD current with PD current and G_{O} is the rate of change of I_{ph0} with LD current. G_{E} and G_{O} represent the relationship of the electrical and optical parameters, respectively.

Comparing the sensitivity of the PBOS system in Eq. (15) with the conventional scheme in Eq. (14), the sensitivity is multiplied by the bracket factor in [9]. Hence, three main factors in the bracket term – T_{f}, G_{E}, and G_{O} – determine the sensitivity enhancement of the PBOS, which are discussed as follows.

3.3.1 Sample Transmittance (T_f)

It should be noted that unlike the conventional measurement system, the transmittance of the sample to be measured affects the sensitivity of the PBOS. This may infer that the range of transmittance to be measured should be determined before tuning the PBOS. In most cases, it can be determined in advance according to the target samples. For example, normal glucose concentration in human blood is in the range of 65–104 mg/dl under the condition of an empty stomach [39]. For water regulation, the standard for water quality also determines the range of transmittance of PBOS sensor.

3.3.2 G_E and Go

G_E is related to the electrical parameters of the pseudo-BJT: how the PD current controls the LED bias. It is tunable by adjusting the electronic parameter such as R_f. Also, G_O is a parameter that is related to the optoelectronic devices and is hardly changeable unless one changes the optoelectronic devices. Practically, G_E and G_O can be extracted by measurement when the optical and electrical parts of the system are determined. To explain the extraction procedure, the examples in [15] are adopted by using the same optical devices (a laser diode as the light source and a photodiode as the detector). The feedback resistance R_f with 560 kΩ and OP-AMP with OPA544 are used for the circuit. A series resistance (180 Ω) is connected to the LD, and its role will be explained in the next section. The system and results are described in [15] in more detail.

To obtain G_O, the photocurrent is measured as the laser diode current without the sample as shown in Fig. 11a. From the circuit, the photogeneration current (I_{ph0}) as the laser diode current is extracted when the reverse bias at the PD is fixed to constant (V_{pd}=5 V). Then, the G_O is obtained by differentiating the result with I_L as shown in Fig. 11a. Since the G_O represents the characteristics of optical pathway, no electrical feedback loop is needed for the G_O extraction measurement.

For the G_E measurement, the electrical feedback loop should be now included, and the current of laser diode is measured as the PD current. The measurement setup of G_E is shown in Fig. 11b. Here, G_E is identical to $R_f/(R_{ld} + R_s)$, which can be approximated to R_f/R_s after the turn-on voltage because $R_{ld} << R_s$. Then, the rate of change of the LED current with the PD current can be adjustable by choosing the resistor values for R_f and R_s. Since G_E depends on the resistor values, this parameter is tunable to find an optimum point. The G_E extraction result is shown in Fig. 11b. For the small PD current, the R_{ld} is large because the LED is not turned on, so that G_E shows a small value. As the PD current increases, G_O saturates to approximately R_f/R_s value since R_{ld} becomes small after the diode turns on voltage. As a remark, the pseudo-BJT without the amplifier in Fig. 7b gives the value of G_E of one that is very small to turn on the light, while pseudo-BJT with the amplifier gives the value of G_E in the order of 10^3. On contrary, the values of G_O with and without the amplifier stage are the same.

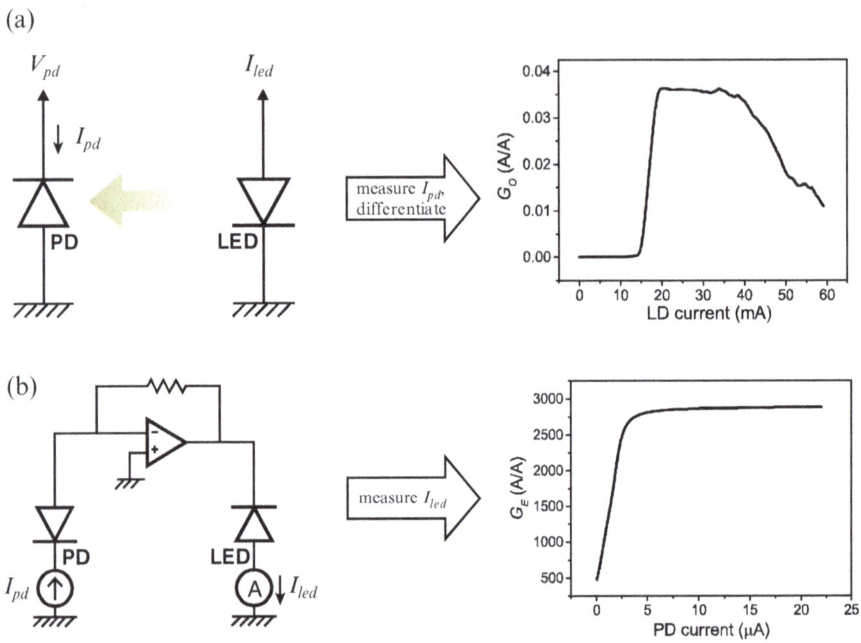

Fig. 11 (**a**) Measurement setup of G_O for the given optical device and the measurement result. (**b**) Measurement setup of G_E for the given optical device and PBOS configuration and the extraction result. Measured data is reprinted from Choi et al., IEEE Trans. Electron Devices 2016;63:2074–2079 [15]

Once the G_E and G_O are extracted, the sensitivity enhancement can be now estimated based on the previous modeling on the pseudo-BJT. Here, starting from the extracted G_E and G_O of sample device [15], the sensitivity of conventional system and the pseudo-BJT can be calculated by Eqs. (14) and (15), respectively.

The sensitivity of conventional absorbance sensor is defined by I_{pho} according to (14) and becomes higher as the light intensity of the light source increases. Therefore, 1.2 mA of the sensitivity can be achieved where the LED current of 60 mA is applied. This value means that the photocurrent of PD is 120uA when T_f changes by 1%. Note that this sensitivity value does not change as the PD current or the sample transmittance varies.

Unlike the conventional case, the sensitivity of PBOS in Eq. (15) is not fixed to a unique value and varies according to the sample condition (T_f), PD current (I_{pho}), and system parameters (G_E, G_O). Hence, the sensitivity is plotted as the T_f and I_{pd} as shown in Fig. 12 where the sensitivity is normalized to that of conventional one (1.2 mA). It clearly shows that when the PD current satisfies a certain condition (14–18 uA), the sensitivity can be enhanced up to two to five times. This will give rise to the larger signal (PD current) changes between the different sample concentrations. For example, if we desire to measure T_f of 0.017 when T_f of blank sample is 0.021, the signal difference of PBOS can be calculated as

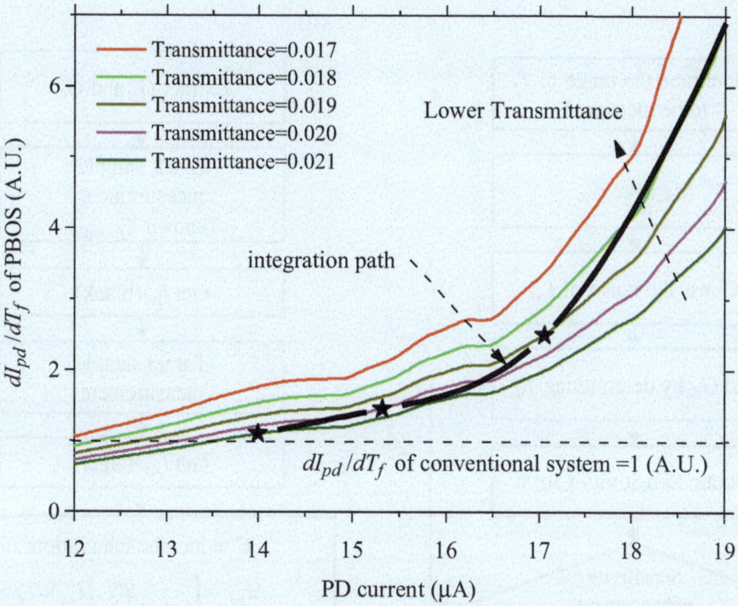

Fig. 12 The sensitivity enhancement compared to that of the conventional system along with the PD current and Transmittance. The integration path for Eq. (16) is shown in the figure as an example. Modified from Choi et al., IEEE Trans. Electron Devices 2016;63:2074–2079 [15]

$$\Delta I_{pd} = \int_{T_f=0.021}^{0.017} S\left(T_f, I_{pd}\right) dT_f \qquad (16)$$

where the integration path is shown in Fig. 12. The result of the integration Eq. (16) gives $\Delta I_{\mathrm{pd}} = 6.06$ uA. In the same manner, the ΔI_{pd} of conventional on is 1.2 mA × (0.019–0.017) = 2.4 uA because the sensitivity S is constant regardless of T_{f} and I_{pd}.

In this way, how the sensitivity of pseudo-BJT sensor can be enhanced is demonstrated. As shown in the PBOS sensitivity curve in Fig. 12, the PBOS gives nonlinear characteristics, whereas the nominal sensor requires the linearity for the data analysis. This means that the sensing signal (ΔI_{pd}) cannot be directly interpreted as the transmittance or the sample concentration with linear relation. Hence, the system tuning and the result interpretation should be done according to the following procedure.

Firstly, we discuss the optimization of PBOS. If the system is not properly tuned, the sensitivity of the PBOS can be lowered as Fig. 12 shows the sensitivity enhancement factor less than one. The optimization process can be shown as the flowchart shown in Fig. 13a which consists of flowing procedures:

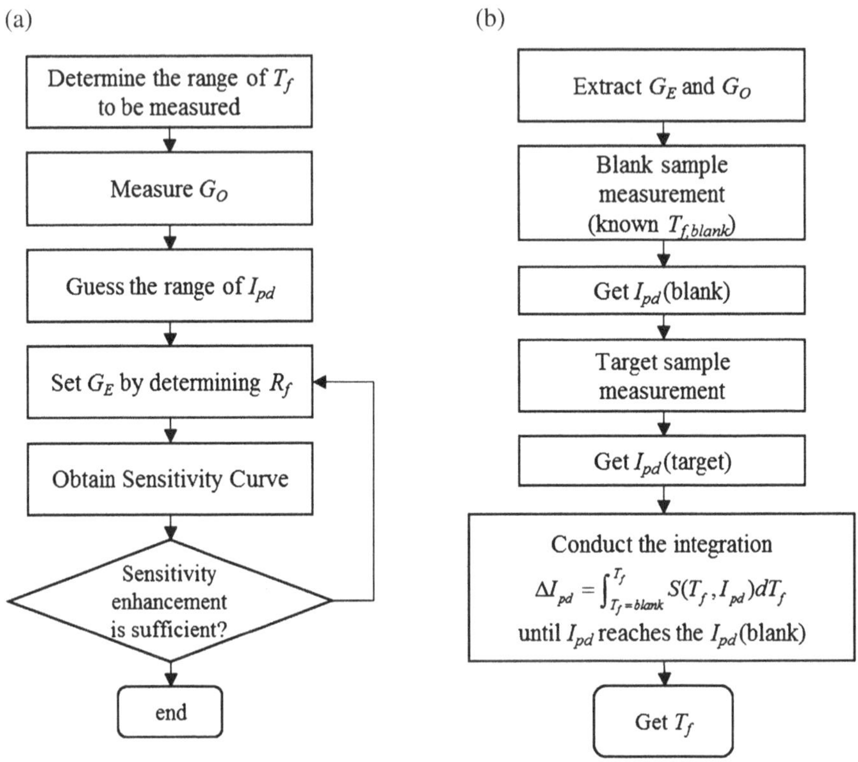

Fig. 13 (**a**) Flowchart of the PBOS optimization and selection rule of the resistances (**b**) Flowchart of the conversion procedure from the measurement result (I_{pd}) to the sample transmittance

- Define the range of a transmittance (T_f) to be measured according to the target material and application.
- Measure the G_O using the measurement setup in Fig. 11a.
- Determine the range of I_{pd} to be measured. Multiplying $G_{O,max}$ and the range of T_f may be an appropriate initial guess.
- Determine the R_s and R_f and measure the G_E using the setup in Fig. 11b.
- Calculate sensitivity matrix from G_O, G_E, and the range of T_f.
- Check the sensitivity is sufficiently enhanced or not.
- Fix the values of R_f and R_s to optimize G_E and thus the sensitivity.

Secondly, the conversion process can be conducted as the following procedures (also refer the flowchart in Fig. 13b):

- G_E and G_O should be extracted after the system setup. From these parameters, $S(T_f,I_{pd})$ can be calculated.
- Obtain the PD current $I_{pd,blank}$ for a blank condition (no sample). $T_{f,blank}$ of blank sample should be known.

- Obtain PD current when target sample is inserted ($I_{\mathrm{pd,blank}}$) and ΔI_{pd} is extracted.
- Conduct the integration in Eq. (16). The integration starts from the point at $S(T_{\mathrm{f,blank}}, I_{\mathrm{pd,blank}})$.
- Find $T_{\mathrm{f,sample}}$ which gives the integration result becomes the same as the measured $I_{\mathrm{pd,sample}}$ value.
- Calculate the sample concentration from T_{f} using the Beer-Lambert law.

As a summary, the pseudo-BJT enhances the sensitivity based on the positive feedback nature of the system in Fig. 9. At the same time, we lose the linearity of the sensor so that the conversion procedure based on Eq. (16) is required. The nonlinear property enhances the gain in the specific range while losing the gain in other range. Hence, the system should be carefully tuned to enhance the sensitivity at the point of interest.

In the next section, an application of the PBOS system to the glucose sensing is demonstrated with the wavelength of 1600 nm. In addition, the sensitivity and the LOD enhancement are demonstrated under the experimental way. Since the same optical devices and system in [15] is used, G_{E} and G_{O} is the same for the glucose sensing.

3.4 Variation of PBOS Circuit

The enhancement mechanism due to the pseudo-BJT seems to be sound but there could be some practical obstacles to be used in real sensor application. In order to resolve the problems during the real implementation, three variations of PBOS circuit are suggested as shown in Fig. 14.

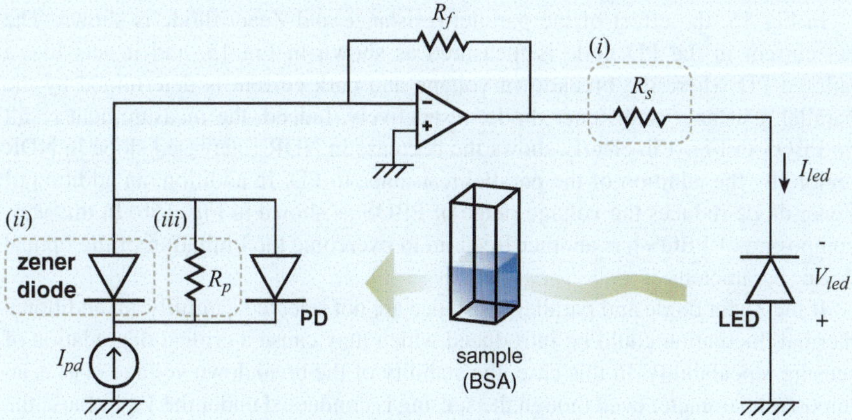

Fig. 14 Variations of the PBOS circuit: (*i*) addition of serial resistance to LED (R_{s}), (*ii*) addition of parallel zener diode to PD, and (*iii*) addition of parallel resistance to PD (R_{p})

Firstly, the inclusion of the feedback resistance can introduce the thermal drift of the system. To suppress the thermal effect, the serial resistance having the same temperature coefficients (%/K) to LED is recommended. In this case, the G_E becomes

$$G_E = \frac{R_f}{R_{ld} + R_s} \approx \frac{R_f}{R_s} \tag{17}$$

The approximation stands for the case when LED is turned on ($R_{ld} << R_s$). Then, if both resistances are placed closely, the thermal drifts of R_f and R_s are canceled out and G_E can be stabilized. Even though the thermal fluctuation due to R_f is neutralized, there is also the thermal drift to be accounted due to the optical device property, and the minimization of this effect is discussed in [15].

Secondly, the system may require too high voltage to drive the PBOS operation. For example, the measurement result of PBOS using a silicon photodiode (fabricated by Advanced photonix) and UV-LED (fabricated by Seoul Viosys) is shown in the following. It requires the applied voltage up to ~80 V because of the small dark current to turn on the LED before the breakdown occurs. The need of high voltage gives rise to the additional cost and may cause an instability problem to the PD. In order to avoid the high voltage bias, a parallel addition of Zener diode in the PD can clamp the voltage without affecting to the pseudo-BJT operation.

Thirdly, the resistance of NDR region is very high so that it may be hard to detect such a high impedance value accurately. Even, in this case, unwanted oscillation can occur, which originated from the diverging property of NDR system. The resistance of NDR region can be controlled by introducing the parallel resistance to the PD. The parallel resistance acts like a dark current component of the PD because the currents are summed up in a parallel connection. Since the dark current is closely related to the NDR, it will result in the decrease of NDR.

In Fig. 15, the effect of the parallel resistance and Zener diode is shown. The net current in the PD node is measured as shown in Fig. 15, and it acts like a tailored PD where the breakdown voltage and dark current is determined by the parallel resistance and Zener diode, respectively. Indeed, the measurement result on PBOS in Fig. 15b clearly shows the decrease in NDR (increased slope in NDR region) by the addition of the parallel resistance to PD. In addition, an addition of Zener diode reduces the voltage range of PBOS as shown in Fig. 15b. In this way, composing of PBOS has another freedom to overcome the limitation of the optical device parameters.

If the Zener diode and parallel resistance are not selected carefully, an additional thermal fluctuation could be introduced which may cause a critical degradation of sensing repeatability. In this case, the stability of the breakdown voltage (V_B) is an important parameter even though the sensing is conducted under the V_B because the V_B determines the LED turn-on point.

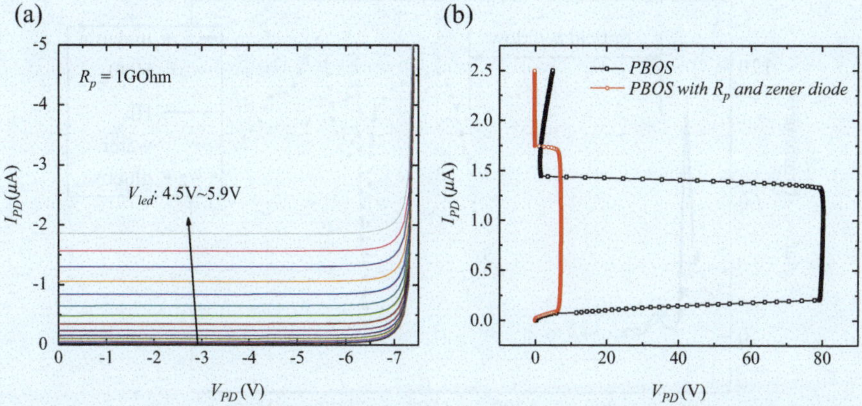

Fig. 15 (**a**) Net current in the PD node under the connection in Fig. 14. The breakdown current of PD itself is ∼80 V. (**b**) The change of I-V characteristics of PBOS by adding the parallel connection of R_p and Zener diode to the PD

4 Application of Smart LED Spectrophotometer

4.1 Glucose Sensing

Monitoring of the glycemic status, as performed by patients and health-care providers, is considered a cornerstone of the diabetes care [40], since the patients should maintain a certain blood glucose level [40–42]. However, most of present glucose-meter uses an invasive method which requires a blood collection so a daily basis usage of the glucose-meter gives a great pain to the patients [43, 44]. To avoid the inconvenience, noninvasive measurements, such as an optical, electrical, and ultrasonic method, have been intensively investigated [45]. As an application of LED spectrophotometer, the optical method of glucose sensing is summarized here.

The optical method for the glucose sensing can be categorized according to the wavelength: near-infrared region and mid-infrared region. For the near-infrared region, the light in the rage of 600–2000 nm is mostly used. In this regime, the light is less absorbed into the skin (due to melanin or so) or other abundant components in the blood such as Hb, HbO_2, and H_2O, as shown in Fig. 16. This is why the regime is called as "optical window." Here, a scattering characteristic of glucose in the blood is use as a sensing mechanism; since the scattering coefficient of blood is affected by a scatter (red blood cell) and surrounding materials (including the glucose), the increase of glucose level gives rise to the decrease of the difference in the reflective index between the scatter and the surroundings. Therefore, the light transmittance increases as the glucose level increases unlike the nominal absorption mechanisms. One of the products using the scattering mechanisms is announced by Orsense Co. with ten LEDs having the wavelength range of 600–100 nm.

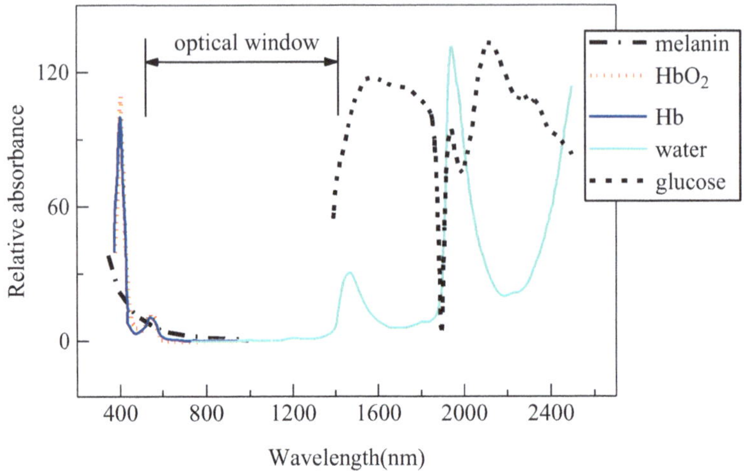

Fig. 16 Absorption spectrum of glucose in the visible and mid-IR region. Absorption spectrums of nonspecific molecules are also plotted [45]

In the mid-infrared region, mostly 1.6–10 um of wavelengths are used. In this case, most light is absorbed in the skin since it is not in the optical window, so the reflected or scattered light is used more than the transmitted light. The advantage in this regime is that the absorption spectrum of glucose is specific because of the C-C, C-H, and O-H bending vibrations of glucose. An application of PBOS to the glucose sensing is demonstrated in [19] based on the bending vibrations in the 1600 nm wavelength. The measurement result of the I-V characteristics of PBOS system is shown in Fig. 17 with various glucose concentrations (20 mg–2 g/dL). A glucose powder is dissolved in the distilled water, and the serial dilution method is used for preparing a low concentration sample. Interestingly, as the sensitivity curve of PBOS in Fig. 12, the sensing signal (ΔI_{pd}) is larger under the application of the pseudo-BJT. As a result, the LOD of PBOS could be enhanced up to 20 mg/dL, while it is hard to distinguish 100 mg/dL of glucose and a blank solution if a conventional LED-PD system is used.

Even though the optical methods are available with the blood sample in a lab, there are many barriers to be resolved for a real implementation. The absorption, scattering, and reflection due to the skin that is different for every person can affect to the result. Moreover, the nonspecific materials such as water, hemoglobin, protein, and fat also affect to the result. Therefore, how to obtain the selectivity of the sensor over the various environments remains as an open question.

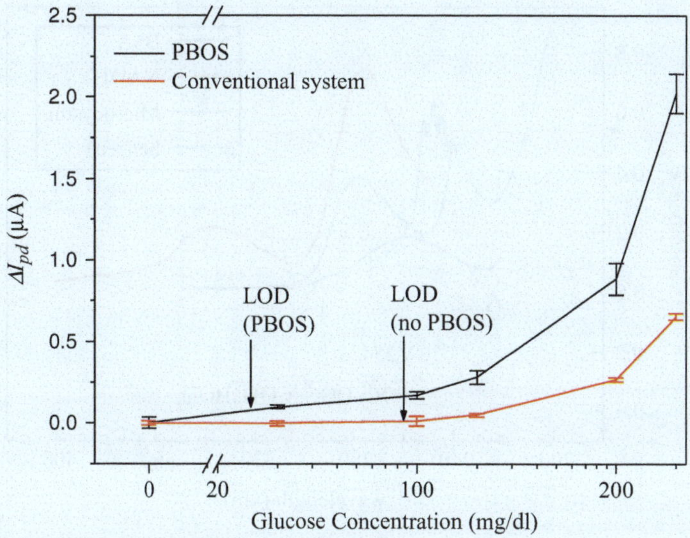

Fig. 17 Sensing signal (ΔI_{pd}) as glucose concentration with and without the pseudo-BJT. An enhanced LOD by introducing the PBOS is <20 mg/dL. Reprinted from Choi et al., IEEE Trans. Electron Devices 2016;63:2074–2079 [15]

4.2 Water Quality Sensor

A real-time monitoring of water quality is an important issue in a modern industrial society. Many parameters especially related to the organic material have an absorption peak in the UV range so that the UV-VIS absorption spectroscopy has been developed as a powerful solution for the online and in situ measurement. (Note that a method using a dye is often called as UV-VIS method, but this is beyond the scope of UV-VIS LED spectroscopy in this chapter.) One of the most important parameters is the biochemical oxygen demand (BOD) and chemical oxygen demand (COD). There have been reported numerous papers to figure out the BOD/COD from the UV-VIS spectroscopy including the absorbance at the one wavelength or the multiple wavelengths ranging from 200 to 300 nm [46, 47]. As another example, in Fig. 18, the spectrum of various materials serving as the pollutants is summarized along with the detection range of water index (TOC, BOD, COD). There are many industrial sensors having a probe-type form factor that is designed to dip in the water. Figure 19 shows the probe-type sensor from s::can co., and they have a Xenon flash lamp and photodetector arrays together with a water flow cell. We expect the LED spectrometer can shrink the form factor of the conventional water probe by capturing several key wavelengths. As an example, S. Choi et al. in SNU released a handheld size protein/DNA meter using LED arrays, as shown in Fig. 19, to replace the conventional lab-based spectrometry.

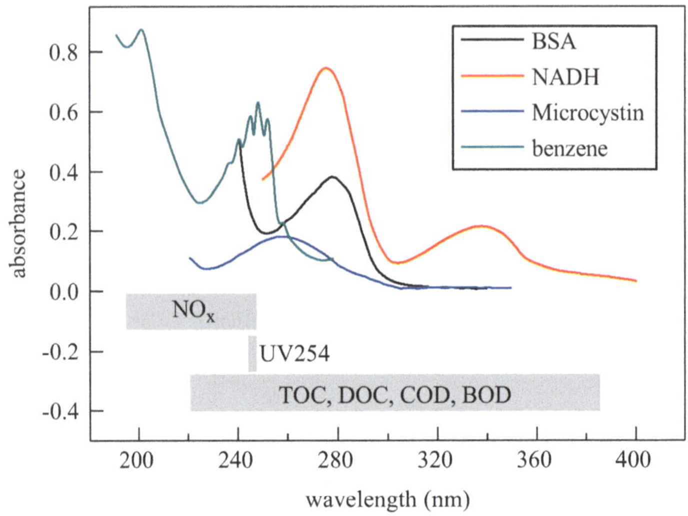

Fig. 18 UV absorption spectrum of various water pollutants including protein, nicotinamide adenine dinucleotide (*NADH*), DNA, microcystine, and benzene. In addition, the region for *BOD*, *COD*, and *TOC* (total organic carbon) measurement is also described [46–47]

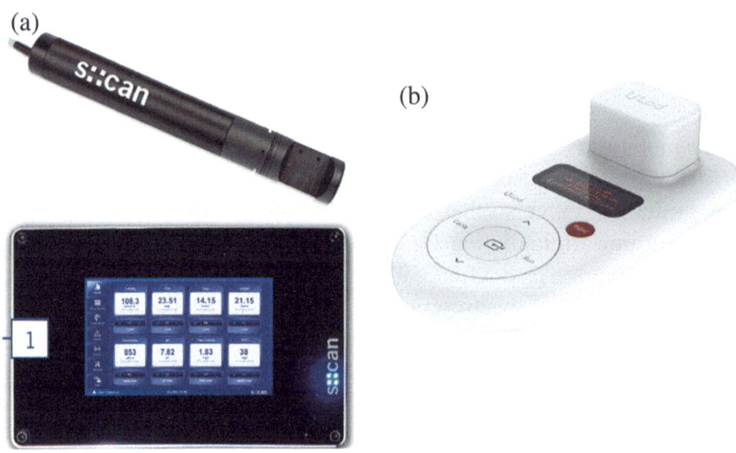

Fig. 19 (**a**) Probe and controllers for UV-VIS spectroscopy in a flowing water, released by s::can; (**b**) protein/DNA meter with a small footprint using LEDs, released by the research group in Seoul National University (SNU)

It is hard to replace the UV-VIS spectroscopy with the LED spectrophotometry completely because the information (number of wavelengths) from LED spectroscopy is less than that from the continuous spectroscopy. Since there are so many parameters to be considered and many unexpected mixtures (e.g., suspended solid or floating matters) in the raw water, the reduction of an information may result in false alert. In order to compensate the problem, the number of LED should be increased with increasing cost especially due to high price in the UV-LED. In addition, the availability of the UV-LEDs having less than 250 nm is limited; even though the information of 200–250 nm is valuable for conducting the BOD, NO_x analysis.

Hence, a clever selection of the wavelength and number of LED well suitable to the situation is needed. For example, the selection of LED for the tap water and wastewater will be different. In addition, the maturity of the UV-LED fabrication will accelerate the adoption of the LED spectroscopy by reducing the cost and enlarging the possible wavelength range for the water quality monitoring.

Acknowledgments This work was supported in part by the Center for Integrated Smart Sensors funded by the Ministry of Science, ICT & Future Planning as Global Frontier Project, and in part by Giparang co.

References

1. Morison, I.: Introduction to Astronomy and Cosmology. Wiley, West Sussex (2008)
2. Colthup, N.B., Daly, L.H., Wiberley, S.E.: Introduction to Infrared and Raman Spectroscopy. Academic Press, New York (1975)
3. Lakowicz, J.R.: Principles of Fluorescence Spectroscopy. Springer, New York (1999)
4. Nilapwar, S.M., Nardelli, M., Westerhoff, H.V., Verma, M.: Absorption spectroscopy. In: Methods in Enzymology, pp. 59. Elsevier, Cambridge (2011)
5. Forster, H.: UV/VIS spectroscopy. Mol. Sieves. **4**, 337–426 (2004)
6. Stuart, B.: Modern Infrared Spectroscopy. Wiley, West Sussex (1996)
7. Griffiths, P.R., Haseth, J.A.: Fourier Transform Infrared Spectrometry. Wiley, New Jersey (2007)
8. Broeke, J., Langergraberb, G., Weingartnera, A.: On-line and in-situ UV/V is spectrocopy for multi-parameter measurements: a brief review. Spectrosc. Eur. **18**(4), 1–4 (2006)
9. Nelson LA, McCann, J., Loepke, A.W., Wu, J., Dor, B.B., Kurth, C.D.: Development and validation of a multiwavelength spatial domain near-infrared oximeter to detect cerebral hypoxia-ischemia. J. Biomed. Opt. **11**, 064022 (2006)
10. Mohammad, K.A., Zekry, A., Abouelatta, M.: LED based spectrophotometer can compete with conventional one. J. Eng. Technol. **4**(2), 399–407 (2015)
11. Malinen, J., Kansakoski, M., Rikola, R., Eddison, C.G.: LED-based NIR spectrometer module for hand-held and process analyser applications. Sensors Actuators B. **51**, 220–224 (1998)
12. Yeh, T.-S., Tseng, S.-S.: A low cost LED based spectrometer. J. Chin. Chem. Soc. **53**, 1067–1072 (2006)
13. Namasivayam, V., Rongsheng Lin, B.J., Brahmasandra, S., Razzacki, Z., Burke, D.T., Burns, M.A.: Advances in on-chip photodetection for applications in miniaturized genetic analysis systems. J. Micromech. Microeng. **14**, 81–90 (2004)
14. Chuang, S.L.: Physics of Photonic Devices. Wiley, New Jersey (2009)
15. Choi, S., Moon, J., Lee, S., Hwang, Y., Park, Y.J.: A pseudobipolar junction transistor for a sensitive optical detection of biomolecules. IEEE Trans. Electron. Devices. **63**(5), 2074–2079 (2016)
16. Fogelman, S., Blumenstein, M., Zhao, H.: Estimation of chemical oxygen demand by ultraviolet spectroscopic profiling and artificial neural networks. Neural Comput. Applic. **15**, 197–203 (2006)

17. Long, D.: Photovoltaic and photoconductive infrared detectors. In: Topics in Applied Physics, pp. 101–147. Springer, Heidelberg (2005)
18. Baumgartner, P., Engel, C., Böhm, G., Abstreiter, G.: Franz–Keldysh effect in lateral GaAs/AlGaAs based npn structures. Appl. Phys. Lett. **70**(21), 2876–2878 (1997) http://dx.doi.org/10.1063/1.119204
19. Kim, H., Choi, S., Yoo, N., Lee, M.J., Park, Y.J.: Near-infrared detection using pulsed tunneling junction in silicon devices. IEEE Trans. Electron. Devices. **63**(1), 377–383 (2016)
20. Zhou, Y., Liu, Y.-H., Lo, Y.-H.: Bias dependence of sub-bandgap light detection for core-shell silicon nanowires. Nano Lett. **12**(11), 5929–5935 (2012)
21. Takeda, K., Hiraki, T., Tsuchizawa, T., Nishi, H., Kou, R., Fukuda, H., Yamamoto, T., Ishikawa, Y., Wada, K., Yamada, K.: Contributions of Franz Keldysh and avalanche effects to responsivity of a germanium waveguide photodiode in the L-band. IEEE J. Sel. Top. Quantum Electron. **20**(4), 64–70 (2014). doi:10.1109/JSTQE.2013.2295182
22. Zhou, Y., Y-h, L., Cheng, J., Lo, Y.-H.: Bias dependence of sub-bandgap light detection for Core–Shell silicon nanowires. Nano Lett. **12**(11), 5929–5935 (2012). doi:10.1021/nl3033558
23. Fair, R.B., Wivell, H.W.: Zener and avalanche breakdown in As-implanted low-voltage Si n-p junctions. IEEE Trans. Electron. Devices. **23**(5), 512–518 (1976)
24. You, J.B., Yu, K.: Near-infrared silicon sub-bandgap photo-detectors for on-chip integrated optical links. In: Lasers and Electro-Optics Pacific Rim (CLEO-PR), 2015 11th Conference on 24–28 2015. Pp. 1–2. (2015) doi:10.1109/CLEOPR.2015.7376068
25. Roucka, R., Mathews, J., Weng, C., Beeler, R., Tolle, J., Menendez, J., Kouvetakis, J.: High-performance near-IR photodiodes: a novel chemistry-based approach to Ge and Ge_xSn devices integrated on silicon. IEEE J. Quantum Electron. **47**(2), 213–222 (2011). doi:10.1109/JQE.2010.2077273
26. Oehme, M., Schmid, M., Kaschel, M., Gollhofer, M., Widmann, D., Kasper, E., Schulze, J.: GeSn p-i-n detectors integrated on Si with up to 4% Sn. Appl. Phys. Lett. **101**(14), 141110 (2012.) doi:http://dx.doi.org/10.1063/1.4757124
27. Su, S., Cheng, B., Xue, C., Wang, W., Cao, Q., Xue, H., Hu, W., Zhang, G., Zuo, Y., Wang, Q.: GeSn p-i-n photodetector for all telecommunication bands detection. Opt. Express. **19**(7), 6400–6405 (2011). doi:10.1364/OE.19.006400
28. Zhang, D., Xue, C., Cheng, B., Su, S., Liu, Z., Zhang, X., Zhang, G., Li, C., Wang, Q.: High-responsivity GeSn short-wave infrared p-i-n photodetectors. Appl. Phys. Lett. **102**(14), 141111 (2013.) doi:http://dx.doi.org/10.1063/1.4801957
29. Tseng, H.H., Li, H., Mashanov, V., Yang, Y.J., Cheng, H.H., Chang, G.E., Soref, R.A., Sun, G.: GeSn-based p-i-n photodiodes with strained active layer on a Si wafer. Appl. Phys. Lett. **103**(23), 231907 (2013.) http://dx.doi.org/10.1063/1.4840135
30. Peng, Y.-H., Cheng, H.H., Mashanov, V.I., Chang, G.-E.: GeSn p-i-n waveguide photodetectors on silicon substrates. Appl. Phys. Lett. **105**(23), 231109 (2014.) doi:http://dx.doi.org/10.1063/1.4903881
31. Colace, L., Masini, G., Assanto, G., Luan, H.-C., Wada, K., Kimerling, L.C.: Efficient high-speed near-infrared Ge photodetectors integrated on Si substrates. Appl. Phys. Lett. **76**(10), 1231–1233 (2000.) doi:http://dx.doi.org/10.1063/1.125993
32. Famà, S., Colace, L., Masini, G., Assanto, G., Luan, H.-C.: High performance germanium-on-silicon detectors for optical communications. Appl. Phys. Lett. **81**(4), 586–588 (2002.) doi:http://dx.doi.org/10.1063/1.1496492
33. Wang, J., Loh, W.Y., Chua, K.T., Zang, H., Xiong, Y.Z., Tan, S.M.F., Yu, M.B., Lee, S.J., Lo, G.Q., Kwong, D.L.: Low-voltage high-speed (18 GHz/1 V) evanescent-coupled thin-film-Ge lateral PIN photodetectors integrated on Si waveguide. IEEE Photon. Technol. Lett. **20**(17), 1485–1487 (2008). doi:10.1109/LPT.2008.928087
34. Feng, N.-N., Dong, P., Zheng, D., Liao, S., Liang, H., Shafiiha, R., Feng, D., Li, G., Cunningham, J.E., Krishnamoorthy, A.V., Asghari, M.: Vertical p-i-n germanium photodetector with high external responsivity integrated with large core Si waveguides. Opt. Express. **18**(1), 96–101 (2010). doi:10.1364/OE.18.000096

35. Kang, Y., Liu, H.-D., Morse, M., Paniccia, M.J., Zadka, M., Litski, S., Sarid, G., Pauchard, A., Kuo, Y.-H., Chen, H.-W., Zaoui, W.S., Bowers, J.E., Beling, A., McIntosh, D.C., Zheng, X., Campbell, J.C.: Monolithic germanium/silicon avalanche photodiodes with 340 GHz gain-bandwidth product. Nat. Photon. **3**(1), 59–63 (2009)
36. Volodymyr, B., Koman, C.S., Martin, O.J.F.: Multiscattering-enhanced absorption spectroscopy. Anal. Chem. **87**, 1536–1543 (2014)
37. Sze, S.M., NG, K.K.: Physics of Semiconductor Devices. Wiley, New Jersey (2007)
38. Grove, A.S.: Physics and Technology of Semiconductor Devices. Wiley, New York (1967)
39. Yoon, J.-Y.: Introduction to Biosensors: From Electric Circuits to Immunosensors. Springer, New York (2013)
40. Goldstein, D.E., Little, R.R., Lorenz, R.A.: Tests of Glycemia in diabetes. Diabetes Care. **27**(7), 1761–1773 (2004)
41. Group UKPDS: Intensive blood-glucose control with sulphonylureas or insulin compared with conventional treatment and risk of complications in patients with type 2 diabetes (UKPDS 33). Lancet. **352**(9131), 837–853 (1998.) doi:http://dx.doi.org/10.1016/S0140-6736(98)07019-6
42. Group TDCCTR: The effect of intensive treatment of diabetes on the development and progression of long-term complications in insulin-dependent diabetes mellitus. N. Engl. J. Med. **329**(14), 977–986 (1993). doi:10.1056/NEJM199309303291401
43. Oliver, N.S., Toumazou, C., Cass, A.E.G., Johnston, D.G.: Glucose sensors: a review of current and emerging technology. Diabet. Med. **26**, 197–210 (2009)
44. Amir, O., Weinstein, D., Zilberman, S., Less, M., Perl-Treves, D., Primack, H.: Continuous non-invasive glucose monitoring technology based on 'occlusion spectroscopy'. J. Diabetes Sci. Technol. **1**, 463–469 (2007)
45. Shen, Y.C., Davies, A.G., Linfield, E.H., Taday, P.F., Arnone, D.D.: The use of Fournier-transform infrared spectroscopy for the quantitative determination of glucose concentration in whole blood. Phys. Med. Biol. **48**, 2023–2032 (2003)
46. Storey, M.V., Bvd, G., Burns, B.P.: Advances in on-line drinking water quality monitoring and early warning systems. Water Res. **45**, 741–747 (2011)
47. Korostynska O, Mason A, Al-Shamma'a AI Monitoring pollutants in wasterwater: traditional lab based vesus modern real-time approaches. In: Smart Sensors for Real-Time Water Quality Monitoring, pp. 1–24. Springer, Heidelberg (2013)

An Air Quality and Event Detection System with Life Logging for Monitoring Household Environments

Hyuntae Cho

1 Introduction

Most people spend a substantial proportion of their time inside buildings. However, many people are exposed to risks, such as air pollution, indoor noise, or property loss. First, the household environment is contaminated by many pollutants. In 2014, the World Health Organization (WHO) reported that around 7 million people died as a result of exposure to air pollution, that is, one in eight of total global deaths, which indicates that air pollution is now the world's largest single environmental health risk [1–4]. WHO also estimates that indoor air pollution in households cooking over coal, wood, and biomass stoves was linked to 4.3 million deaths. In the case of outdoor air pollution, WHO estimates that 3.7 million deaths occurred worldwide as a result of sources of pollution from both urban and rural areas and that 1 million deaths were affected by both of indoors and outdoors. A number of contamination sources can affect the indoor environment. Cooking, such as frying and roasting, generates hazardous gases that include carbon oxide (CO), carbon dioxide (CO_2), nitrogen dioxide (NO_2), volatile organic compounds (VOCs), and particulate matters (PMs). These dangerous gases cause oxygen insufficiency in the lungs and increase the risk of lung cancer in nonsmoking women. In particular, nitrogen oxide and dioxide (NOx), as well as CO leaked from indoor boilers, are the most dangerous gases in the household environment [5–7]. People can also be exposed to sick house syndrome. Dangerous gases that can induce sick house syndrome (sick building syndrome) include VOCs like toluene (C7H8), benzene (C6H6), and formaldehyde (HCHO) [8, 9]. They can be contained in furniture, wallpaper, electronic devices, etc.

H. Cho (✉)
Center for Integrated Smart Sensors, N1-306, Building N1, KAIST, Daehak-ro 291,
Yuseong-gu, Daejeon 305-701, Republic of Korea
e-mail: phd.marine@kaist.ac.kr

H. Yasuura et al. (eds.), *Smart Sensors at the IoT Frontier*,
DOI 10.1007/978-3-319-55345-0_10

Second, indoor noise caused by upper-floor neighbors is a serious social problem in apartment houses and buildings [10]. The noise sources are various. Running, walking, appliances such as TV and laundry machine, musical instruments, etc. are noise sources that can cause severe stress and even result in revenge, fighting, and murder. Third, some people lose their property due to external events, such as fire or malicious intruders. Indoor fire occurs by short-circuiting of electronic devices, gas stoves, lit candles, etc. In addition, it can take place during the winter season by heaters and electric blankets. Fire can break out in people's property, such as their house, furniture, electronic devices, and clothes. Malicious intruders like thieves steal people's luxury items, such as jewelry, cash, and costly electronic devices.

The aim of this study is to develop an air quality and event detection system to measure environmental contamination and to save people's property from fire and malicious intruders. In order to detect environmental contamination, the proposed system uses multiple environmental sensors detecting temperature, humidity, ambient light, gas sensors, and particulate matters. Microphone and camera module are used for fire and intruder detection purposes. All measured and detected data are simultaneously stored in internal storage and an Internet server. The proposed system uses many sensors and components. So, it basically consumes much energy. We also propose a low-power consumption algorithm to prolong the system's lifespan. The proposed system consists of a full-function device (FFD) and a reduced-function device (RFD). They can be installed in a house to accurately measure ambience and communicate with each other via Wi-Fi and bluetooth low energy (BLE). Basically, ad hoc on-demand distance vector (AODV) based on BLE is used for the network-routing protocol. This chapter also includes how they communicate with each other.

This paper is organized as follows. First, we present an air quality and event detection system with life logging for the household environment. Section 3 describes the network architecture and network protocol for air quality and event detection systems. Section 4 deals with performance evaluation, consisting of power consumption and network performance. Finally, Sect. 5 concludes this paper and proposes future work.

2 Air Quality and Event Detection System for Household Environments

The proposed system consists of a full-function device (FFD) and reduced-function devices (RFD). As aforementioned, the air quality and event detection system detects indoor air quality, noise, fire, and intruders. The FFD contains all functions, operates independently, and acts as a gateway. The RFDs have limited functions and operate alone. However, an RFD cannot directly connect to the Internet, because it does not contain a Wi-Fi module. In order to connect to the Internet, the RFD should go through the FFD.

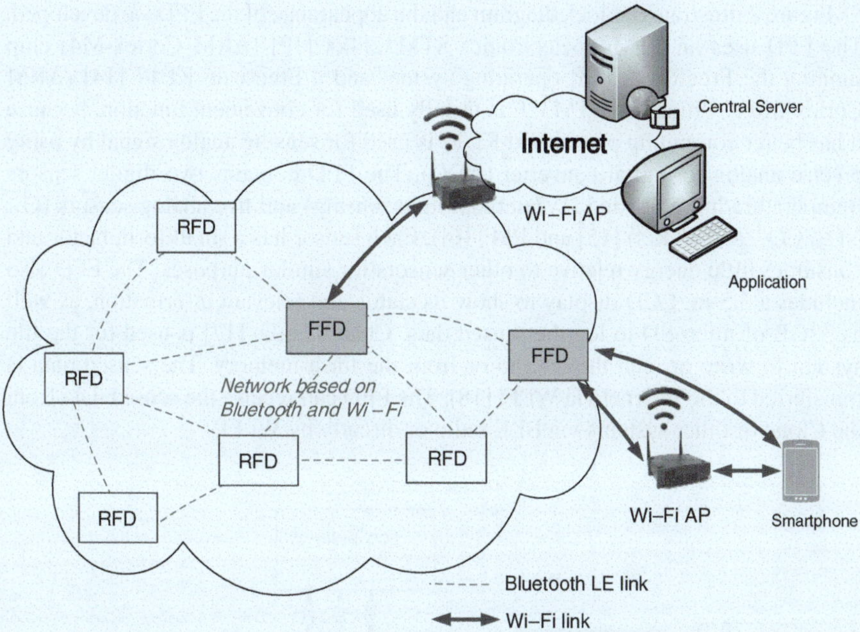

Fig. 1 Conceptual overview of the air quality and event detection system

Figure 1 shows a conceptual overview of the air quality and event detection system. The FFD contains more functions than the RFDs and is connected to the Internet via Wi-Fi. The RFDs having less functions than the FFD are connected to the FFD via BLE. All devices construct a network topology on BLE and then communicate with each other in multi-hop pattern [11].

2.1 *Full-Function Device (FFD)*

The FFD refers to the personal environmental monitoring system (PEMS). This can measure indoor air pollution, such as CO, NO_2, PM, and VOCs, as well as temperature/humidity. It also includes a microphone and audio CODEC to measure the indoor noise level and a camera to detect visual events. The FFD periodically measures indoor air pollution to save energy because gas sensors consume a lot of energy. A camera and a proximity sensor detect whether people are in front of the device or not. If they detect people, the FFD will turn on and start to measure the environment and display it on the screen. Snapshots and environmental data are also stored in the Cloud via Wi-Fi for security purposes.

Figure 2 illustrates a block diagram and the appearance of the FFD we developed. The FFD uses an STMicroelectronics STM32f4xx [12] (ARM Cortex-M4) chip running the FreeRTOS [13] operating system and a Freescale KL17 [14] (ARM Cortex-M0+) chip. The STM32F is usually used for convenient function, because it has better computing power, and KL17 is used for sensing analog signal by using precise analog to digital converter (ADC). The FFD contains two digital sensors (temperature/humidity and UV/ambient light sensing) and five analog sensors (O_3, NO_2, CO_2, and VOCs) [15] and PM [16]). Each sensor has a small form factor and consumes little energy relative to other sensors for similar purposes. The FFD also includes a 3.5 in. LCD display to show its status and relevant information, as well as 32GB of microSD to log the sensed data. Chan's FatFs [17] is used for the file system to write or read the data to or from the flash memory. The sensed data is transferred to the Internet via Wi-Fi [18]. The FFD can receive the sensed data from the Cloud or other systems via BLE radio or directly by Wi-Fi.

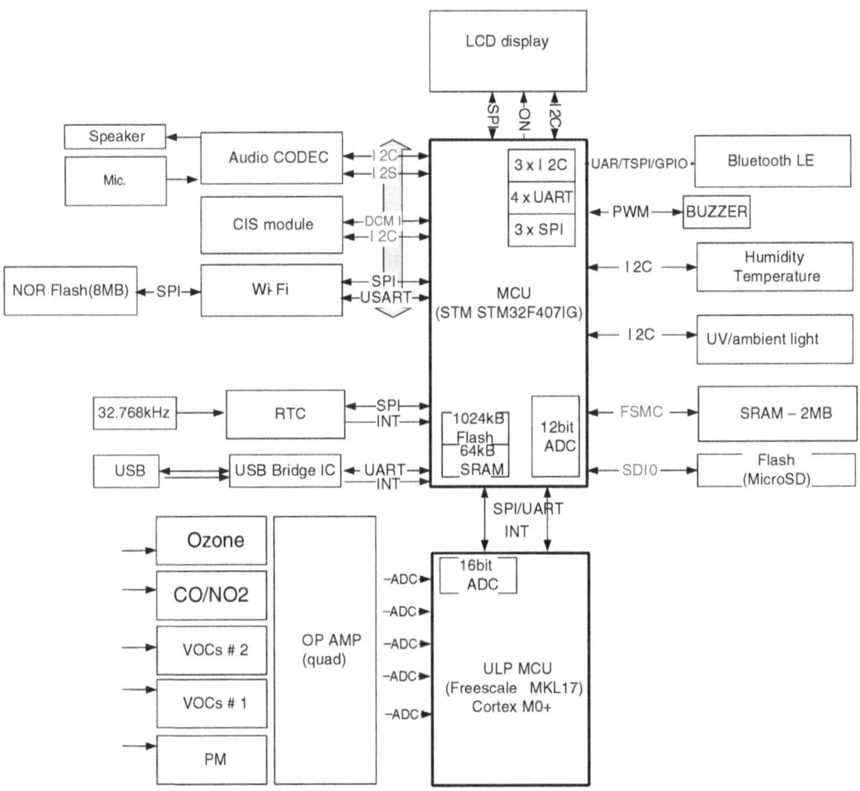

(a)

Fig. 2 Full-function device: (**a**) block diagram and (**b**) appearance

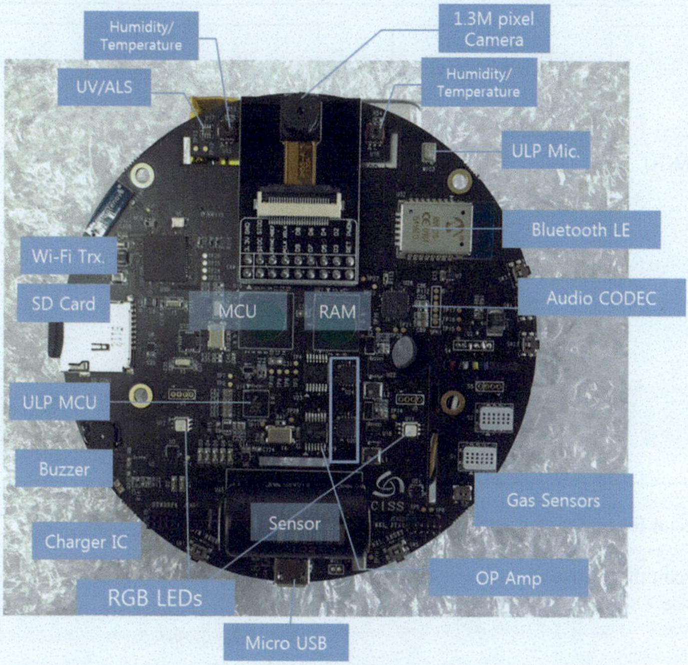

(b)

Fig. 2 (continued)

2.1.1 Gas Sensors

As earlier mentioned, in this system, we use multiple sensors to measure air pollution. The PM sensor is an optical sensor. We also use two types of gas sensors. One is based on the MEMS gas sensor, and the other is an electrochemical sensor. First, O_3, NO_2, and TVOC sensors are semiconductor-based gas sensors. In general, the MEMS sensors use a heater to increase the temperature of the sensor to approximately 300 °C or higher [19–21]. For example, when the sensor temperature reaches 300 °C, contaminants such as VOCs can easily attach to the sensor. The heating function is driven by using different voltages for the MEMS sensors. We designed a broad voltage range for general use by using a 5 V output and then adjusting it to the proper levels. Since the system uses a 3.7 V Li-Pol battery, we use a step-up DC/DC converter. Figure 3 shows that the sensors share this voltage through an analog switch to provide different heating voltages [22].

After heating, the resistance of the MEMS gas sensors varies according to the concentration of contaminants. The FFD then measure the resistance of a sensor between two electrodes to evaluate the environmental contamination. Figure 4 depicts how the system measures the resistance of the gas sensors. For a given sensor and load resistance, the output voltage can be calculated as:

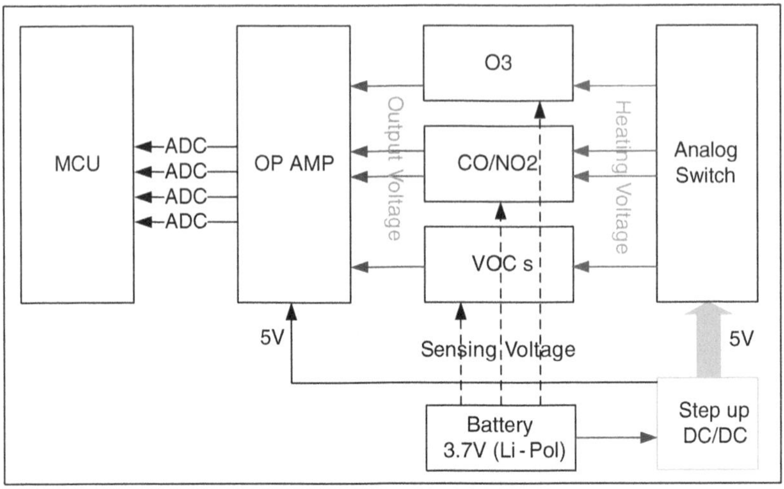

Fig. 3 Circuit for heating the sensor

Fig. 4 Circuit that measures the resistance

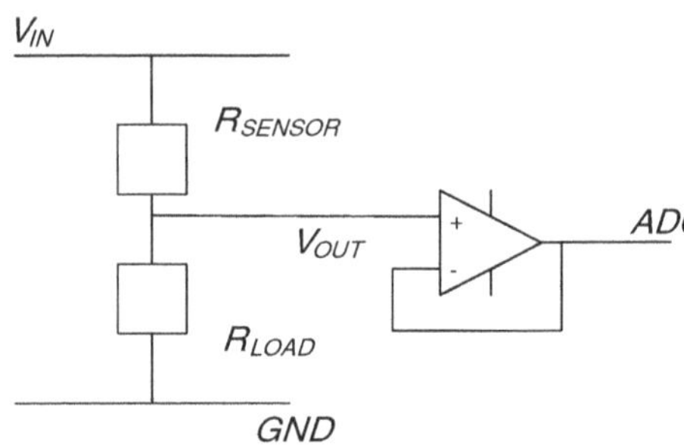

$$V_{\mathrm{OUT}} = \frac{R_{\mathrm{LOAD}} \times V_{\mathrm{IN}}}{R_{\mathrm{LOAD}} + R_{\mathrm{SENSOR}}} \tag{1}$$

where V_{OUT} is *the measured ADC value/ADC resolution* * *VIN*. From Eq. 1, the sensor resistance can be calculated as in Eq. 2.

$$R_{\mathrm{SENSOR}} = R_{\mathrm{LOAD}} \times \left(\frac{V_{\mathrm{IN}}}{V_{\mathrm{OUT}}} - 1 \right) \tag{2}$$

The FFD has a 16-bit ADC that can take highly accurate measurements of the resistance. The accuracy of the resistance measurements is related to the number of ADC bits and also to the difference between R_{SENSOR} and R_{LOAD}. The accuracy is highest when R_{SENSOR} is close to R_{LOAD}. The FFD also contains a quad-operational amplifier buffer that can prevent the ADC input impedance loading the sensor circuit at very high values of R_{SENSOR} (e.g., >1 M). The nature of the MOS gas sensors is such that they are often used for cost-sensitive applications where high measurement accuracy is not required.

The value for R_{LOAD} is important because it determines the accuracy of the measurement. As earlier mentioned, the accuracy is highest when R_{SENSOR} is close to R_{LOAD}. However, R_{SENSOR} varies according to the concentration of the air pollution. Therefore, we have implemented a simulator that can determine the optimal reference resistance value as follows. The users want to see the actual concentration of gases. The measured resistance is converted into PPM as follows: $PPM = R_{air}/R_{gas}$ or R_{gas}/R_{air}. However, semiconductor-based gas sensors do not have selectivity whereby the sensor reacts to a specific gas. So many works used principle components analysis (PCA) [23–25] or independent components analysis (ICA) [26]. Both of them also cannot help solve the selectivity problem. The selectivity is very important in some applications. Machine and deep learning approaches [27] can be used to obtain sensing selectivity. But, the proposed system does not use them yet in this chapter.

In order to detect fire, the selectivity for carbon monoxide is very critical. We use Figaro TGS5342 CO sensor [28], which is an electrochemical gas sensor. This sensor always works with an ultra-low-power microprocessor to detect fire and carbon monoxide.

2.1.2 PM Sensor

The system uses a Sharp GP2Y1010 sensor [16] to measure particulate matters. The PM sensor is operated by 5 V power and pulse signals. The FFD periodically measures information from the PM sensor, because the PM sensor consumes a lot of power. In order to read the signal, the PM sensor requires a 100 Hz pulse signal, where a pulse consists of 0.32 ms high signal and 9.68 ms low signal. The processor reads the output signal from the PM sensor 0.28 ms after the start of the high signal. Then the processor converts the ADC value into amperes and compares the look-up table provided by the datasheet. However, it cannot read signals for the initial three pulse periods. Figure 5 shows that after three pulse signals, the system measures ten output signals from the fourth to fourteenth signals and then extracts the maximum signal as PM data. After measuring the sensor data, it returns to power-down mode.

2.2 Operation

Figure 6 shows the software architecture of the FFD. It contains the main features of the main microcontroller. FreeRTOS 8.1.x is used for the operating system. The device driver, file system, and cortex microcontroller software interface standard (CMSIS) are located under RTOS. It also includes task management, power management, communication, and sensor middleware. In order to extend the battery life, the FFD operates according to the duty cycle and sporadically transmits sensing

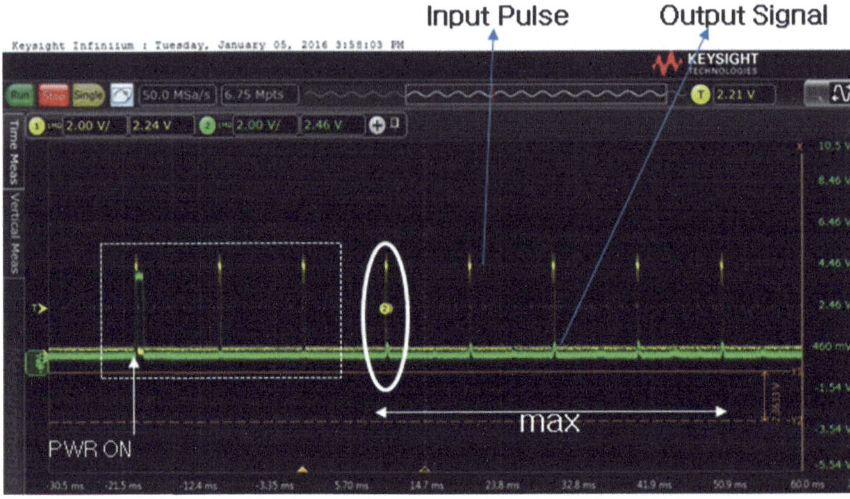

Fig. 5 Output signal of the particulate matter sensor

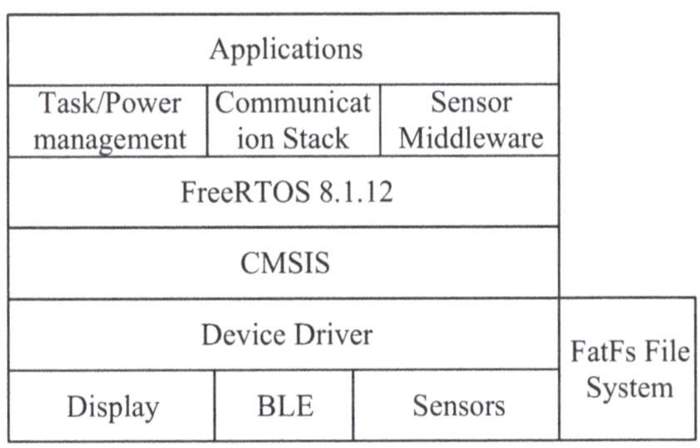

Fig. 6 Software architecture of the full-function device

data in a low-power mode. The FFD can change the duty cycle according to day/night cycles, time, or season, as well as the amount of remaining energy. The program of the coprocessor, MKL17, is implemented by firmware to extremely reduce the power consumption.

Figure 7 illustrates the sensors used in the system and applications. Seven sensors are used to monitor indoor air quality, five sensors are used to detect fire, a microphone array is used to measure indoor noise, and the combination of microphone and camera is used to detect malicious intruders. All data are stored in the internal storage for further processing.

We describe the operating procedure of each component. Figure 8 shows that first, for environmental sensing, the procedure consists of resistance measurements for the gas sensors followed by data conversion to PPM (or PPB) and then further conversion to an air quality index (AQI), followed by showing the values on the dis-

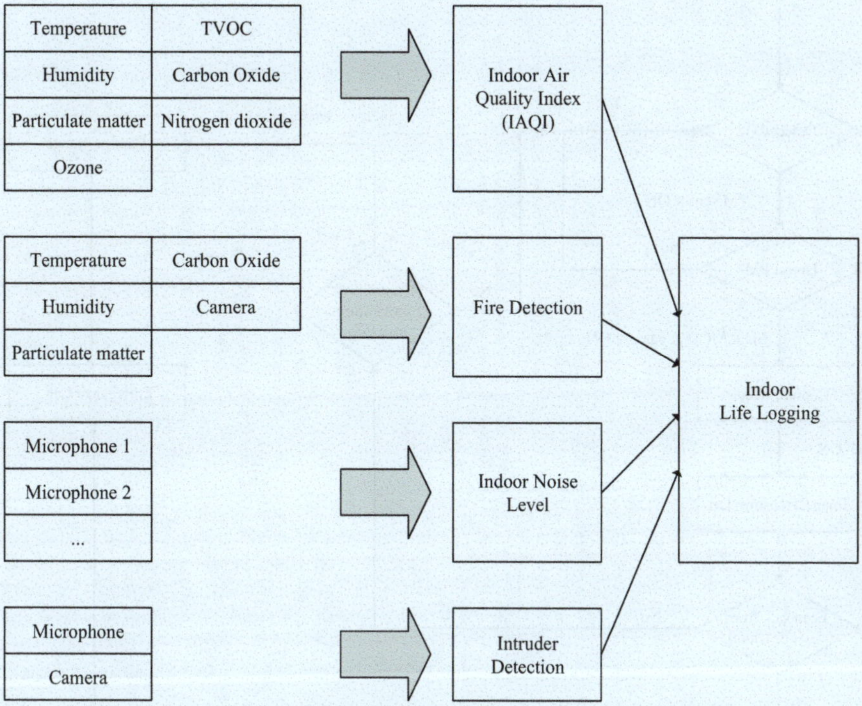

Fig. 7 Flow chart of the system operation for the full-function device

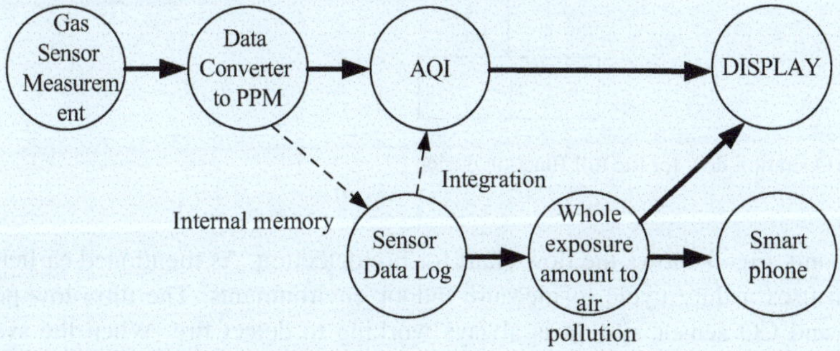

Fig. 8 Operation flow for the full-function device

play. The AQI is a number that is used by government agencies to communicate to the public how polluted the air currently is, or how polluted it is forecast to become. Different countries use their own air quality indices, which correspond to different national air quality standards. In this paper, we use a simple AQI instead of a governmental recommendation. The AQI also has a parameter that requires continuous measurement over a period of 24 h. As a result, the FFD records a log in its internal memory. This log file can also be used to assess the total exposure to air pollution.

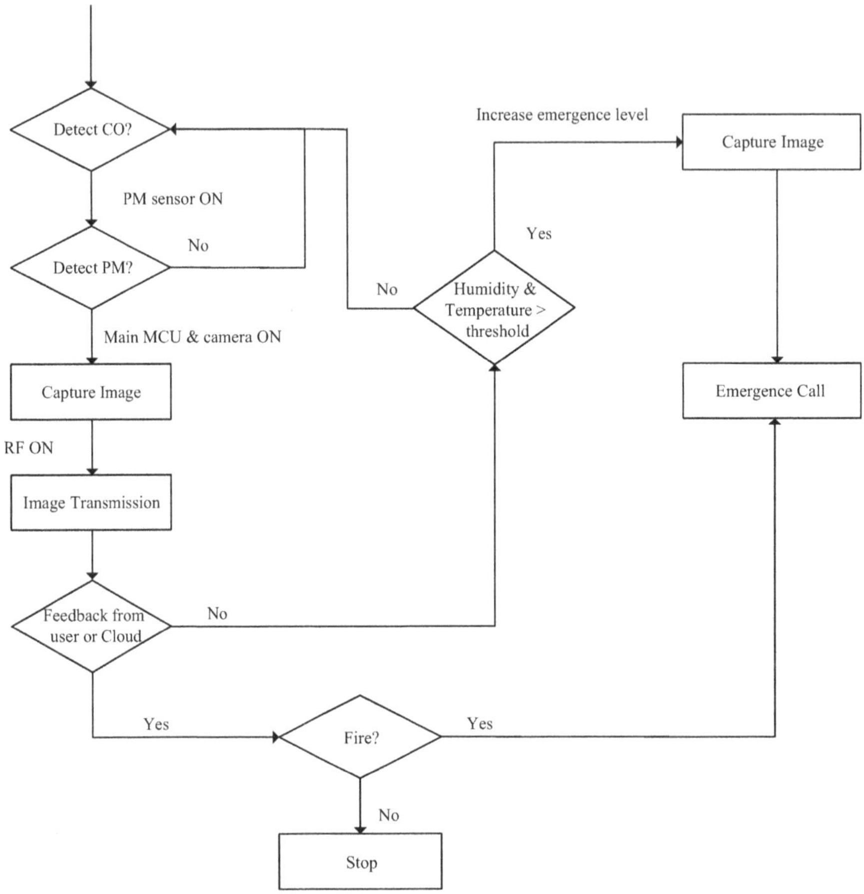

Fig. 9 Operation flow for the full-function device

Second, Fig. 9 shows the flow chart for fire detection. As mentioned earlier, the system uses a duty cycle to measure indoor environments. The ultra-low-power MCU and CO sensor system is always working to detect fire. When the system detects CO gas, it turns on the PM sensor and checks the density of particulate matter. If it also detects excessive density, it then turns on the camera, takes a photo, and then transmits the photo to the user or the Cloud. The user or Cloud gives the system feedback by recognizing the photo. If it is fire, the system sends an emergency message to a call center. If the system does not receive any response from the user or Cloud, it additionally checks the temperature and humidity. Then, it makes a decision on whether it is an actual fire or not.

Third is indoor noise measurement. In order to measure noise, the system has the following procedures as shown in Fig. 10. It:

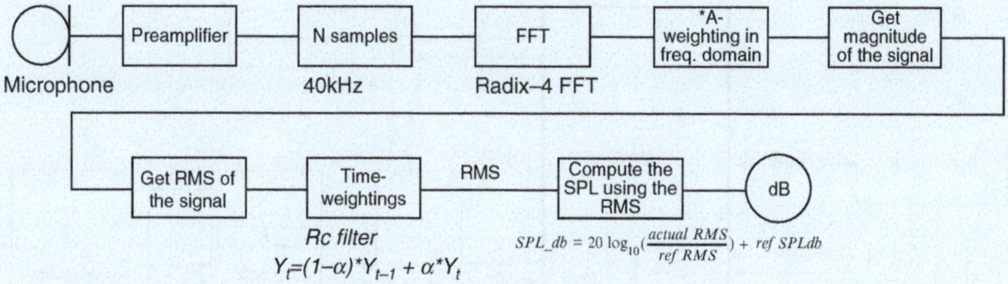

Fig. 10 Process for the measurement of sound pressure level

1. Collects N samples from the microphone
2. Calculates a fast Fourier transform (FFT) for the N samples collected
3. Applies A-weighting in the frequency domain to apply different weight
4. Gets the magnitude of the signals
5. Gets the root-mean-square (RMS) value of the signal
6. Applies a time-weight filter to the RMS value
7. Computes the sound pressure level, using the RMS value

Fourth, a microphone and camera are used for intruder detection. The microphone listens to the external signal. If the signal exceeds some threshold, the system takes a photo and then records it in storage. The camera also detects external events. The camera detects a scene change using two photos: a photo at time $t - 1$ and a photo at time t.

2.3 Reduced-Function Device

Figure 11 contains a block diagram and the appearance of the RFD. The RFD uses a Freescale KL17 (ARM Cortex-M0+) chip. The RFD basically contains an ultra-low-power MCU, a humidity/temperature sensor, and a bluetooth module. It can include environmental sensors, such as O_3, NO_2, CO, PM, or VOCs. The RFD has the combination of a few sensors, even though it has enough sensor interfaces. It depends on the application. The sensed data is transferred to the FFD or smartphone via BLE.

3 Networking Among Devices

Figure 1a in Sect. 2 shows a conceptual overview of the environmental monitoring system and the network platform used in this study. Environmental monitoring systems can be placed either indoors or outdoors. The device sends the sensed data to the Cloud and receives useful data or other data from other devices from the Cloud service via Wi-Fi connectivity. However, a Wi-Fi-based platform is too expensive,

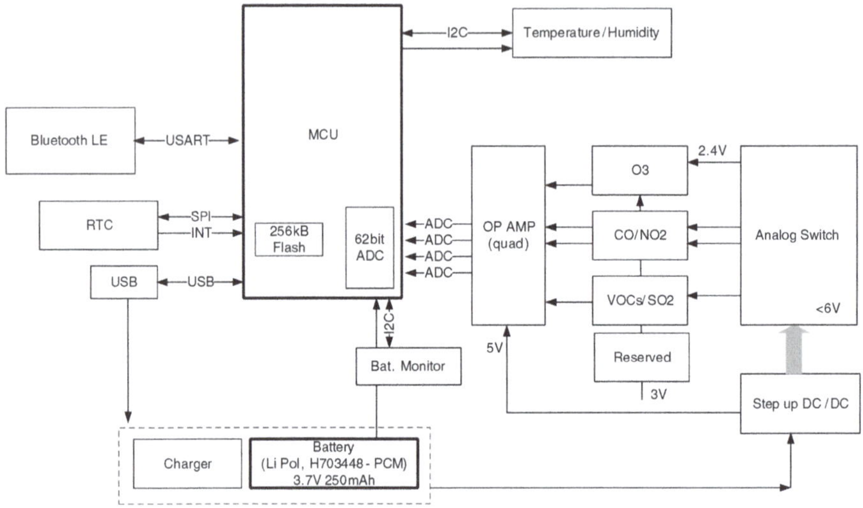

(a)

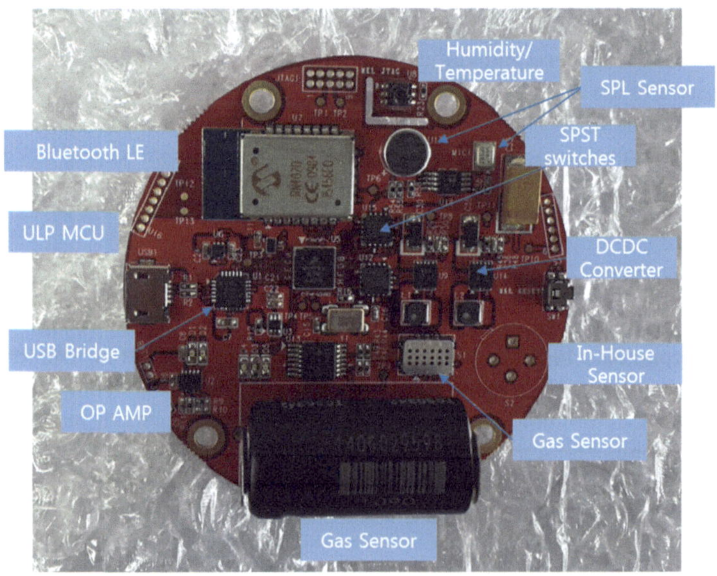

(b)

Fig. 11 Reduced-function device: (**a**) block diagram and (**b**) appearance

in terms of the energy and cost required. So, our device can use a secondary communication channel, BLE. The secondary radio is useful and can be included at low cost to enable multi-hop communication, inter-device communication, and wake-up radio functionality based on Wi-Fi. The device also uses one radio to connect to the Internet or an environmental monitoring network.

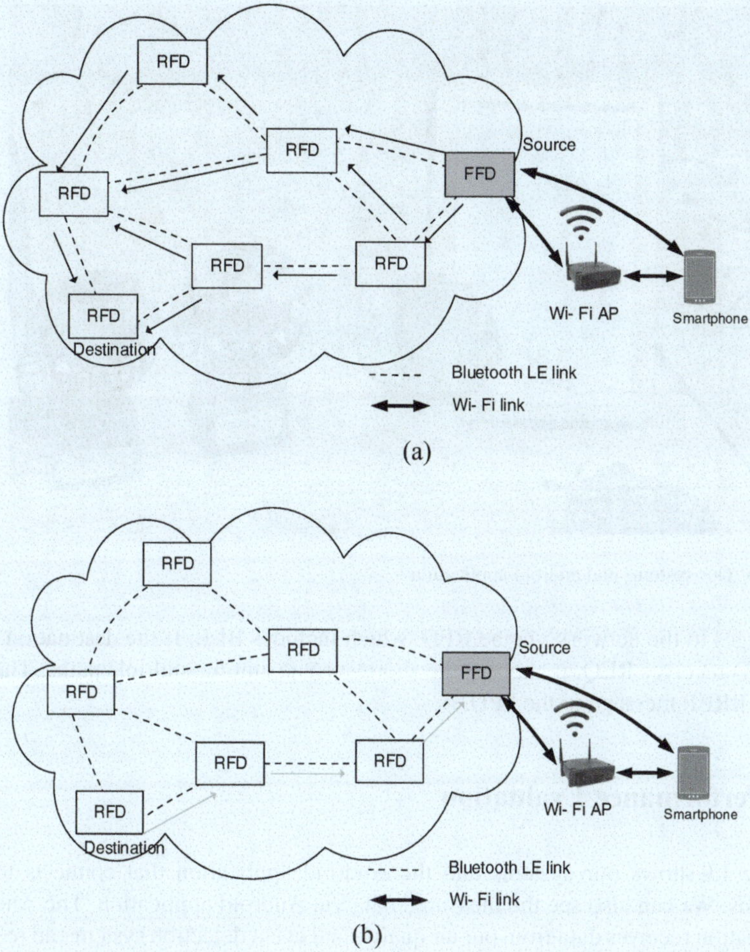

Fig. 12 Ad hoc on-demand distance vector routing protocol in the PEMS network: (**a**) REQ message and (**b**) RREP message

Our device establishes the network topology based on AODV [29–31]. AODV is a well-known routing protocol in wireless sensor networks based on a ZigBee radio. Because we use BLE, this paper redefines the protocol. BLE uses services and characteristics to send data or other information. We also define services and characteristics for our specific application, which transmits and receives information about toxic gases and environmental sensing. Figure 12 shows an example of the network protocol where the smartphone acts as a sink node. The smartphone based on Android OS establishes and maintains session with the FFD and RFD. If the smartphone sends a Route Request (RREQ) message to discover all environmental monitoring systems, the FFD forwards that message to the network. In this case, the FFD acts as a level 1 node. If there is no FFD device, the smartphone directly sends

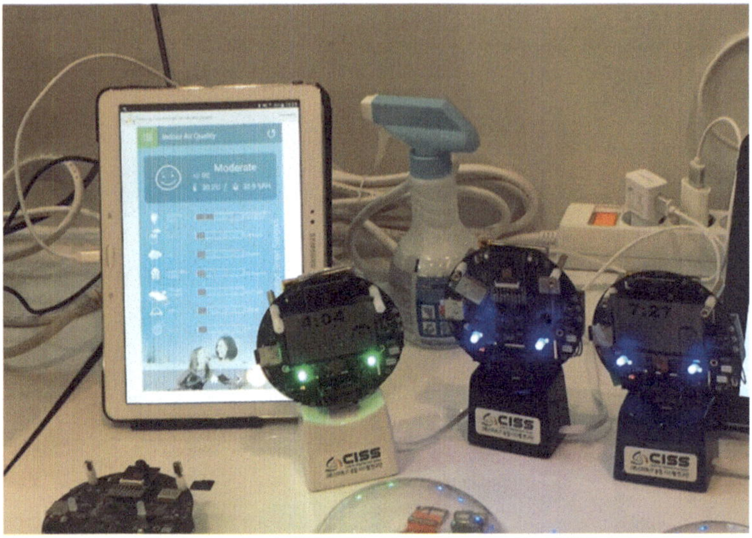

Fig. 13 Our systems and android application

messages to the network via the RFD, which includes BLE. If the destination FFD or RFD receives REQ, it chooses the shortest route among multiple paths. Then, it sends RREP message to the FFD.

4 Performance Evaluation

Figure 13 shows our systems and the Android application that connects to the systems. We can also see the measured data via Android application. The Android application receives data from our air quality and event detection system and records the received data in memory. The application shows a summary, total air quality index, and information about each sensor. In addition, when the air quality and event detection system detects some event, such as fire or intruder, it sends an emergency message to the application, and the application notifies the user.

4.1 Evaluation of Gas Sensors

This section deals with the performance of conventional semiconductor gas sensors and engineering techniques to improve the performance. Four semiconductor sensors are used in the FFD. So, first we evaluate R_{air}, which is the resistance when the ambient air is clean and there is no gas. Figure 14 shows the results. We use TVOC, NO_2, and O_3 sensors. We compared eight sensors for each gas. The left-hand side graphs in the figure indicate that sensors are exposed to the dried condition, while the right-hand side graphs indicate that sensors are exposed to the humid condition. Both of them represent that not all of the semiconductor

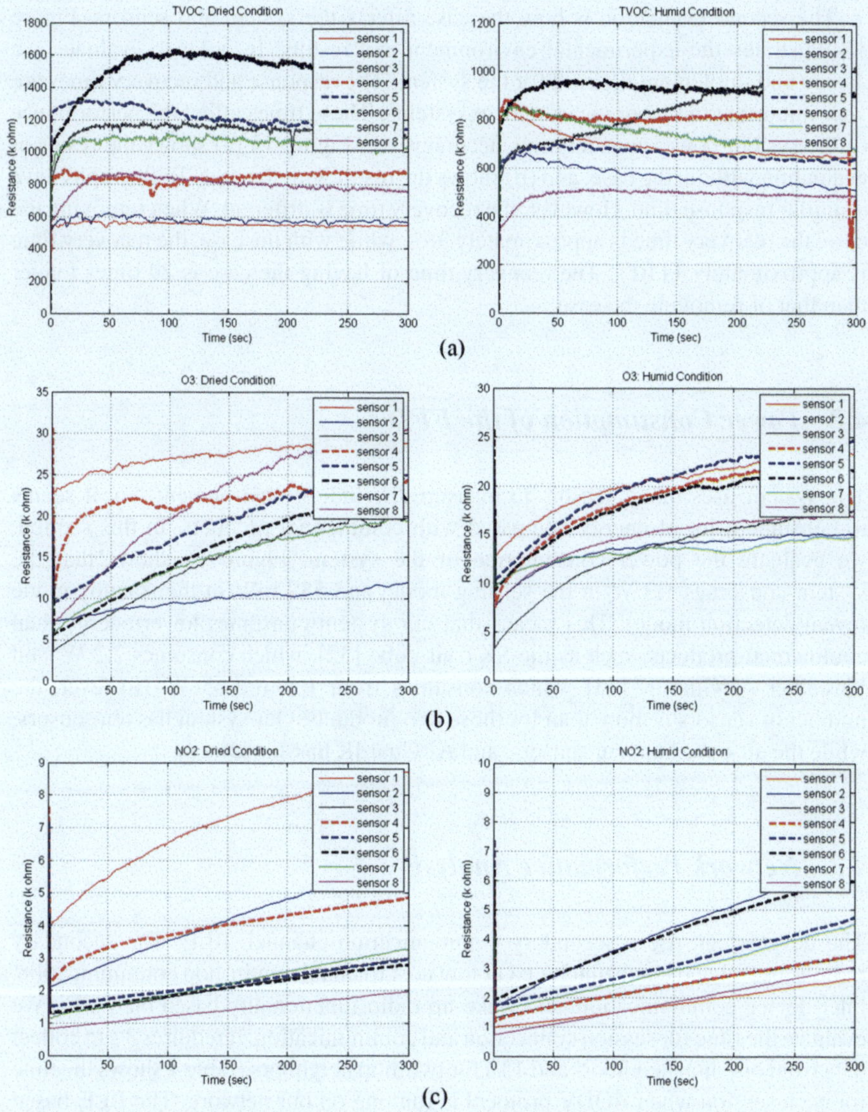

Fig. 14 R_{air} of semiconductor-based gas sensors: (**a**) TVOC, (**b**) O_3, and (**c**) NO_2

gas sensors we used have the same R_{air} value. This means that sensors do not have reproducibility. This problem makes it complicated for the system to measure gases. Unfortunately, the graphs have different deviations and maxima and minima values. When the ambient condition is dry, the deviation of sensors is larger than when the condition is humid. In order to correct this problem, the system measures this R_{air} value and records it in memory in the initial time.

The second evaluation is how the case affects the system and sensors. Figure 15 illustrates the experimental environment and results. In order to evaluate, we design and implement the case for the system. The response and recovery times are very important in sensor measurement systems. These times affect the performance of the system power consumption, accuracy, etc. Figure 15 (a) shows the result of evaluation without the case, and (b) shows the result with the case. Both graphs have a similar response time. However, the recovery time is different. When removing the case, the recovery time is approximately 38 s, while with the case, the recovery time is approximately 1110 s. The recovery time of having the case is 29 times longer than that of removing the case.

4.2 *Power Consumption of the FFD*

The system uses many sensors to measure the indoor environment. So, it seems to consume a lot of energy compared with commercial products. In this section, we evaluate the power consumption of the system. Figure 16 shows that the system consumes 1.3 W in the sensing mode, and 588 mW in the standby mode (event detection mode). This means that our system consumes lower energy than commercial products, such as the SKT air cube [32], which consumes 2.2 W, and Kweather AirGuardK [33], which consumes from 1.5 to 3.65 W. However, the number of sensors is more than for those two products. Our system has ten sensors, while the air cube has four sensors, and AirGuardK has six sensors.

4.3 *Network Performance Analysis*

The devices can use a secondary communication channel, BLE. The secondary radio is useful and can be included at low cost to enable multi-hop communication, inter-device communication, and wake-up radio functionality based on Wi-Fi. We evaluate the time for session connection and communication. It requires 3 s to collect data from one-hop neighbors and 15 s for two-hop neighbors. Table 1 shows the time for each session when AODV protocol is running on our network. The BLE-based routing protocol requires some delay or latency, because it is basically based on a pairing mechanism. It consumes much time to send data to the smartphone.

5 Conclusion

These days, people spend a substantial proportion of their time inside buildings. However, many people are exposed to risks, such as air pollution, indoor noise, and property loss. In particular, indoor air pollution is critical. Air pollution, which

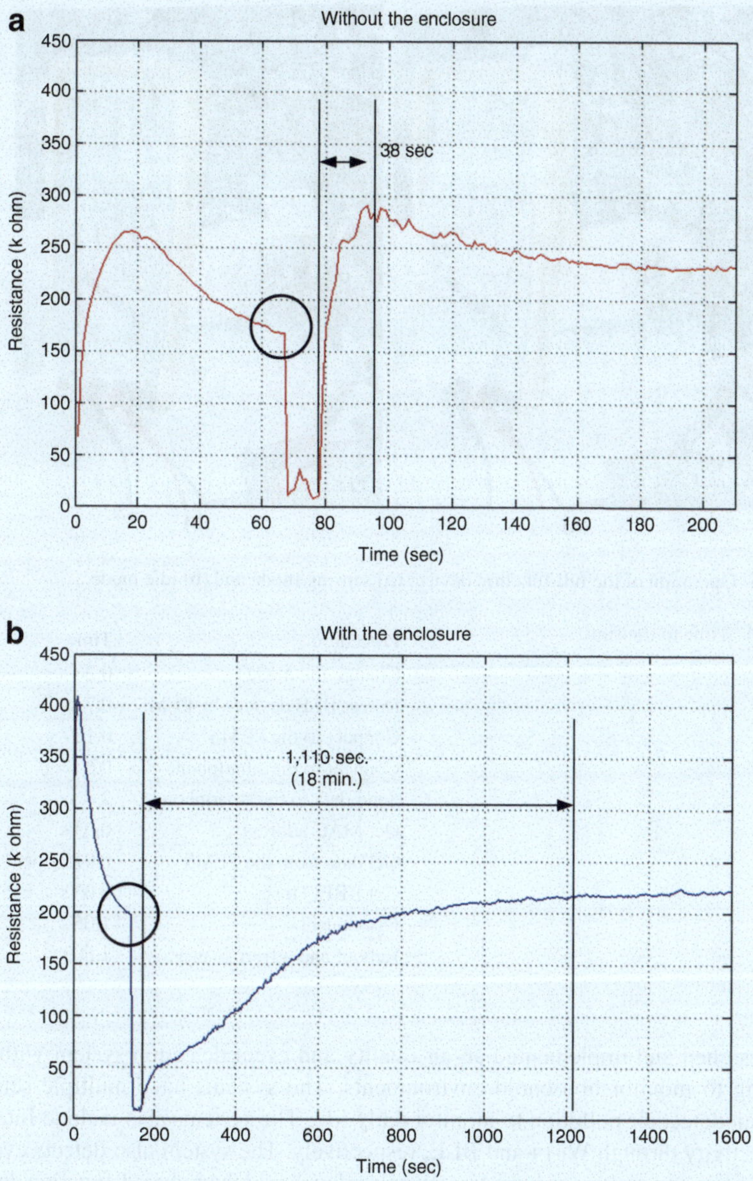

Fig. 15 Response time and recovery time of the system: (**a**) without case and (**b**) with case

includes dangerous materials like nitrogen, carbon, particulates, and toxic gas, is a global problem. Individuals suffering from lung disease, such as asthma, and those who work or exercise outside are particularly susceptible to the adverse effects of smog, such as damage to lung tissue, and reduction in lung function. In this chapter,

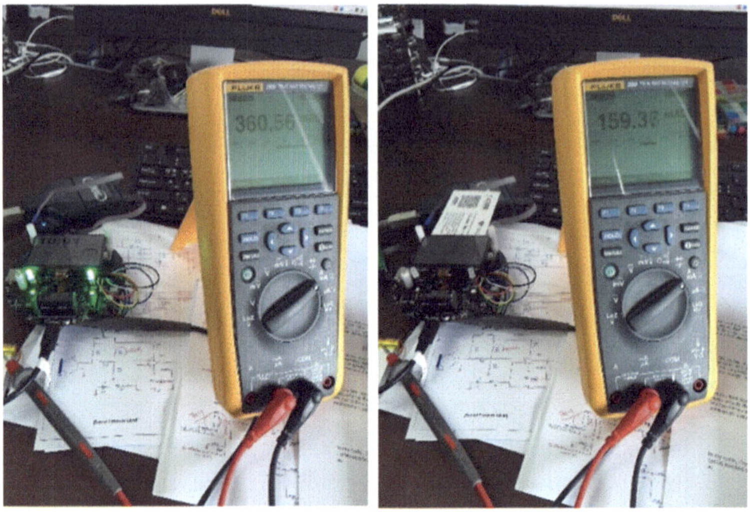

(a) (b)

Fig. 16 Operation of the full-function device: (**a**) sensing mode and (**b**) idle mode

Table 1 Time to transmit data

Procedure	Time
Change the mode	0.2 s
Detect BLE devices by PEMS	0.7 s
Connect to the PEMS	0.189 s
Connect to the smartphone	0.6 s
Send data to the smartphone	2.1 s
Get MAC address	0.03 s
Get data from the PEMS	0.015 s per char
Get RREQ msg.	0.03 s
Disconnect	1.05 s
Reboot and advertise msg.	1.609 s

we designed and implemented an air quality and event detection system with life logging to monitor household environments. The systems have multiple sensors that can detect air pollution to monitor daily life. The systems also include Internet connectivity through Wi-Fi and BLE, respectively. The system also detects events, such as fire and malicious intruders. We also developed the reduced-function device to reduce cost and to enhance the lifespan of the system. Therefore, we hope that devices that detect air pollution and events could save millions of lives and preserve people's property.

Acknowledgments "This work was supported by the Center for Integrated Smart Sensors funded by the Ministry of Science, ICT and Future Planning as Global Frontier Project" (CISS-2011-0031870).

References

1. WHO: Burden of Disease from Household Air Pollution for 2012. World Health Organization, Geneva (2014)
2. WHO: Public Health, Environmental and Social Determinants of Health (PHE), World Health Organization, Geneva. http://www.who.int/phe/health_topics/outdoorair/databases
3. WHO: WHO Guidelines for Indoor Air Quality: Selected Pollutants. WHO, Copenhagen (2010)
4. Pope III, C.A., Burnett, R.T., Thun, M.J., Calle, E.E., Krewski, D., Ito, K., Thurston, G.D.: Lung cancer, cardiopulmonary mortality, and long-term exposure to fine particulate air pollution. J. Am. Med. Assoc. **287**(9), 1132–1141 (2002)
5. Goldsmith, J.R., Friberg, L.T.: Effects of air pollution on human health. Air Pollut. **2**, 457–610 (1977)
6. Seinfeld, J.H., Pandis, S.N.: Atmospheric Chemistry and Physics: From Air Pollution to Climate Change. Wiley, Hoboken (2016)
7. Dockery, D.W., Arden Pope, C.: Acute respiratory effects of particulate air pollution. Annu. Rev. Public Health. **15**(1), 107–132 (1994)
8. Lee, D.-D., Lee, D.-S.: Environmental gas sensors. IEEE Sensors J. **1**(3), 214–224 (Oct. 2001)
9. George, F., Fine, L.M., Cavanagh, A.A., Binion, R.: Metal oxide semi-conductor gas sensors in environmental monitoring. Sensors. **10**, 5469–5502 (2010.) 2010
10. Houtgast, T.A.M.M.O.: Indoor speech intelligibility and indoor noise level criteria. In: Noise as a Public Health Problem, vol. 10, pp. 172–183. ASHA Reports, Rockville (1980)
11. Banerjee, S., Misra, A.: Minimum energy paths for reliable communication in multi-hop wireless networks. Proceedings of the 3rd ACM international symposium on Mobile ad hoc networking & computing. ACM (2002)
12. STMicroelectronics: STM32F407 datasheet. http://www.st.com (2014)
13. FreeRTOS. http://freertos.org (2014)
14. Freescale: Kinetis KL17 datasheet. http://www.freescale.com (2015)
15. SGX Sensortech. http://www.sgxsensortech.com
16. Sharp GP2Y1010. https://www.sparkfun.com/datasheets/Sensors/gp2y1010au_e.pdf (2006)
17. FatFs Generic, F.A.T: File system module. http://elm-chan.org/fsw/ff/00index_e.html (2015)
18. TI. CC3100 datasheet. http://www.ti.com (2015)
19. Mo, Y., et al.: Micro-machined gas sensor array based on metal film micro-heater. Sens. Actuators B. **79**(2), 175–181 (2001)
20. Suehle, J.S., et al.: Tin oxide gas sensor fabricated using CMOS micro-hotplates and in-situ processing. IEEE Electron. Device. Lett. **14**(3), 118–120 (1993)
21. Hwang, W.-J., et al.: Development of micro-heaters with optimized temperature compensation design for gas sensors. Sensors. **11**(3), 2580–2591 (2011)
22. Cho, H.: Personal environmental monitoring system and network platform. 2015 9th International Conference on Sensing Technology (ICST). IEEE (2015)
23. Dunteman, G.H.: Principal components analysis. No. 69. Sage (1989)
24. Price, A.L., et al.: Principal components analysis corrects for stratification in genome-wide association studies. Nat. Genet. **38**(8), 904–909 (2006)
25. Syms, C.: Principal components analysis. In: Jorgensen, S.E., Fath, B.D. (eds.) Encyclopedia of Ecology, pp. 2940-2949. Elsevier, Oxford 2008. ISBN: 9780444520333
26. Vasilescu, M. Alex O., Terzopoulos, D.. Multilinear independent components analysis. 2005 IEEE computer society conference on Computer Vision and Pattern Recognition (CVPR'05), vol. 1. IEEE, Machine learning (2005)
27. Goldberg, D.E., Holland, J.H.: Genetic algorithms and machine learning. Mach. Learn. **3**(2), 95–99 (1988)
28. Figaro. http://figarosensor.com. TGS5342 (2013)
29. Akyildiz, I.F., Su, W., Sankarasubramaniam, Y., Cayirci, E.: Wireless sensor networks: a survey. Comput. Netw. **38**, 393–422 (2002)

30. Perkins, C., Belding-Royer E., Das S.: Ad hoc on-demand distance vector (AODV) routing. No. RFC 3561 (2003)
31. Royer, E.M., Perkins, C.E.: An implementation study of the AODV routing protocol. Wireless communications and networking conference, 2000. WCNC. 2000 IEEE, vol. 3. IEEE (2000)
32. SKT Air cube. http://www.sktworld.kr (2014)
33. Kweather. http://airguard.com. Airguard K (2014)

Mobile Crowdsensing to Collect Road Conditions and Events

Kenro Aihara, Hajime Imura, Bin Piao, Atsuhiro Takasu, and Yuzuru Tanaka

1 Introduction

Cyber-physical systems (CPSs) seek to provide users with optimal control of the world in which they live by modeling physical space in cyberspace, coupled with the use of related databases. More than big data systems, a social CPS is the operating system of urban society. It provides a user environment that supports people's agency in decision-making. The need for social CPSs to assist in building sustainable, safe, and secure urban societies is growing, and the prerequisite technologies are maturing rapidly. Challenges that remain include opening data silos maintained by both the private sector and the government and analyzing massive, complex datasets that cannot be completely described by a single monolithic model. The field of social CPSs provides tantalizing challenges for researchers and developers.

This paper provides an overview of the ongoing social CPS project, which aims to develop a mobile sensing framework to collect sensor data reflecting personal-scale, or microscopic, roadside phenomena using crowdsourcing and big data, such as traffic and climate data, as well as the contents of social networking services such as Twitter.

K. Aihara (✉) • B. Piao • A. Takasu
National Institute of Informatics, 2-1-2 Hitotsubashi, Chiyoda-ku, Tokyo 101-8430, Japan
e-mail: kenro.aihara@nii.ac.jp; piaobin@nii.ac.jp; takasu@nii.ac.jp

H. Imura • Y. Tanaka
Hokkaido University, N-13, W-8, Sapporo, Hokkaido 060-8628, Japan
e-mail: hajime@meme.hokudai.ac.jp; tanaka@meme.hokudai.ac.jp

H. Yasuura et al. (eds.), *Smart Sensors at the IoT Frontier*,
DOI 10.1007/978-3-319-55345-0_11

2 Background

2.1 Driving Problems: The Situation in Sapporo

Sapporo is the fifth largest city in Japan, with a population of about 1.91 million. The city receives an average of about 6 m snowfall annually, with an average maximum snow depth of about 1 m in February. The city spends more than 15 billion yen every winter on road management activities such as snowplowing and snow removal.

The snowfall in winter causes significant changes to road conditions. For example, Fig. 1 shows various situations in Sapporo. Four photographs were taken from the same location using a camera mounted on the dashboard of a vehicle. In summer, the climate in Sapporo is more stable than in other areas in Japan and the road is very clear (Fig. 1a). However, in winter, the conditions can change drastically. After heavy snowfall, the roads are covered by a thick layer of snow (Fig. 1b). Snowplows are deployed along all main roads administered by Sapporo city and other public sectors after heavy snowfalls. After snowplowing, the roads are clear but wet (Fig. 1c).

Figure 2 shows some typical scenes in Sapporo in winter. The plowed snow forms huge heaps on the roadside, which causes other serious problems. One of these problems is dead angles. Huge snow heaps create a lot of dead angles, which can cause accidents, especially when the roads are slippery (Fig. 2a). Another problem is the narrower lanes created by the plowed snow. Snow heaps become higher

(a) summer

(b) winter: heavy snowfall

(c) winter

(d) winter: surface freezing

Fig. 1 Comparison of road conditions in summer (**a**) and winter (**b**, **c**, and **d**). The road conditions vary from hour to hour, especially in winter

Fig. 2 Various snow heaps and related problems. (**a**) Dead angle. (**b**) Narrower lanes. (**c**) Huge snow heap (1). (**d**) Huge snow heap (2)

after snowplowing and are sometimes taller than people (Fig. 2c, d). Further, the higher the snow heap, the longer its tail. Long tails often result in narrower lanes, thereby blocking traffic movement (Fig. 2b). Therefore, it is important to detect road segments or intersections where the number of available lanes has been reduced as a result of snow heaps.

The photograph shown in Fig. 1d was taken the day after the photograph shown in Fig. 1c. It can be seen that just 1 day of snowfall can make a significant difference, and the situation can vary from day to day in a given location.

Driving in Sapporo in winter is affected by the amount of snowfall, and thus the depth of the snow, the temperature, the road surface, the volume of traffic, the amount of snowplowing, and other road conditions. The road surface often becomes frozen as a result of the extremely low temperatures (Fig. 1d).

Sapporo appears to be one of the most "challenged" cities in the world because its citizens demand good facilities and services even though the climate is severe. Table 1 shows the population and average annual snowfall of several major cities around the world. Sapporo has almost twice as much snowfall as Quebec City, while its population is nearly four times larger and is increasing. Therefore, the authors

Table 1 Population and average annual snowfall of major world cities [2, 4, 12]

City	Population (k)	Average annual snowfall (cm)
Sapporo	1,914	597
Tokyo (23 wards)	8,947	11
Innsbruck	125	99
Vienna	1,767	67
Moscow	11,794	136
Montreal	1,718	218
Quebec City	532	316
Ottawa	883	236
Toronto	2,615	133
Vancouver	641	48
Boston, Massachusetts	646	111
Buffalo, New York	259	241
Chicago, Illinois	2,719	93
Cleveland, Ohio	390	173
Denver, Colorado	649	137
Detroit, Michigan	689	109
Milwaukee, Wisconsin	599	119
New York, New York	8,406	64
Pittsburgh, Pennsylvania	306	106
Salt Lake City, Utah	191	143
Washington, DC	646	37

believe that a case study set in Sapporo provides useful information that can be used to help solve problems relating to the management of big, smart cities.

2.2 *Goal-Directed Sensing with Active Participants: Crowdsourcing*

The term "crowdsourcing" was first described in 2006 [10] and was subsequently defined as the act of taking a task traditionally performed by a designated agent and outsourcing it by making an open call to an undefined but large group of people [11]. This can take the form of peer production, but is also often undertaken by individuals [9].

The concept of smart cities can be viewed as recognition of the growing importance of digital technologies in the search for a competitive position and a sustainable future [16]. The smart city agenda, which uses information and communications technology to achieve strategic urban development goals such as improving the quality of life of citizens and creating sustainable growth, has gained a lot of momentum in recent years.

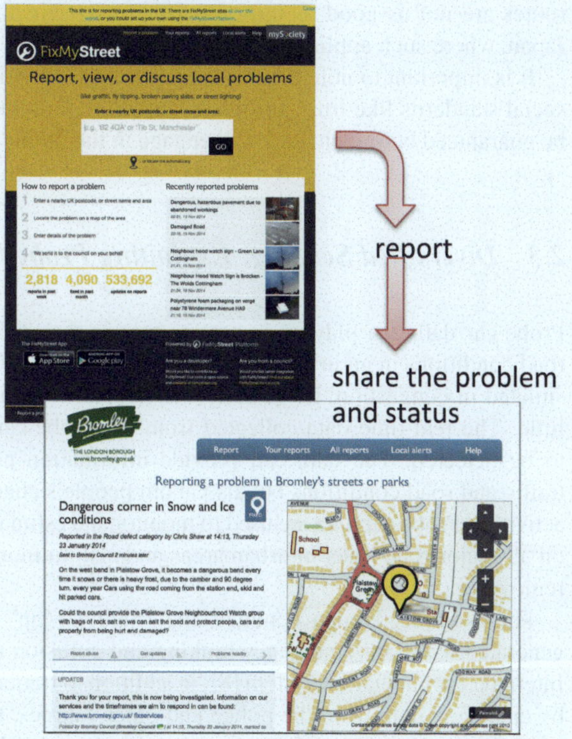

Fig. 3 FixMyStreet

Tools such as smartphones provide the opportunity to facilitate co-creation between citizens and authorities. Such tools have the potential to organize and stimulate communication between citizens and authorities and to allow citizens to participate in the public domain [1, 17]. One example is FixMyStreet,[1] which enables citizens to report broken streetlights and potholes in roads [13].

FixMyStreet, which was launched in early February 2007, is a website that enables people to report problems they have found to their local council by simply locating them on a map (Fig. 3). FixMyStreet is used to report items that are damaged, dirty, or dumped and which need fixing, cleaning, or clearing, respectively. The reports are emailed to the relevant council, which then takes action to resolve the problem. Alternatively, citizens can discuss a problem with other people on the website and then either lobby the council to act or take direct action themselves.

Waze[2] is another crowdsourcing service that is used to collect traffic data. Even though Waze provides traffic information collected from users and a navigation function, this seems to be insufficient to attract users because the recommended

[1]https://www.fixmystreet.com/

[2]https://www.waze.com/

routes are not as good as those provided by navigation appliances, especially in Japan, where such appliances are well developed.

It is important to note that these approaches will not succeed automatically, and social standards like trust, openness, and consideration of mutual interests need to be guaranteed before citizens will engage in the public domain.

2.3 *Diversified Sensing: Exploiting Probe Car Data*

Probe car data can play an important role in the analysis of changing traffic and road conditions in an urban area. Probe cars act as mobile sensors. If a probe car is situated in congestion, its velocity will decrease and its position will change very little. The real-time data collected from the probe car will enable the congestion to be detected. The data can provide information not only about the changing traffic and road conditions but also about people's changing mobility demands and activities. Probe car data are used to monitor and estimate traffic and road conditions on all major road links in urban areas and also to monitor snowplowing and snow removal operations.

For example, Honda conducts the "Safety Map" project, whereby maps are generated based on emergency braking and collision data collected from Honda's Internavi car navigation system [8], in addition to frequent collision points identified by analyzing data from the police and other sources, as well as information about areas that local residents find dangerous. Frequent hard braking points are identified using probe car data collected from vehicles equipped with Honda's Internavi system.

There are some issues in relation to the use of probe car data in estimating road conditions. One is the density of the data. Only a small number of vehicles can operate as a probe car. In addition, other types of data are required in addition to location and velocity. Finally, probe car data are often collected by various car manufacturers, who consider the data too sensitive to be released for public analysis.

3 Crowdsourced Mobile Sensing and Its Applications

3.1 *Overview*

CPSs are a promising new class of systems that deeply embed cyber capabilities into the physical world, either in relation to humans, infrastructure, or platforms, to transform interactions with the physical world [3, 15]. CPSs facilitate the use of information that is available from the physical environment. Advances in the cyber world in relation to such fields as communications, networking, sensing,

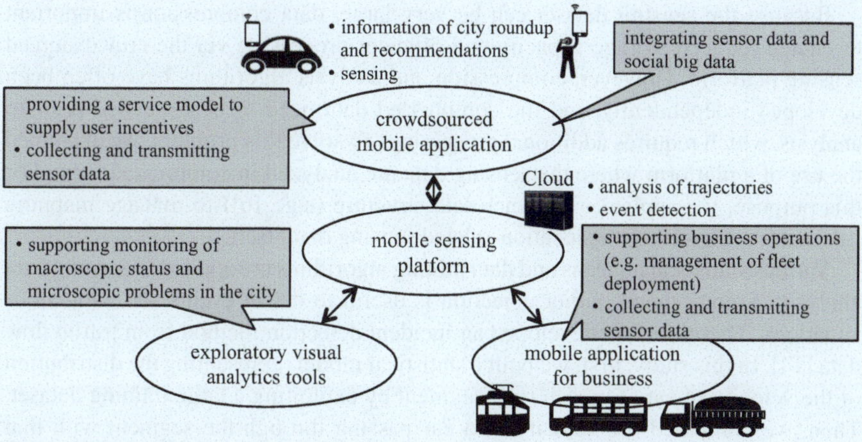

Fig. 4 Overview of the proposed mobile sensing system

computing, storage, and control, as well as advances in the physical world in relation to materials and hardware, are rapidly converging to realize a new class of highly collaborative computational systems that rely on sensors and actuators to monitor and effect change. In this technology-rich scenario, real-world components interact with cyberspace via sensing, computing, and communications elements.

Social CPSs focus on human aspects in the parallel world because humans are not only able to exploit such systems but are also able to be observed and affected by those systems. Information flows from the physical to the cyber world, and vice versa, thereby adapting the resulting converged world to reflect human behavior and social dynamics. Indeed, humans are at the center of this converged world because information about the context in which they operate is the key to adapting CPS applications and services.

Figure 4 shows the proposed system for crowdsourced mobile sensing. The cloud, which is the service platform, is located at the center. Several applications are then developed using this platform.

3.2 *Service Platform*

The service platform facilitates applications not only to receive data from other applications but also to provide ordinary functions for location-based services, such as nearest and up-to-date places. The platform also plays a role in integrating the data that has been collected using big data such as traffic and climate information, as well as the contents of social networking services, and then analyzing these data to identify specific phenomena in the city, especially relating to roads.

Because the sensing dataset can be very large, data compression is important to enable reduced storage capacity and efficient processing via the crowdsourced sensing platform. However, compression and analysis algorithms have often been developed independently, and the compressed data need to be expanded prior to analysis, which requires additional processing. To solve this problem, we examined the use of a platform where the sensing data are analyzed in compressed form. For this purpose, we applied a succinct data structure (e.g., [6]) to manage mapping information, as well as the location-related sensing data itself.

Various statistical analysis and data mining algorithms are applied to sensing data analysis. Among them, outlier detection is useful to detect events and anomalous situations. Therefore, we developed an incident detection method from traffic flow data [14]. In this study, first, we built a statistical model representing the distribution of the velocity of cars for each road segment by exploiting a large training dataset. Then, we compared the velocity of a car passing through the segment with that estimated by the model. If the velocity was an outlier with respect to the velocity distribution model, we judged the road segment to be an anomalous situation. Because of the large amount of sensing data, we used a complex model and achieved a high detection rate [14].

Crowdsourced sensing data can be biased because the people who provide the data are not always typical users. Therefore, when using a statistical model for the analysis, as in our incident detection method, bias correction is necessary.

3.3 Mobile Applications for End Users

For citizens who are end users, a mobile application has been developed based on the cloud platform. Although the main target users are drivers, the application can be used by pedestrians using public transport such as the subway and buses.

For drivers, a drive recording function, or video event data recorder, is provided. Users mount the recording appliance on the dashboard or attach it to the windshield to record the behavior of the car during the journey, such as the trajectory (a sequence of locations with time stamps), acceleration, and speed. One of the strongest motivations for using such appliances is that they can provide evidence in relation to an accident if necessary. Therefore, drivers should use an appliance of some sort whenever they drive. We expect that the proposed smartphone application can replace existing appliances. Table 2 shows the main features of various drive recorders. Ordinary appliances, such as the Garmin Dash Cam, are commercial products that usually work automatically. The appliance begins recording when the driver starts his/her engine and stores the data it records. PAYD is used not to record data relating to accidents but to reduce insurance premiums. It monitors and records the driver's behavior by assessing speed, braking, acceleration, and cornering and the time of day when journeys are made. The data are transmitted to the insurance provider via mobile networks.

Table 2 Drive recorders

Name	Type	Sensors (input)	Monitor	Storage	Motivation
Garmin Dash Cam TM	Appliance	Camera (movie, photo), GPS, integrated microphone, incident detection sensor	Image (2.3-inch LCD display)	Local (4–32 GB microSD)	Lifelogging, recording incidents
Telematics auto insurance, or PAYD (pay as you drive), e.g., progressive insurance	Appliance	Speed, time of day, number of miles, (via Onboard diagnostic (OBD-II) port)		Cloud	Reduce insurance fee
iSymDVR 2	Smartphone app	Camera (movie), location, G-sensor	Image, map, speed	Local	Lifelogging, recording incidents
Safety sight (free)	Smartphone app	Camera (10 s. movie), location, G-sensor	Image, map, speed, distance to the vehicle ahead, safe driving diagnosis	Local	Recording incidents, ecological driving
PROPOSED (free)	Smartphone app	Camera (movie), location/heading/speed/course, accelerometer/gyro/magnetic	Image, map, speed, real-time events in the city	Cloud	Lifelogging, recording incidents, getting real-time information

iSymDVR 2 and Safety Sight[3] are smartphone applications for drive recording. Although Safety Sight is provided by an auto insurance company, it only assesses the driver's behavior and provides feedback. It does not transmit data back to the insurance company. It also provides a warning to drivers by estimating the distance to the vehicle ahead, which it calculates by analyzing an image of the scene to detect the shapes and sizes of objects (Fig. 5a). It automatically records a 10-s video of the scene in front of the vehicle before and after impact when the app detects the possibility of an impact, such as from sudden braking (Fig. 5b).

The advantages of applications compared with appliances are as follows:

1. Ordinary appliances are stand-alone, which means that local storage is limited, whereas the application is connected to the cloud.

[3] http://www.sjnk.jp/app_pc/safetysight/

(a)

Fig. 5 Safety sight by Sompo Japan Nipponkoa Insurance Inc. (**a**) Approaching forward vehicle warning. (**b**) Event data recorder

2. Appliances are not cheap, whereas some applications, including the one we have developed, are free.
3. Appliances only store driving records, whereas the application can provide feedback.

We believe that these advantages provide citizens with the incentive necessary to use the application.

The drive recorder is able to collect data reflecting roadside situations whenever drivers are on the road. Power consumption is not critical, because power can be supplied by the car's electrical system. Further, drivers are not required to manipulate their smartphones while they are driving. The details of the drive recorder application are presented in Sect. 4.

3.4 *Applications for Civil Administration*

In our prior study on the CPS-IIP Project, we implemented smartphone-based mobile sensing applications in city buses and snowplowing vehicles to investigate the influence of snowfall and snow removal operations on traffic. The experimental field is shown in Fig. 6d. The mobile sensing system for buses is an important tool for collecting both periodic and continuous road traffic information. We implemented 20 sensing systems in cooperation with Hokkaido Chuo Bus, one of the bus companies operating city bus services in Sapporo (Fig. 6c). The bus sensing data are useful for both periodic and continuous monitoring of major traffic routes. Meanwhile, the sensing system used in snowplowing and snow removal vehicles is useful for monitoring snow removal operations. Both sensing systems collect data relating to vehicle speed, bearing, and latitude and longitude, which are determined by GPS. However, each system also provides additional information. The bus sensing system provides three-axis acceleration, route information, and other operating information, while the snowplowing vehicle sensing system provides operating information (e.g., snowplowing or snow removal information). Therefore,

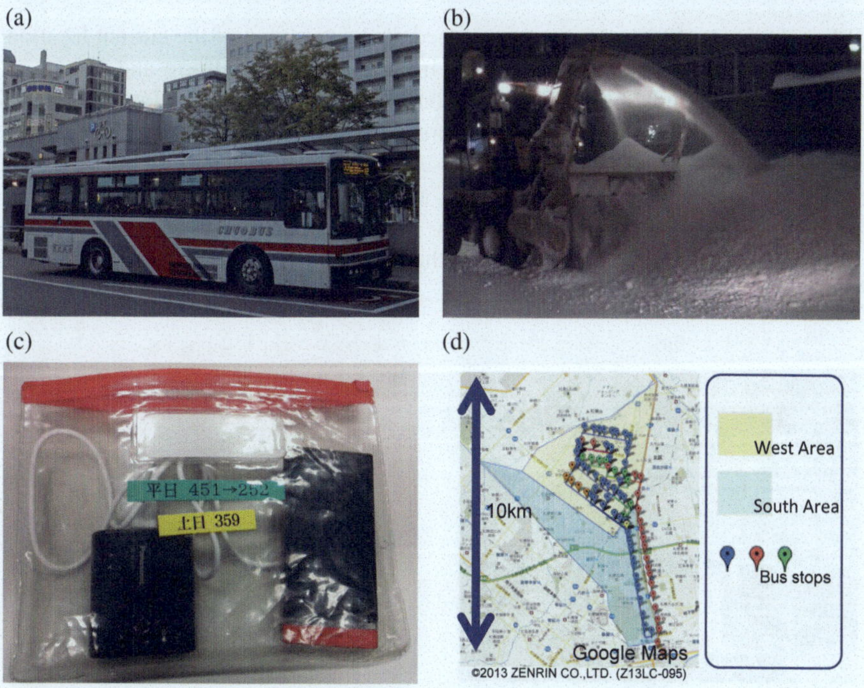

Fig. 6 A Chuo bus, a snowplowing vehicle, the smartphone-based mobile sensing terminal, and the experimental field. (**a**) Hokkaido Chuo bus. (**b**) Snow plowing vehicle. (**c**) Smartphone based mobile sensing terminal for operators. (**d**) The experimental field

it is important to develop a unified platform for mobile sensing that facilitates the deployment of several applications for operational monitoring.

4 "Drive Around-the-Corner.": A Drive Recorder Application

In February 2015, we developed a drive recorder application called "Drive around-the-corner (Drive ATC)." This application was made available to the public in February 2016.[4] Drive ATC collects driving behavior logs, records events, and delivers information regarding the vehicle's current position.

The service can be accessed via the iOS application. Before commencing a journey, users mount their smartphone in a holder and connect a power supply cable if necessary (Fig. 7) and then open the application (Fig. 8). The application records driving behavior logs and videos and uploads them to the service platform.

4.1 Map with Event Information

When the Drive ATC application is opened, it shows a map of the current location (Fig. 8a). Roadside events are retrieved from the service platform and shown on the map. For example, the traffic sign icon located at the center of Fig. 8a denotes road construction. This information has previously been posted by other Drive ATC users.

In addition, footprint markers, which are placed in locations that the user has passed previously, are shown as triangles. The size of the markers varies according to the speed of the vehicle. The shorter the triangle marker, the slower the vehicle was traveling.

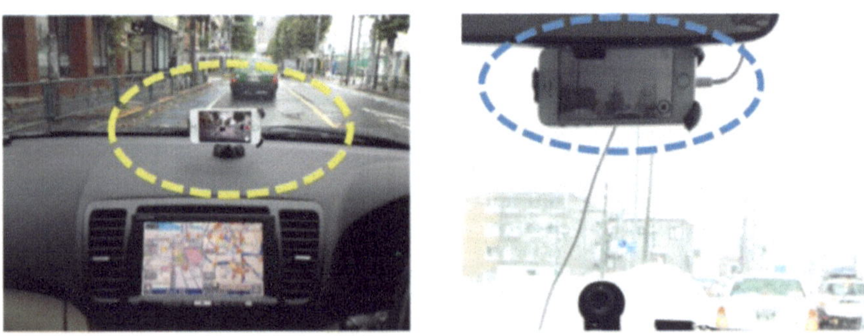

Fig. 7 Smartphones mounted in the vehicle

[4]https://itunes.apple.com/app/drive-around-the-corner./id1053216595

(a)

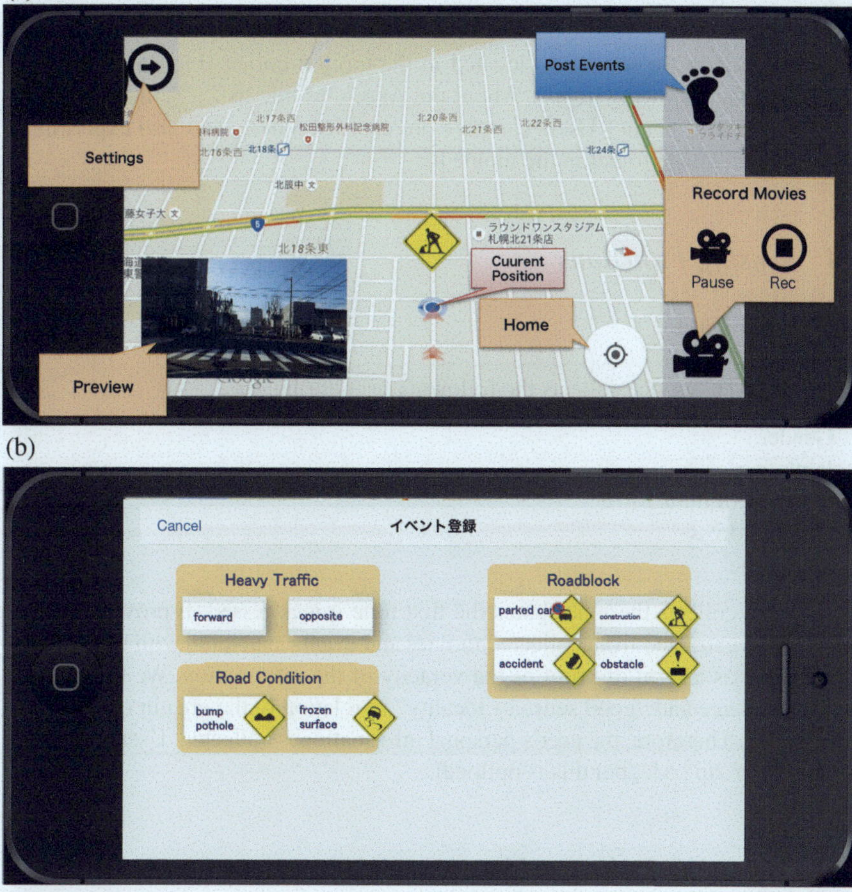

Fig. 8 "Drive around-the-corner" application. Traffic information, events posted by other users, events extracted from sensor data, and footprints are shown on the map on the main screen. (**a**) Main screen. (**b**) Post events

4.2 *Posting Event*

To enable users to report a roadside event to others while they are stationary, the application provides them with the ability to post event information. After tapping the footprint marker in the top right corner, users are requested to select an event that they recognize (Fig. 8b). There are eight possible events grouped into three categories: heavy traffic, road condition, and roadblock. The selected event is posted to the service platform with details of the current time and location.

4.3 *Settings*

The menu button for settings is located in the top left corner (Fig. 8a). The menu consists of the following items: "About the App," "Movie list," "Settings," "Event list," and "User account." Users can play prerecorded movies and also export them to the general image folder in the movie list.

4.4 *Sensing Functions*

4.4.1 User Data

The Drive ATC service collects the following user attributes:

- Gender
- Birth year
- Zip code of hometown
- Email address
- Nickname

The service collects these attributes the first time it is accessed. It provides shared information extracted from collected movies and events annotated with the sender's information as a form of proof of the veracity of the information. We believe that the sender's age and "good sense of locality" in relation to the area in question are informative. Therefore, the user's personal information is collected. Users are asked to enter their zip code, but this is optional.

4.4.2 Onboard Location and Motion Sensors

The Drive ATC application obtains location and motion data from onboard sensors. While users are driving and using the application, behavior logs and movies are recorded. The data that are collected are pooled in the local data store and then transmitted to the service platform. The data that are collected are shown in Table 3.

Table 3 Data collected by drive around-the-corner

Type	Attributes
Location	Latitude, longitude, and altitude with accuracy
Heading	True_north with accuracy
Move	Speed, course
Acceleration	x, y, z
Rotation rate	x, y, z

It is important that the data timestamp is accurate. For example, location data can be timestamped using GPS-adjusted time. This ensures that the time is precise and credible. However, motion values and movies are usually timestamped with the clock time of the terminal, and the clock time of computing devices is generally incorrect. To enable processing and integration of the various kinds of sensor data, they must be aligned in the same timeframe. One solution is to obtain an offset of two timestamps so that location data are recorded using both GPS-based time and clock time.

4.4.3 Movies

The Drive ATC application records two types of movies, one to be uploaded and the other to be saved locally. To reduce traffic to the service platform, uploaded movies are transferred intermittently, the frame rate being adjusted in accordance with the speed of the vehicle.

Because these movies are uploaded via a mobile network such as 3G or LTE, they should be compressed. The movies that are saved locally are of higher quality and can be used as evidence in the event of an accident (Fig. 9).

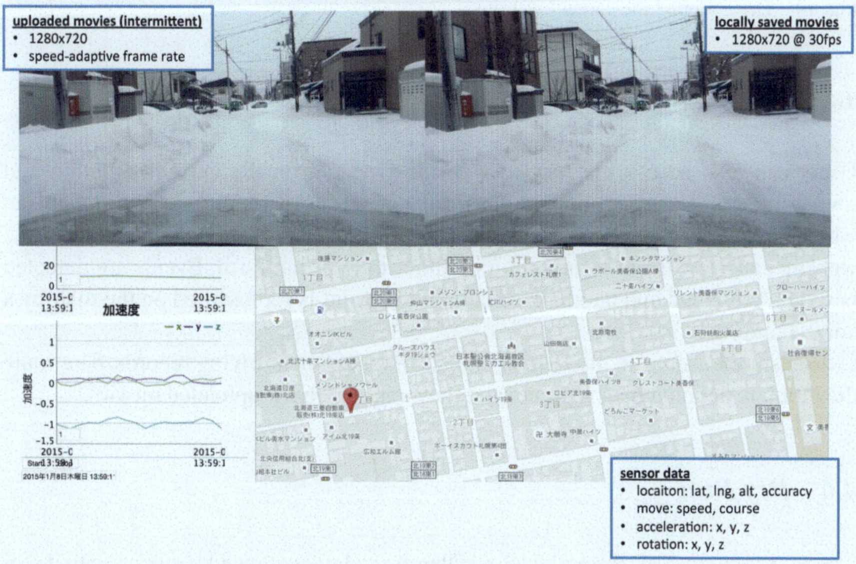

Fig. 9 Data collected using the "Drive around-the-corner" application

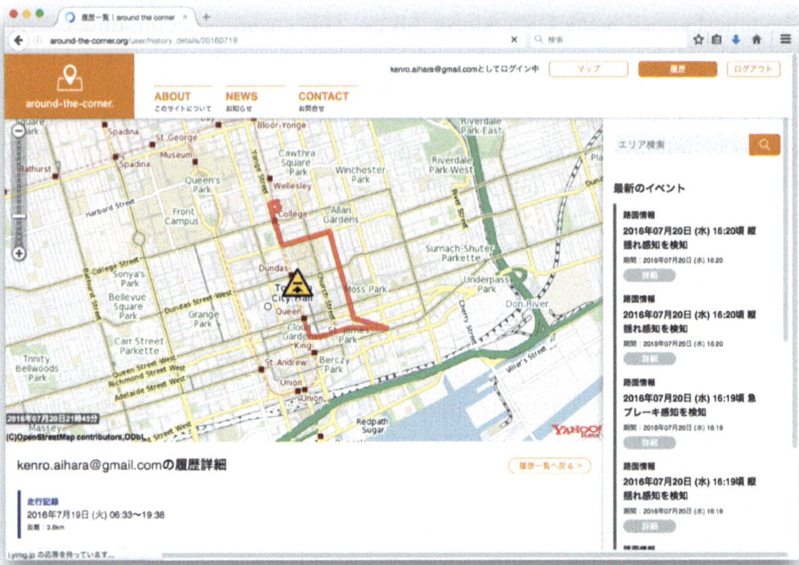

Fig. 10 “Drive around-the-corner” website

4.5 Website

Users can also access the service website to check on the current situation and review their journey and driving performance.[5] Figure 10 illustrates the website, which mainly consists of a map and a list of detected events. All visitors to the website can view event information in either map or list form. Events are detected based on collected and anomalized data. Each event is represented on the map by a corresponding icon.

Registered users can also log in and access their own driving records. The route they took is denoted by a red line, and they can play back uploaded movies.

4.6 Dry Run

Figure 11 shows an example of acceleration data collected from a vehicle in Sapporo. The z-axis (vertical) denotes offset acceleration value for gravity and the y-axis is the heading direction of the car. Figure 12 illustrates scenes from the journey shown in Fig. 11.

[5]http://around-the-corner.org/

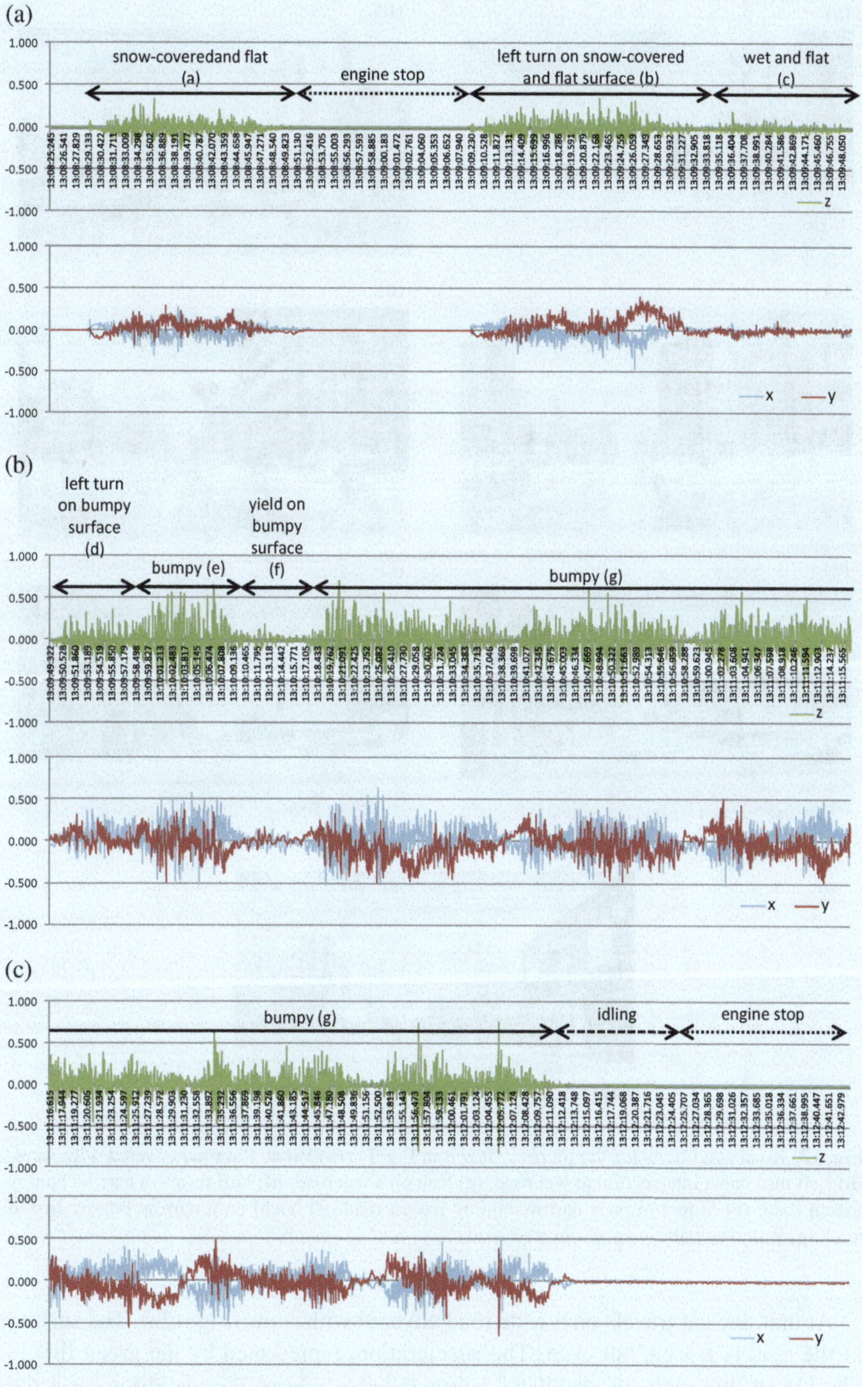

Fig. 11 An example of recorded acceleration data. (**a**) Section 1/3. (**b**) Section 2/3. (**c**) Section 3/3

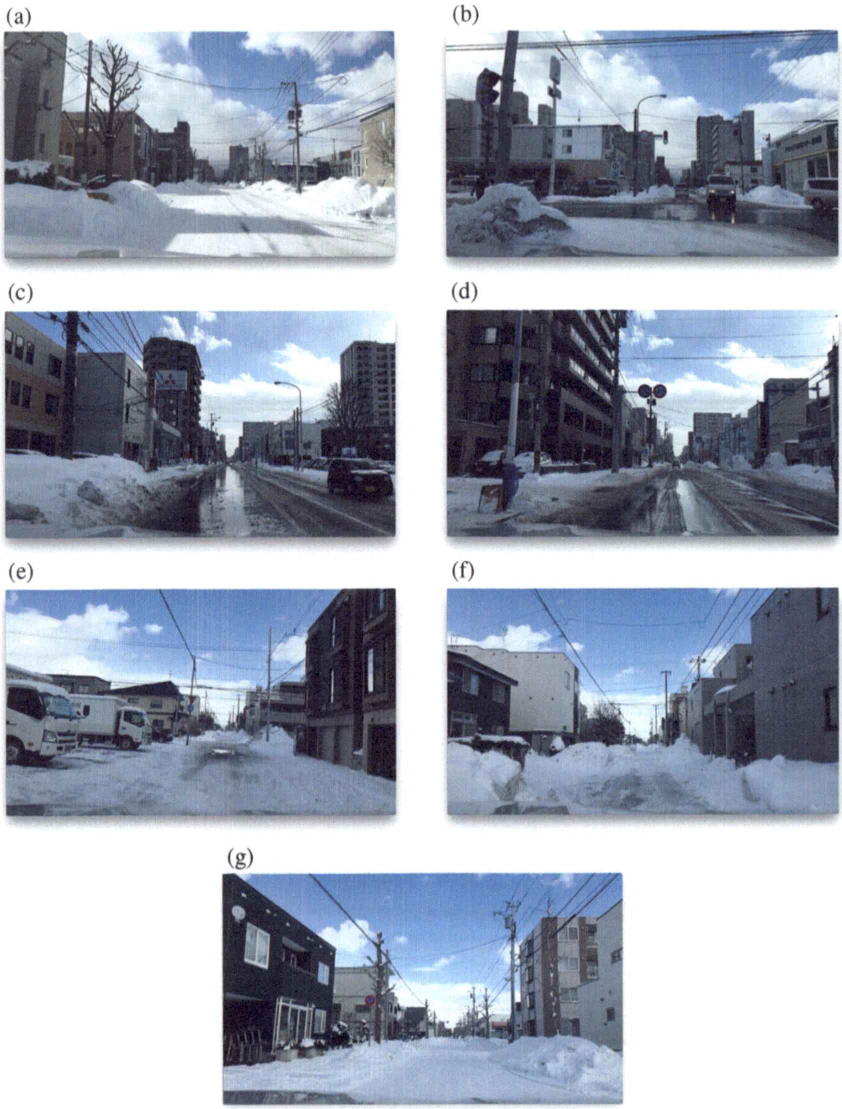

Fig. 12 Road conditions for the journey shown in Fig. 11. (**a**) Broad flat road covered with snow. (**b**) Left turn on an intersection to wet road. (**c**) Run on a wet road. (**d**) Left turn to a narrow bumpy frozen road. (**e**) Slow run on a narrow bumpy frozen road. (**f**) Yield on a narrow bumpy frozen road. (**g**) Run on a narrow snow-covered road

At first, the car travels on a wide road covered with snow (Fig. 12a). The surface of the road is frozen, but even. The acceleration, represented by the green line in Fig. 11a of this segment, oscillates below $0.3\,\mathrm{m/s^2}$. Next, the car stops for a red light. The car's engine is automatically shut down by the start–stop system, and the oscillation falls to the minimum level. Then, the car moves to the next intersection and turns left onto a wet road (Fig. 12b). The increase in transverse acceleration represented by the red line in Fig. 11a indicates the left-hand turn. The acceleration increases significantly because this wet road is more stable than the previous frozen road (Fig. 12c).

The car then turns left onto a narrower road (Fig. 12d). The road is frozen, but is uneven because the ice is thawing. The car travels very slowly and pitches wildly (Fig. 12e). Its acceleration reaches more than $0.5\,\mathrm{m/s^2}$, and even rolling and yawing are recognized because the car slips and drifts on the road (Fig. 11b). The acceleration is greater than on the wet and even road of (c), even when the car travels over the bumps in this road (Fig. 12f).

The car finally stops at the end of the bumpy road section (Figs. 11c and 12g) for a red light. The engine idles for a while and then stops.

4.7 *Survey*

To understand how user functions such as drive recording influence users in their decision on whether to purchase the application, we conducted a survey of users.

Fifty participants were instructed to use the application at least three times in February 2015. They all lived in Sapporo or its vicinity.

At the end of the designated period, they were asked to complete a questionnaire. Twenty-seven (19 males and 8 females) of the 50 participants responded. A summary of their responses is shown in Table 4.

In relation to motivation, 20 of the 27 subjects advised that they would select the application if it was cheaper than appliances. This supports the idea that the attempt to motivate users to adopt the service by giving them a drive recording function is effective.

Table 4 Results of the survey

Question		# of answers
Motivation		
	"Do you use the application instead of appliances if the application is cheaper than appliances?"	20
Attractive functions		
	Real-time information related to traffic	23
	Route navigation to the destination	19
	Automatic recording lifelogs that can be reviewed on the cloud	16
	Sharing up-to-date posts from users	15
Requests		
	Reducing size of locally saved movies	21
	Reducing traffic for uploading data	17

Based on the answers to the questions, the respondents seem to feel that real-time traffic information collected from other users and route navigation to their destination are attractive features. These functions are not included in general drive recorder appliances. Conversely, drive logging and social functions such as sharing posts were not deemed to be attractive features.

Therefore, we believe that the types of functions that are included in the application can serve as an incentive to attract users.

There are some issues that need to be resolved. The biggest problem is the size of locally saved movies. Twelve of the 27 respondents wanted the data storage to be less than 1 GB per week, even though it often required more than 100 MB every 3 min. Although users can increase the storage capacity for locally saved movies, we may have to consider increasing the amount of compression.

4.8 Data Analysis

As mentioned in the previous section, motion values, such as acceleration and gyro, may reflect the road surface conditions. Road surface conditions have long been a concern in society because they have a significant impact on transport safety and driving comfort, especially in snowfall areas. Figure 13 shows the traffic accident rates on various road surfaces in snowfall areas in Japan. It can be seen that about 50% of accidents have occurred on frozen road surfaces. Therefore, it is important to detect frozen road surfaces effectively. In previous studies, many road monitoring systems have used image processing techniques [5, 18] whereby the video cameras are usually placed at representative points on the main roads or mounted on the dashboards of vehicles. However, the effectiveness of the cameras may be impacted by factors such as low light levels at nighttime or snowfall. Further, this method cannot identify the surface conditions when a frozen road surface has been covered by a layer of snow.

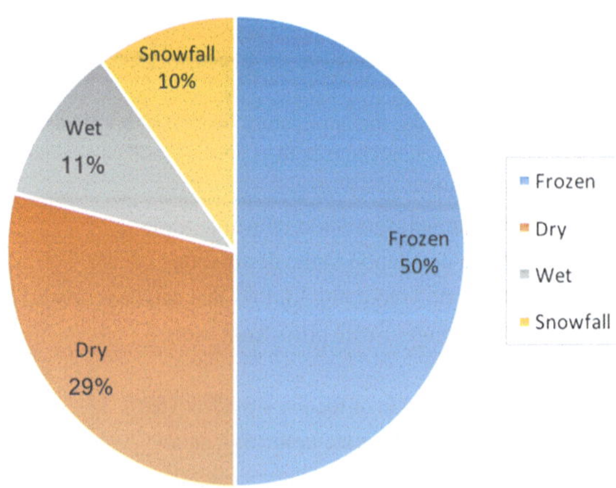

Fig. 13 Traffic accident rates under different road surface conditions

In addition to the condition of the road surface, acceleration depends on the car, the driver, the smartphone, and how it is mounted. All of these elements must be considered when estimating the surface condition of each road segment from a collection of motion values.

We believe that some assumptions must be addressed when developing a suitable model. One assumption is consecutiveness. That is, data collected from the same route are recorded using the same smartphone mounted in the same car driven by the same driver. Therefore, any difference in motion should depend on road surface conditions and driving behavior. The second assumption is concurrentness. If the weather is stable, the surface condition of the same road segment over the same time range should be similar. Here, the time range may be a few hours or even longer. Then, the motion values during stationary periods must depend solely on the structural characteristics of the car, such as the eigenfrequency and the mounted smartphone, because road surface conditions and driving behavior can be ignored. We need to pay attention to two kinds of motion at zero velocity, especially if the car is equipped with a start–stop system.

We assume that driving behavior does not have a high level of influence on Fourier analysis of motion values. Therefore, the analysis must depend on other factors such as road surface conditions and the vehicle.

To enable the detection of road surface conditions from the collected data, we studied a methodology for estimating the condition of snow-covered roads using the collected motion values. This method extracts features from both the time domain and the spectral domain of data collected from accelerometers and gyroscopes. Then, it uses principal component analysis (PCA) to reduce the dimensions of various features and extract characteristics enabling the classification of driving behavior in relation to road conditions. Finally, we apply the support vector machine (SVM) to the data to estimate the road surface conditions.

This method focuses on the following three road surface conditions:

- Frozen
- Sherbet
- Normal

Figure 14 illustrates these three road surface conditions. For the examples of frozen roads shown in Fig. 14a, b, the motion values should reflect surface features such as ruggedness and potholes. Figure 14c also shows an example of a frozen road surface. However, this surface looks like a mirror and does not have any apparent ruggedness or potholes. It is difficult to reflect these kinds of road surface conditions in motion values. Therefore, in this paper, a frozen road surface refers to either of the first two scenarios.

A sherbet road surface means that the road surface, which is covered by snow, contains water or ice. Examples of sherbet road surfaces are shown in Fig. 14d, e. A normal road surface means that the road is flat, regardless of whether snow has fallen. Examples of normal road surfaces are shown in Fig. 14f, g.

Fig. 14 Three kinds of road surfaces. (**a**) Frozen road surface with ruggedness. (**b**) Frozen road surface with potholes. (**c**) Mirror-like frozen road surface. (**d**) Sherbet road surface with ruggedness. (**e**) Sherbet road surface with potholes. (**f**) Flat road surface with asfalto. (**g**) Flat road surface with snowfall

4.8.1 Feature Extraction and Selection

In addition to road conditions, the acceleration and gyro values depend on the car, the driver, the smartphone, and the way it is mounted. To reduce the number of variables, we consider the same car driven by the same driver using the same smartphone mounted in the same way. In this case, the different motions should only depend on road conditions and driving behavior. According to He [7], human activity often occurs at low frequency. Therefore, we may assume that driving behavior is also centralized at low frequency. Based on this assumption, features extracted from the spectral domain should be divided into two groups, high frequency and low frequency.

In the snowfall area, the variation in the road surface shapes is large, even under the same kinds of road conditions. The series of acceleration and gyro values from the road segments cannot assume fixed time series patterns. In contrast to the series of motion values, the statistical features such as the root-mean-square, standard deviation, and correlation may be more important for estimating the road surface conditions.

Features were extracted from each of the three axes of the accelerometer and gyroscope values using a 2-s sliding window. In addition, the mean speed over the sliding window is a feature. The extracted features were:

- Root-mean-square
- Standard deviation
- Correlation
- High-frequency energy
- Low-frequency energy
- Mean speed

The root-mean-square and the standard deviation can be extracted directly from the raw data of three axes of the accelerometer and gyroscope values. Correlation is calculated between each pair of axes as the ratio of the covariance and the product of the standard deviations.

Energy is the sum of the squared discrete FFT component magnitudes of the signal. If $x_1, x_2, \ldots x_n$ are the FFT components of an axis of the 2-s sliding window, the energy is defined as Eq. (1).

$$\text{Energy} = \frac{\sum_{i=1}^{n} |x_i|^2}{n} \tag{1}$$

We use a rate parameter r to divide the FFT components into high-frequency energy and low-frequency energy. In other words, high-frequency energy is defined as Eq. (2), and low-frequency energy is defined as Eq. (3).

$$\text{Energy}_H = \frac{\sum_{i=r \cdot n}^{n} |x_i|^2}{(1-r) \cdot n} \tag{2}$$

$$\text{Energy}_L = \frac{\sum_{i=1}^{r \cdot n} |x_i|^2}{r \cdot n} \tag{3}$$

In this paper, we specify the parameter r as 0.5.

The extracted features have a total of 31 dimensions. The PCA algorithm will be used to evaluate these dimensions. PCA is a mathematical algorithm that reduces the dimensionality of the data while retaining most of the variation in the dataset. It accomplishes this reduction by identifying directions, called principal components, along which the variation in the data is maximal.

4.8.2 Classification

In this paper, we use the SVM to classify the road conditions based on the one-versus-one (OVO) method. The SVM is one of the most popular classification methods in the field of machine learning. However, because the SVM was originally designed for binary classification, it cannot deal with multi-class classification directly. The multi-class classification problem is usually solved by decomposition of the problem into several two-class problems. In the OVO method, a set of binary classifiers is constructed using corresponding data from two classes. While testing, we used the voting strategy of "Max-Wins" to produce the output. Because the training dataset is relatively limited here, the generalization capability of the classifier is more important in terms of recognition. We used the leave-one-subject-out validation test to evaluate the ability of the classifiers to estimate the road conditions.

4.8.3 Experimental Results and Discussion

This section describes the experiment using the proposed road estimation method. In this experiment, the training data were generated using hand-labeling based on the drive recording videos. The total size of the training dataset is 1,364 items, including 129 frozen road surfaces, 235 sherbet road surfaces, and 1,000 normal road surfaces. Each of these data items is segmented into two-second lengths.

Figure 15 shows the classification results with different numbers of PCA components. We know that the first 21 components are sufficient to provide high accuracy. The definitions of precision, recall, and the F-measure are explained in Eqs. (4), (5), and (6). In addition, the means of TP, TN, FP, and FN are shown in Table 5. We can see that the estimation results for the frozen road surfaces in the different PCA components always maintain both high precision and high recall. However, the estimation results for the sherbet road surfaces are maintained at a relatively low level. A possible reason for this is that the amount of water in the sherbet road surface is excessive. In this case, the motion values for the sherbet road surface will be similar to those for the normal road surface.

This result shows that the proposed method can estimate the frozen road surface condition effectively. This means that it is possible to use motion values to estimate the condition of a frozen road surface.

$$\mathrm{Precision} = \frac{\mathrm{TP}}{\mathrm{TP} + \mathrm{FP}} \tag{4}$$

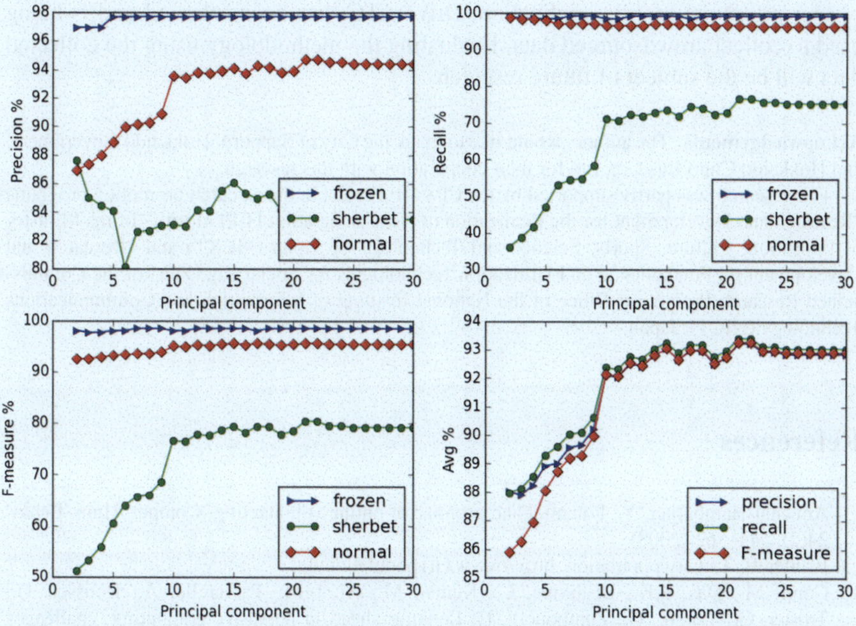

Fig. 15 Experimental results using different numbers of PCA components

Table 5 Confusion matrix

	Actual positive	Actual negative
Predicted positive	TP	FP
Predicted negative	FN	TN

$$\text{Recall} = \frac{\text{TP}}{\text{TP} + \text{FN}} \tag{5}$$

$$\text{F} - \text{measure} = \frac{2 \ \text{Recall} \cdot \text{Precision}}{\text{Recall} + \text{Precision}} \tag{6}$$

5 Conclusion

This paper provides an overview of an ongoing project that aims to develop a mobile sensing framework to collect sensor data reflecting personal-scale, or microscopic, roadside phenomena using crowdsourcing as well as big data, such as traffic and climate conditions, and the contents of social networking services such as Twitter.

To make this framework effective, it is important that the system is able to deal with a large volume of data reflecting the daily lives of citizens. Therefore, we propose a service model that involves citizens.

Our prototype mobile application, Drive ATC, has been released and is being used to collect crowdsourced data. Evaluating the methodology using the collected data will be the subject of future research.

Acknowledgements The authors would like to thank the City of Sapporo, Hokkaido Government, and Hokkaido Chuo Bus Co., Ltd for their cooperation with this research.

This research was partly supported by the CPS-IIP Project in the research promotion programs "Research and Development for the Realization of Next-Generation IT Platforms" of the Ministry of Education, Culture, Sports, Science and Technology of Japan (MEXT) and "Research and Development on Fundamental and Utilization Technologies for Social Big Data" of the Commissioned Research Promotion Office of the National Institute of Information and Communications Technology (NICT), Japan.

References

1. Amichai-Hamburger, Y.: Potential and promise of online volunteering. Comput. Hum. Behav. **24**(2), 544–562 (2008)
2. Brinkhoff, T.: City population. http://www.citypopulation.de/
3. Conti, M., Das, S.K., Bisdikian, C., Kumar, M., Ni, L.M., Passarella, A., Roussos, G., Tröster, G., Tsudik, G., Zambonelli, F.: Looking ahead in pervasive computing: challenges and opportunities in the era of cyber-physical convergence. Pervasive Mob. Comput. **8**(1), 2–21 (2012). doi:10.1016/j.pmcj.2011.10.001. http://www.sciencedirect.com/science/article/pii/S1574119211001271
4. Current Results Publishing Ltd.: Annual average snowfall for cities in the United States. http://www.currentresults.com/Weather/US/annual-snowfall-by-city.php
5. Feng, F.: Winter road surface condition estimation and forecasting. Ph.D. thesis, University of Waterloo (2013)
6. Grossi, R., Gupta, A., Vitter, J.S.: High-order entropy-compressed text indexes. In: Proceedings of the 14th Annual ACM-SIAM Symposium on Discrete Algorithms (SODA), pp. 841–850. Society for Industrial and Applied Mathematics, Philadelphia (2003). http://dl.acm.org/citation.cfm?id=644108.644250
7. He, Z., Jin, L.: Activity recognition from acceleration data based on discrete cosine transform and SVM. In: The 2009 IEEE International Conference on Systems, Man, and Cybernetics, pp. 5041–5044 (2009). http://ieeexplore.ieee.org/xpl/downloadCitations
8. Honda Motor Co., L.: A traffic safety map made by everyone. http://world.honda.com/safety/hearts/2013/03/index.html
9. Howe, J.: Crowdsourcing: Rise of the Amateur. (2006) http://www.crowdsourcing.com/cs/2008/02/chapter-two-ris.html
10. Howe, J.: The rise of crowdsourcing. Wired Mag. **14**(6), 1–4 (2006)
11. Howe, J: Crowdsourcing: How the Power of the Crowd is Driving the Future of Business, Random House Business, New York (2009).
12. Japan Meteorological Agency: Open data. http://www.jma.go.jp/jma/menu/report.html
13. King, S.F., Brown, P.: Fix my street or else: using the internet to voice local public service concerns. In: Proceedings of the 1st International Conference on Theory and Practice of Electronic Governance, pp. 72–80 (2007). doi:10.1145/1328057.1328076, http://doi.acm.org/10.1145/1328057.1328076
14. Kinoshita, A., Takasu, A., Adachi, J.: Traffic incident detection using probabilistic topic model. In: the Workshop Proceedings of the EDBT/ICDT 2014 Joint Conference, pp. 323–330 (2014). http://ceur-ws.org/Vol-1133/paper-52.pdf

15. Poovendran, R.: Cyber-physical systems: close encounters between two parallel worlds. Proc. IEEE **98**(8), 1363–1366 (2010) doi:10.1109/JPROC.2010.2050377
16. Schuurman, D., Baccarne, B., De Marez, L., Mechant, P.: Smart ideas for smart cities: investigating crowdsourcing for generating and selecting ideas for ICT innovation in a city context. J. Theor. Appl. Electron. Commer. Res. **7**(3), 49–62 (2012)
17. Stembert, N., Mulder, I.J.: Love your city! an interactive platform empowering citizens to turn the public domain into a participatory domain. In: International Conference Using ICT, Social Media and Mobile Technologies to Foster Self-Organisation in Urban and Neighbourhood Governance (2013). http://resolver.tudelft.nl/uuid:23c4488b-09e1-4b90-85e3-143e4a144215
18. Yamada, M., Ueda, K., Horiba, I., Tsugawa, S., Yamamoto, S.: A study of the road surface condition detection technique based on the image information for deployment on a vehicle. IEEJ Trans. Electron. Inf. Syst. **124**(3), 753–760 (2004). doi:10.1541/ieejeiss.124.753

Sensing and Visualization in Agriculture with Affordable Smart Devices

Takashi Okayasu, Andri Prima Nugroho, Daisaku Arita, Takashi Yoshinaga, Yoshiki Hashimoto, and Rin-ichiro Tachiguchi

1 Introduction

Agriculture is a highly complex system that depends on the climate, weather, soil conditions, plant types, and so on. Thus, farmers have attempted to modify cultivation techniques to fit the ambient weather conditions and geographical factors using

T. Okayasu (✉)
Faculty of Agriculture, Department of Agro-environmental Sciences, Kyushu University, 6-10-1, Hakozaki, Higashi, Fukuoka 812-8581, Japan
e-mail: okayasu@bpes.kyushu-u.ac.jp

A.P. Nugroho
Faculty of Agricultural Technology, Department of Agricultural and Biosystems Engineering, Universitas Gadjah Mada, Jl. Flora No 1 Bulaksumur, Yogyakarta 55281, Indonesia
e-mail: andrew@ugm.ac.id

D. Arita
Faculty of Information Systems, Department of Information Systems, University of Nagasaki, 1-1-1, Manabino, Nagayo, Nishisonogi, Nagasaki 851-2195, Japan

Institute of Systems, Information Technologies and Nanotechnologies, 2-1-22, Momochihama, Sawara, Fukuoka 814-0001, Japan
e-mail: arita@sun.ac.jp

T. Yoshinaga
Institute of Systems, Information Technologies and Nanotechnologies, 2-1-22, Momochihama, Sawara, Fukuoka 814-0001, Japan
e-mail: yoshinaga@isit.or.jp

Y. Hashimoto
Department of Advanced Information Technology, Graduate School of Information Science and Electrical Engineering, Kyushu University, 744 Motooka Nishi, Fukuoka 819-0395, Japan
e-mail: hashimoto@limu.ait.kyushu-u.ac.jp

R. Tachiguchi
Faculty of Information Science and Electrical Engineering, Department of Advanced Information Technology, Kyushu University, 744 Motooka Nishi, Fukuoka 819-0395, Japan
e-mail: rin@kyudai.ac.jp

H. Yasuura et al. (eds.), *Smart Sensors at the IoT Frontier*,
DOI 10.1007/978-3-319-55345-0_12

long-standing experience. Over the past century, agricultural machinery such as tractors, planters, and harvesters, as well as agricultural equipment in greenhouses, has developed considerably in support of farm work. These developments have evidently contributed to an improvement in productivity as well as a reduction in working time. Further, in Japanese agriculture, the average age of the core people engaged in farming now exceeds 65 years. This presents a very serious problem, because their knowledge and techniques are likely to disappear in the near future. Therefore, it is vital to consider how to collect this information and use it to both improve agricultural practice in Japan and train new farmers. Additionally, it should be recognized that consumer demands are shifting to fresh, high-quality, and high-security fruits and vegetables. In response to these issues, various research and development that takes advantage of new technologies and services has been performed. In particular, these studies are expected to advance the use of Information and Communication Technologies (ICT) in Japanese agriculture.

To date, ICT has been applied to improve agricultural productivity and cultivation skills in several fields of agriculture, e.g., precision agriculture including spatial data collection, precision irrigation and data supply, facility automation including greenhouse control, and animal-feeding facilities [26]. In addition, various affordable devices such as low-price microcomputers, sensors, and open-source software are being developed. These technological advances have changed conventional views on the use of ICT in agriculture. For instance, the reduction in cost of ICT devices means that small- and medium-scale farmers have been able to introduce field monitoring and farm management systems. Various trials and challenges on ICT-based faming have started all over Japan. The authors have also developed and released several devices and software to support agricultural production using affordable devices and open-source software, because our main focus is the advancement of small- and medium-scale farms [5–7, 19–21].

The effectiveness and possibilities of ICT in agriculture are introduced with reference to the affordable smart devices and services developed by the authors. In particular, this chapter discusses a framework for environmental monitoring and control, plant image measurement to evaluate growth and environmental stress responses, and the recording of farm work to optimize operations.

2 Field Environmental Monitoring and Control Framework

Long-term field monitoring of factors such as air temperature, humidity, and solar radiation is important for managing plant growth, pests, and disease, as well as for optimizing cultivation procedures. Further, under the likely effects of global warming and frequency of abnormal weather, it is increasingly important to establish sustainable agriculture.

Various field monitoring systems have been developed by many researchers. Among them, Hirafuji and Fukatsu [2, 8] have proposed a Web-based field monitoring system called "Fieldserver," which is equipped with sensors to measure air temperature and humidity, solar radiation, and other variables relevant to agriculture and includes switches to control heaters and water sprinklers. It is possible to construct a sensor network by connecting a number of Fieldserver systems under the de facto standard network protocol (i.e., TCP/IP). No special software or additional hardware is needed to collect the monitoring data and actuate the various tools installed on the server. All of the commands are fully transferred by typical Web browsers. Based on Global System of Mobile (GSM) communication, Jiang et al. developed a wireless automatic monitoring system that records both environmental variations and pest population dynamics [11]. This monitoring system also provides monitoring data to users through a Web-based application. The authors have developed a simple field monitoring system (FMS) for agricultural production and management using the Fieldserver technology [5]. However, an agent program that collects the measured data is needed at each monitoring site, because the monitoring system does not include a function for the self-transfer of data.

To monitor and control the field environment to achieve suitable agricultural production, a monitoring and control framework [19] was developed based on the client–server architecture shown in Fig. 1. The framework is composed of environmental monitoring nodes as local management subsystems and Web applications as the global management subsystem, which conduct several communication and data exchange functions via the Internet. Details of this framework are explained below.

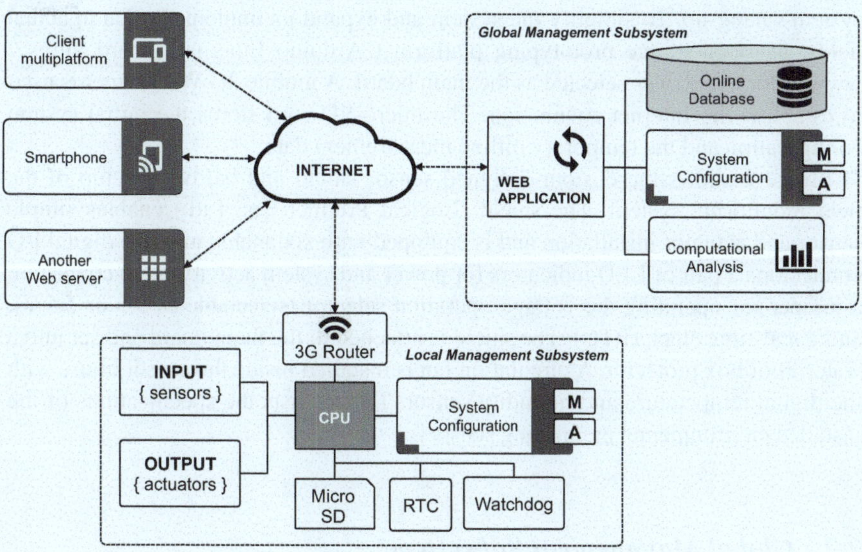

Fig. 1 Architecture of field environmental monitoring and control framework

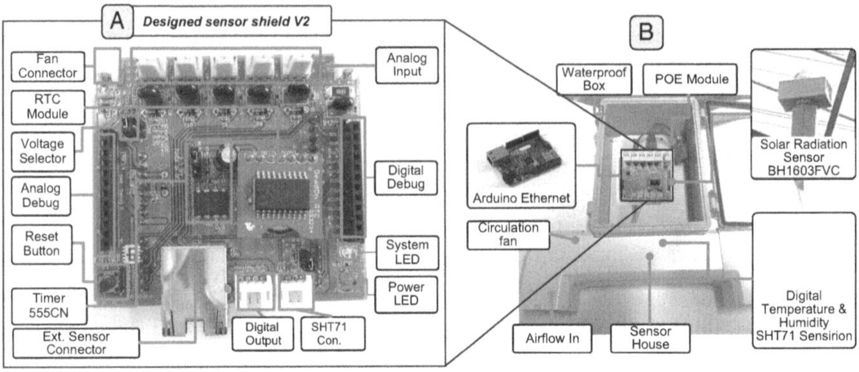

Fig. 2 Custom-designed sensor shield (**a**) and hardware setup of the monitoring system (**b**)

2.1 Local Management Subsystem

The local management subsystems work as "on-field" devices to monitor and control the field environment. These can operate as both the client and local manager depending on the availability of an Internet connection. The components are a CPU, analog/digital inputs for environmental sensors, digital outputs for actuators, additional modules such as a micro-SD card, and a watchdog timer to eliminate system's hang-up. To simplify the system and expand its implementation in actual fields, an open-source prototyping platform ("Arduino Ethernet" board, https://www.arduino.cc/) was selected as the main board. A mobile 3G WiFi router is used to establish the Internet connection. The micro-SD card stores the initial system configuration and the temporary offline measurement data.

Figure 2 shows the custom-designed sensor shield and hardware setup of the field monitoring system. The shield (Logical Product Co., Ltd.) enables simple sensor and actuator installation and is equipped with six analog and two digital I/O connectors, a pair of LED indicators for power and system activity, a fan connector, a jumper for operating the voltage operation selector (selectable as 3.3 or 5.5 V), and a real-time clock (RTC). The shield is attached to the main board and set into a waterproof box protector. A circulation fan is installed inside the sensor house with the digital temperature and humidity sensor. Table 1 lists the specifications of the installed environmental monitoring sensors.

2.2 Global Management Subsystem

The global management subsystem manages the overall system through the Internet. This is provided as Web-based applications running on an Apache Web server and MySQL database server, which are programmed by PHP and JavaScript. The

Table 1 Specifications of the installed environmental monitoring sensors

Sensor type	Hardware specification
Temperature humidity sensor	SHT71 Sensirion
	V input: 2.4–5.5 V
	Temperature accuracy: ±0.4 °C
	Humidity accuracy: ± 3.0% RH (20%–80% RH)
	Humidity range: 0%–100%
Solar radiation sensor	BH1603FVC
	V input (max.): 7 V
Soil moisture content sensor	WD-3-W-5E
	V input: DC +4.5–+15 V
	Accuracy: ±5% F.S (VWC 0%–50%)
	±15% F.S (VWC 50%–100%)

interaction between the local management subsystems and the global management subsystem is established by API programs based on HTTP, which allows for effective field measurement and control, data provision, and system management.

Figure 3 shows the Web interface for local subsystem management provided by the global management subsystem. System management is an important role in the implementation of a cloud-based environmental measurement and control system. The configuration parameters for the local management subsystem in the micro-SD card are synchronized at specified time intervals with those stored in the global management subsystem, which can be managed by using the Web interface. Thus, the local management subsystem maintains the latest working conditions.

2.3 *Application of Framework to Field Environmental Monitoring and Irrigation Control*

Irrigation is an important part of plant cultivation. The framework described above can be applied to control drip irrigation. Also known as trickle irrigation or micro-irrigation, this technique reduces water consumption by slowly supplying water to the root of the plant. The control system uses data from a soil moisture content sensor as a reference for controlling the actuators. If the soil moisture content is lower than some minimum value, the control system opens the water valve to supply water for a set duration. The water flows onto the soil surface or directly to the root zone through a network of valves, pipes, tubing, and emitters. Several scenarios were tested with and without a network connection.

Figure 4 shows the drip irrigation setup for tomato (*Solanum lycopersicum*) cultivation in an experimental greenhouse at Kyushu University. The experiment commenced in June 2015. The soil moisture sensor was installed 5 cm below the soil surface to monitor the soil moisture content in the root zone of the tomato

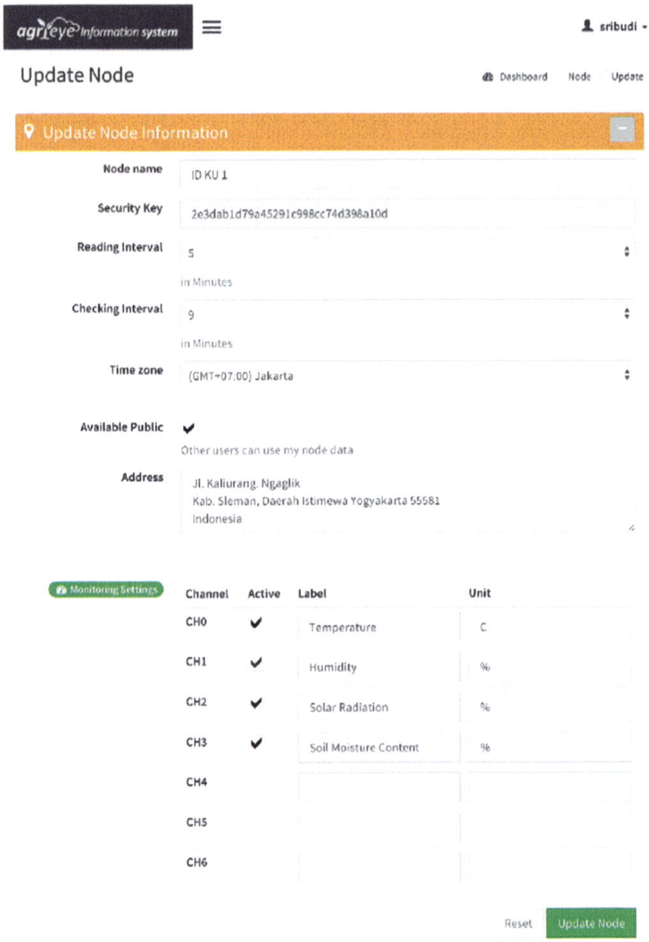

Fig. 3 Web interface for local subsystem management

plants. The output signal port of the system was connected to a relay unit with a solenoid valve. Irrigation water was supplied to each plant pot from the water tank. The specifications of the irrigation system are listed in Table 2. During the experiment, the system also measured the air temperature, relative humidity, and solar radiation.

The response and error in the irrigation control were evaluated under several irrigation scenarios (see Table 3). Increasing the duration of water supply is expected to increase the soil moisture content. The errors E_U and E_L are given by

$$\begin{cases} E_U = s_t - s_{\max} \\ E_L = s_{\min} - s_t \end{cases}, \tag{1}$$

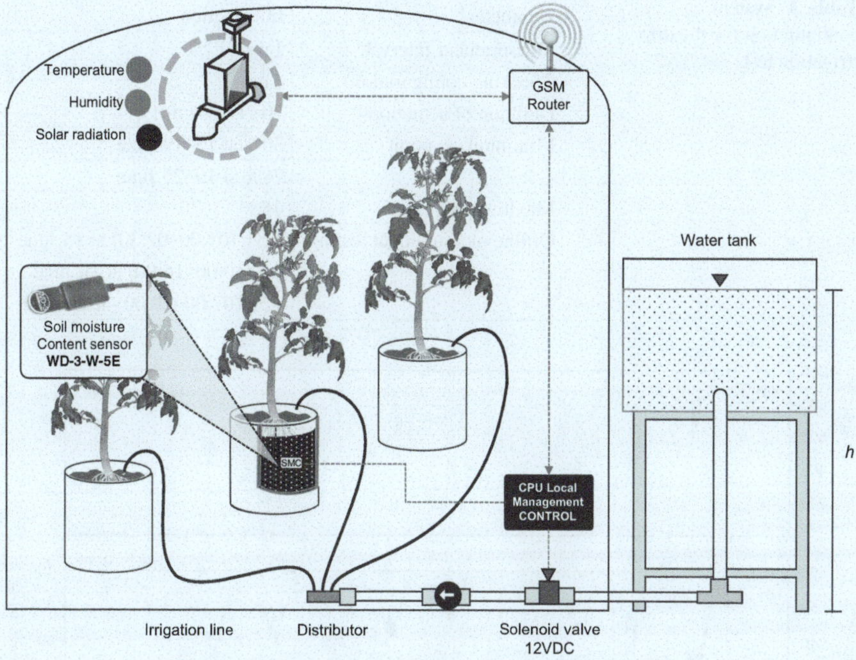

Fig. 4 Irrigation control setup for tomato cultivation

Table 2 Specification of the drip irrigation system

Parameter	Information
Flow rate	3.59 cc/sec
Irrigation type	Gravity flow drip irrigation system
Controller type	Soil moisture sensor
Soil moisture content sensor	WD-3-W-5E (ARP Co., Ltd., Japan)
Number of emitter	12
Height of tank	1 m
Diameter of main line	1.5 cm
Diameter of branch irrigation line	0.4 cm
Length of irrigation line	1 m

where s_t is the soil moisture content measured at time t. $s_{\max}$ and $s_{\min}$ are the maximum and minimum values of soil moisture content, at which points the water supply is controlled by switching the irrigation system off or on.

Figure 5 shows the environmental data and experimental results from automatic drip irrigation with specified testing scenarios over a 10-day observation period. The air temperature ranged from 18.7 to 33.1 °C with an average of 22.7 °C, as displayed by the solid line. The relative humidity varied from 42% to 96%, denoted by the dotted line. The solar radiation attained a maximum value of 890 W/m^2 (solid gray

Table 3 System configuration for the drip irrigation test

Parameter	Information
Measurement interval	1 min
Checking configuration	4 min
Duration of actuation	70 s (1.16 min)
Minimum set point	30% at 15–19 June
	28% at 19–25 June
Maximum set point	45%
Offline measurement testing	1.5 h (01:00–02:30) at 15 June
	6 h (10:00–16:00) at 16 June
	12 h (07:00–19:00) at 22 June

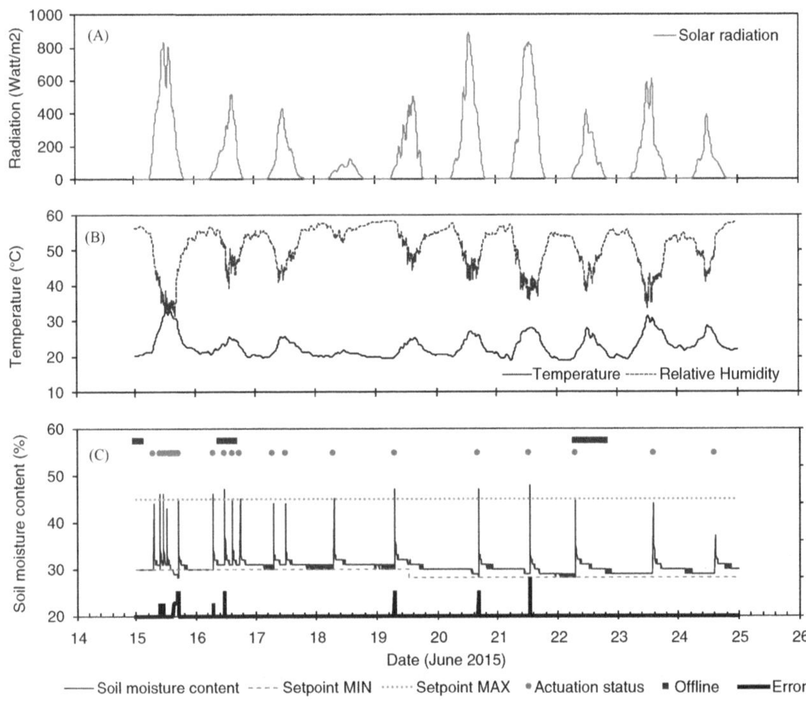

Fig. 5 Environmental data and experimental result of automatic drip irrigation control

line). The error between the monitoring value and the set maximum and minimum values is shown by black rounded markers in the lower chart. In this experiment, the entire irrigation event could be performed according to the scenarios considered.

On 19 June, the minimum moisture content was changed from 30% to 28% via the Web application in the global management subsystem. The soil moisture content was maintained according to the new minimum set point. In real horticultural farming, the minimum and maximum set points must be carefully determined, because the plant water requirements change dynamically with the plant growth stage and environmental condition.

Finally, the offline management was also tested. The offline condition was realized by disconnecting the connection between the local and global management subsystems. In this condition, the irrigation control again performed appropriately.

3 Plant Growth and Motion Measurement

Plant growth, color, and shape are influenced by the weather, soil type, and nutritional factors. Researchers have attempted to extract characteristic values from plants during the cultivation process. Red–green–blue (RGB) and near-infrared (NIR) images are frequently used to measure the leaf area index and plant height. Hyperspectral and multispectral imaging techniques have been adopted to extract leaf color features with the aim of estimating the plant canopy and the effect and impact of fertilizer [3, 24]. Recently, there has been a sharp increase in research on plant phenotypes, focusing on the comprehensive assessment of complex plant features (growth, tolerance, resistance, structure, physical property, and yield) [22]. However, the speed and resolution of the measurement and analysis have many limitations [14]. High-throughput plant phenotyping is being extensively studied in the US and EU countries, resulting in precise and large-scale measurements and analyses on plant phenotypes. This research is largely in response to concerns about food and biomass production under drastic climate change and the increase in the global population. We believe that high-throughput plant phenotyping will be a key technology in developing new cultivar and sustainable agriculture. However, the cost of this research is very high, and thus the development of low-cost measurement systems based on affordable devices is desirable. In this section, two applications are introduced that use affordable image-capture systems to estimate plant growth and identify plant motion based on optical flow.

3.1 *Plant Growth Measurement*

Figure 6 shows the field environmental monitoring system for plant growth measurement. The monitoring system is composed of a microcomputer (Raspberry Pi Type B+), a five-megapixel RGB camera (2592×1944 pixels, Raspberry Pi Camera Board 775–7731, Raspberry Pi Foundation), an air temperature and humidity sensor (SHT-25, Sensirion), an illuminance sensor (AEH11, Holly & Co., Ltd.), and a 3G WiFi router as a data transmitter. The air temperature and humidity sensor is placed in a ventilator and the illuminance sensor and camera are installed in a waterproof box, as shown in Fig. 6a. The monitoring device and the data transmitter are connected using an RJ45 network cable. The environmental data and plant images are sent to the database via an affordable 3G network (250 kbps, ServersMan SIM, Tone

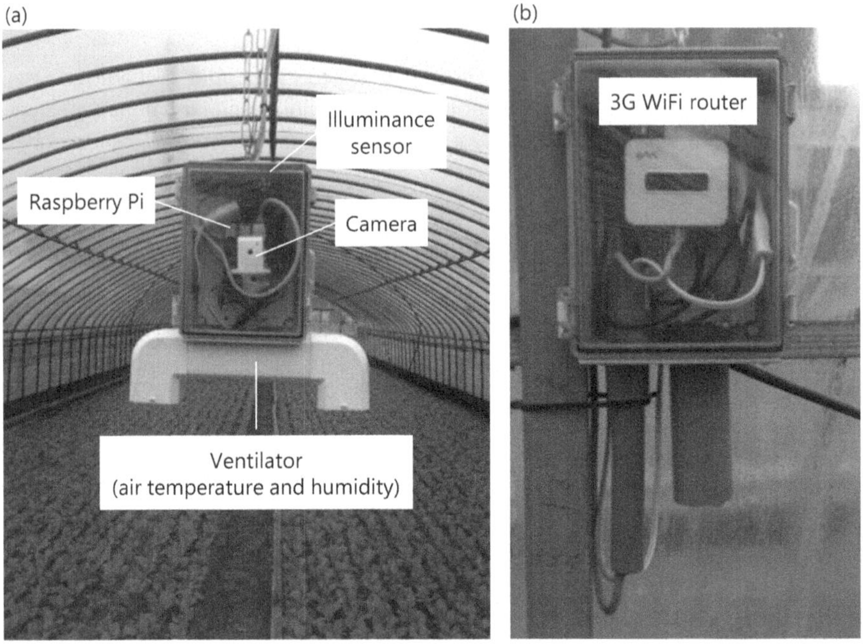

Fig. 6 Field environmental monitoring system for plant growth measurement. (**a**) Monitoring node. (**b**) 3G WiFi box

mobile Inc.) to reduce the management and running costs. All the environmental information and images stored in the database can be accessed using a Web browser on any PCs or smartphones.

Figure 7 illustrates the method of measuring plant growth characteristics. A plastic ruler made of black and white square markers is used to measure plant height. The actual height is directly calculated from the segments of the ruler in view.

A feasibility study for a plant growth prediction method from time-lapse images of the plant was conducted in a Komatsuna (*Brassica rapa var. perviridis*) greenhouse. The feasibility test commenced on 24 Dec., 2015. The air temperature, humidity, and illuminance inside the greenhouse were measured every 5 min, and the plant was photographed every hour by the monitoring system. The resolution of the captured images was reduced to 1440 × 960 pixels to enable efficient transmission over the 3G network. The plant growth characteristics were calculated from the plant image and recorded against the measured environmental information.

Figure 8 shows the plant images at different dates. Farmers often check plant size using the packing film for sale, but such information is not recorded. The plant images contain important information for checking the size of plants to be sold, as well as for evaluating the plant growth stage and situation. In this study, the plant heights were directly calculated using the ruler in each image.

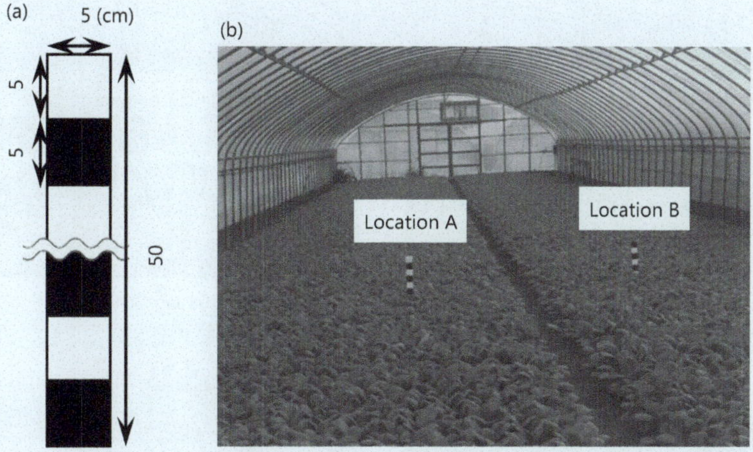

Fig. 7 Growth measurement method. (**a**) Plastic ruler. (**b**) Set up in ground

Figure 9 compares the plant height with the accumulated mean air temperature at the two locations described in Fig. 7. The mean air temperature for that day was calculated at midnight and used as a parameter to determine plant growth. The relation between plant height and accumulated mean air temperature exhibits a clear linear correlation, although the relation is slightly different in each location because of individual or environmental differences. The plant height distribution can be obtained by increasing the number of rulers. However, the current result is sufficiently accurate to estimate a short-term harvesting date in leafy plants' production.

3.2 Plant Motion Measurement

Like other living things, plants display regular motion with a constant period of approximately 24 h, called the circadian rhythm. The internal activities of plants such as the stomatal aperture, flower-bud formation, and growth are triggered by the circadian clock, which helps to adapt the organism to environmental changes in light and temperature. Ever since it provided the first evidence for the existence of circadian rhythm, the physical indicator of plant motion has been used to investigate the clock activity of plants, even under constant environmental conditions [13]. Continuous time-lapse photography has been used to establish effective and efficient leaf motion measurements [1]. Most studies on image-based leaf motion analysis have monitored individual leafy plants to determine their motion in the vertical direction from a side-view projection. Under this approach, it is difficult to estimate the leaf movement of mature plants.

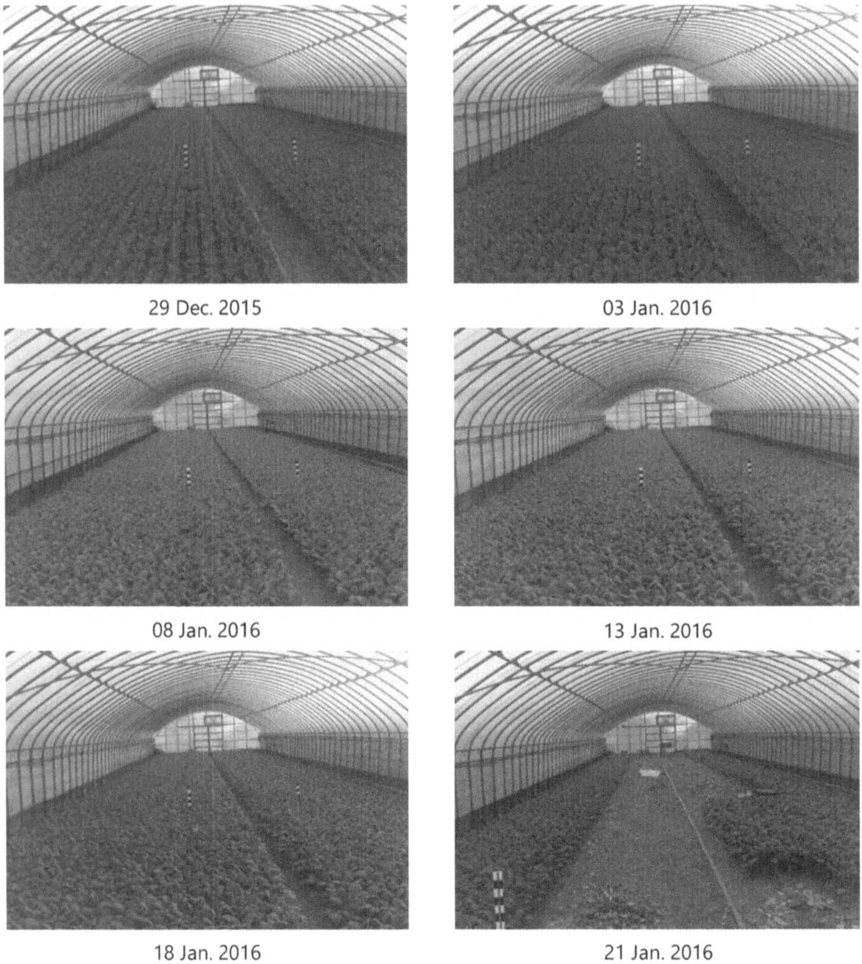

Fig. 8 Plant growth images at different dates

Figure 10 shows a schematic diagram of the automatic plant image-capture system. The system uses a Raspberry Pi B+ as the main computer and includes an infrared camera (Pi NoIR, Raspberry Pi Foundation), a pair of infrared LED light modules to support night vision, and a USB webcam (1280 × 720 pixels, BSW20K07HBK, Buffalo Inc.) to capture RGB images. An environmental monitoring system was also installed to measure the air temperature, humidity, and illuminance. All captured images and environmental data were transferred automatically to the local server via the WiFi connection. The plant motion calculation was implemented using Python and OpenCV.

The plant motion is calculated from the optical flow based on the Pyramid Lukas–Kanade method [12], which is a motion-tracking technique for analyzing the motion

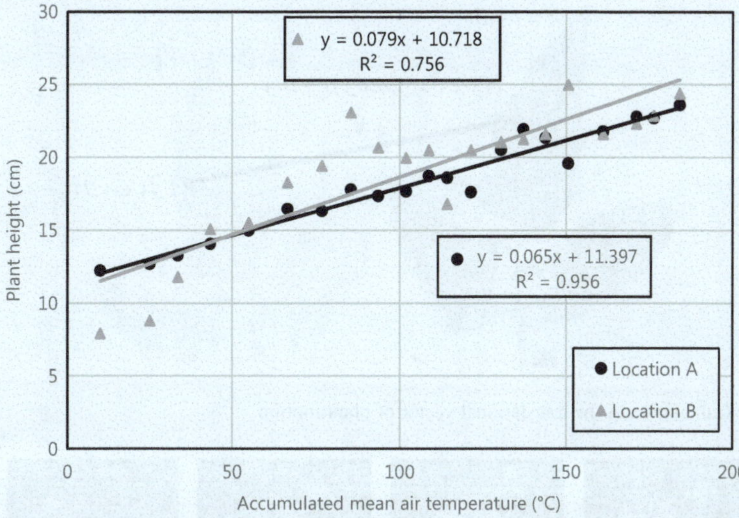

Fig. 9 Comparison between plant height and accumulated mean air temperature at two locations

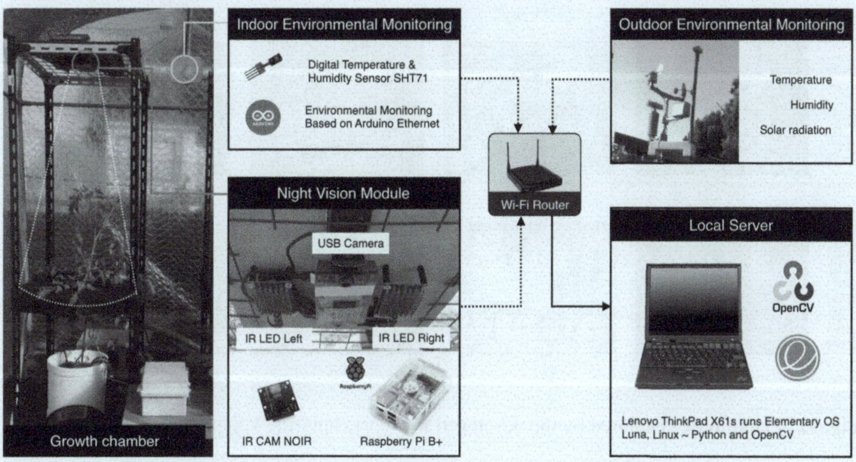

Fig. 10 Schematic diagram of the automatic plant image-capture system

between two consecutive grayscale images. Figure 11 illustrates the process of calculating the translational motion of the plant, where the brightness of an arbitrary point (x, y) at time t and $t+\delta t$ are defined as $I(x, y, t)$ and $I(x+\delta x, y+\delta y, t+\delta t)$. The Shi–Tomasi corner detection algorithm [23] is adopted to obtain suitable tracking points in the optical flow.

A plant motion measurement experiment was conducted in the tomato greenhouse at Kyushu University. The tomato plant had been planted about 30 days prior to the start of the experiment and measured 120 cm in height. The image-capture

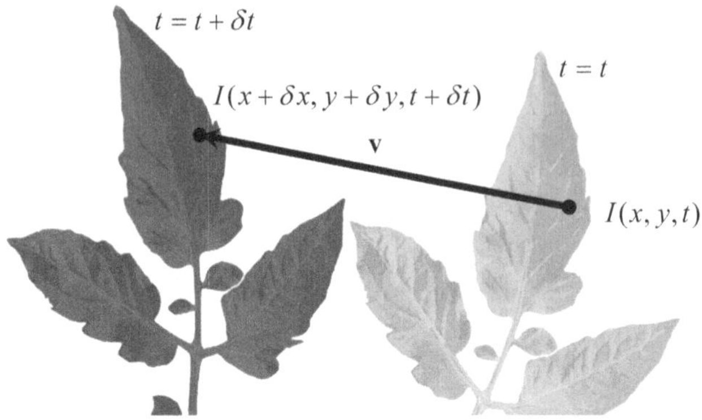

Fig. 11 Calculation of the translational vector of plant motion

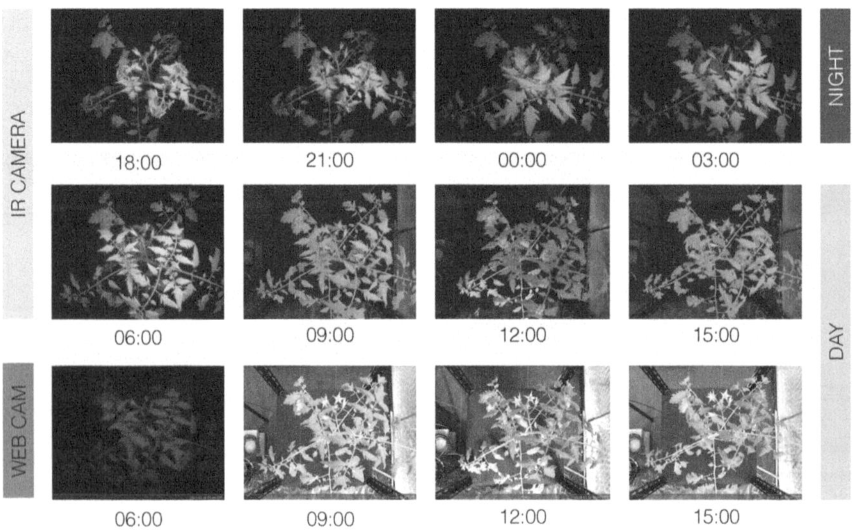

Fig. 12 Plant images captured by the developed low-cost capturing system

system was arranged at the top of the tomato plant to capture day and night time-lapse images every 30 min from 16 to 26 Nov., 2015. Figure 12 shows the day and night tomato plant images captured by the developed system. Color images can be used to observe plant growth, water stress, and damage by calculating the percentage of green area over time. Images from the IR camera are suitable for extracting plant motion under both day and night conditions. Table 4 gives the input parameters for the Pyramid Lukas–Kanade method and the Shi–Tomasi corner detection algorithm for the optical flow.

Figure 13 shows the raw images used as the input for motion estimation and the calculated plant motion obtained by the Lucas–Kanade method. The visualized

Table 4 Input parameters for the optical flow

Aspect	Information
Capturing time interval	30 min.
Lukas–Kanade parameter	
Window size	20 px.
Pyramid level	3
Block size	2
Shi-Tomasi parameter	
Maximum number of corner	500 pts.
Minimum distance	5 px.
Quality level	0.01

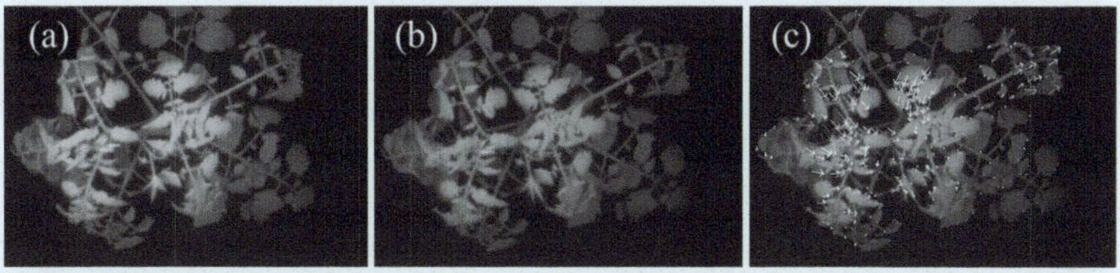

Fig. 13 Captured images at different times (**a**) and (**b**), plant motion calculated by the Lucas–Kanade method (**c**)

results clearly show that the leaf motion of the tomato plant can be extracted (length of the arrowed lines represents the magnitude of the translational motion). This confirms that the plant motion can be obtained using the optical flow.

Figure 14 displays the change in the translation motion of the tomato plant. The mean value of the magnitude of the translational vector is given by

$$\bar{v} = \frac{1}{\bar{n}} \sum_{i=1}^{\bar{n}} ||\bar{v}_i||, \tag{2}$$

where $\bar{n}$ is the number of calculated translational vectors at time t. The calculated translational vector v generally contains some errors. Here, we adopted a 2-h moving average for the calculated vectors to clarify the plant motion. The trend and behavior of translational motion displayed a regular pattern resembling the diurnal cycle. A Fourier transform of the plant motion suggests a 24.55-h cycle. The circadian rhythm of the tomato plant can be clearly measured using the affordable image-capture system. In future work, we will investigate the influence on plant growth and behavior of stress caused by changes in the ambient environment.

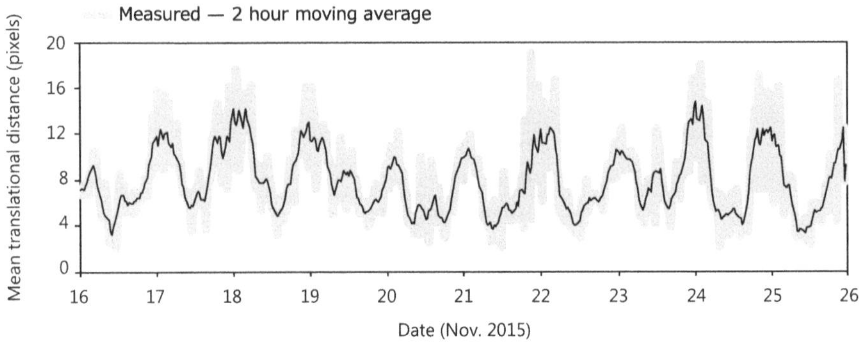

Fig. 14 Change in the translational motion of the tomato plant

4 Farm Work Information Recording

Recording the work conducted on a farm is an important task, as it is not only used to find and solve current problems by referring to previous farm work information but also enables whole work processes and flows to be checked and controlled in order to stabilize farm management. Some farmers record farm work information in notebooks to improve their own knowledge and skills. However, handwritten farm work information cannot always identify which information is suitable for a certain purpose, and such information cannot easily be shared with other farmers. In Japanese agriculture, the average age of the core people engaged in farming exceeds 65 years. Thus, establishing a sustainable system of agriculture is a serious problem, because the expertise and skills of experienced farmers are not being handed down to young or new farmers. Hence, the collection of current expertise and skills for training new farmers is of vital importance. Various studies on the collection of farm work information have been performed. Guan et al. [4] developed a work recording application by which farmers can input information on devices such as mobile phones or personal digital assistants. Murakami et al. [17] and Okayasu et al. [20] proposed work recording systems using barcodes and QR codes. Nanseki et al. [18] developed a recording system using Radio Frequency Identification (RFID) tags, allowing farm work information to be recorded by simply holding an RFID reader to the tags. However, the arrangement of tags in each field must be considered. In this section, a manual farm work information recording system and an automatic farm work recording system using smart devices are introduced.

4.1 Manual Farm Work Information Recording System

Figure 15 shows an overview of the manual farm work recording application. This was developed as a Web-based application using devices such as PCs and mobile

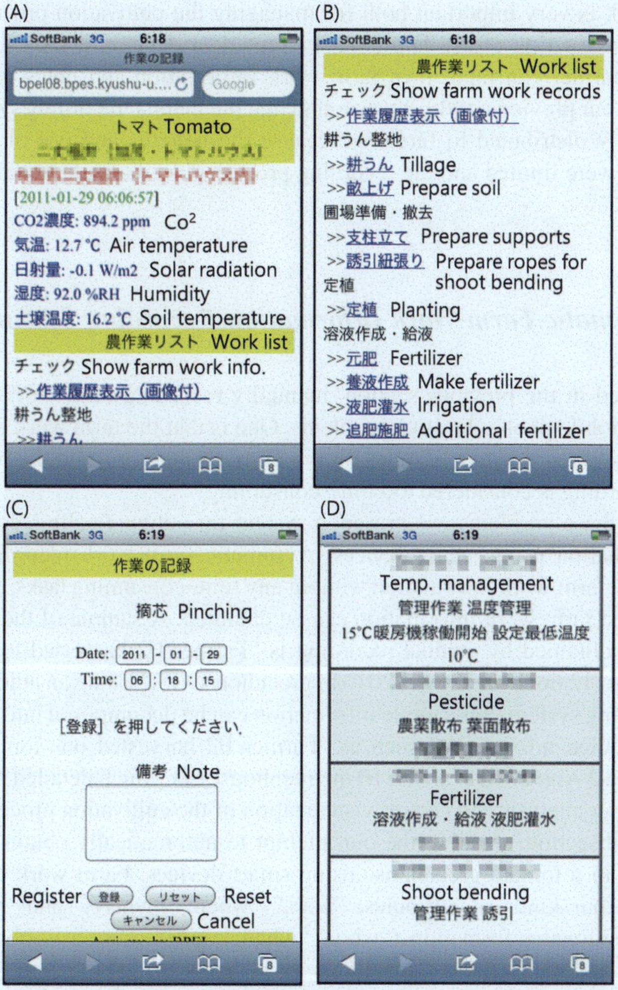

Fig. 15 Overview of manual farm work recording application. (**a**) Current environmental data. (**b**) List of farm work. (**c**) Registration window. (**d**) List of farm work information

devices. The current field environmental information measured by the monitoring system mentioned in Sect. 2 is displayed in the top of the application window. The optimal farm work can be determined and performed considering the environmental information and the farmer's experience and intuition. The farm work is listed below the field environmental information. By selecting an objective item of farm work from the list, the farm work information can be recorded. Additional information and notes for the selected item of farm work are also recorded in the registration window. The recorded farm work information is also displayed. Farmers can add new items of farm work and fields as necessary using the system. The recording of farm work is very important both for managing the cultivation process and for estimating production costs. Thus, we attempted to develop a recording system so that the farmer can input own farm work information and notes more easily. However, in our previous study, it was found that the manual recording system could not be widely distributed to farmers, because the benefits and use of farm work information were limited and the recording process placed an additional burden on the farmers.

4.2 *Automatic Farm Work Information Recording System*

As mentioned in the previous section, manually recording farm work with notebooks and mobile phones has two problems. One is that the farm work information is not precisely recorded, as it relies on the farmer's memory, and the other is that manual recording is considered too time-consuming.

To solve these problems, we present a system for automatically recording farm work information using smart devices. Automatic farm work recording obtains more precise farm work information without any time-consuming tasks. In addition, more detailed farm work information can be obtained. A sample of the farm work information obtained by manual recording is "Farmer F2 harvested tomato fruits from 8 a.m. to 12 noon on 12 Apr. 2016 in greenhouse G3." Using an automatic farm work recording system, this sample information can be decomposed into a sequence of more detailed information, such as "Farmer F2 harvested one tomato fruit at 08:15:22 on 12 Apr. 2016 at area A10 in greenhouse G3." Such detailed information enables deeper analysis and clearer visualization of the cultivation process.

In this subsection, we describe our attempt to automatically obtain farm work information in a tomato greenhouse using smart devices. Farm work information consists of four kinds of attributes: "who," "when," "where," and "what." The farmer's positional information ("who," "when," and "where") is obtained by a smartphone and small radio transmitters or beacons [6]. The action information ("who," "when," and "what") is obtained by smartwatches [7]. Once obtained, the positional and action information is combined into farm work information by considering "when" and "who" as common attributes.

4.2.1 Farmer Position Information

To obtain the farmer's position information, we used beacons placed in a tomato greenhouse, which broadcast their own IDs via Bluetooth, and a smartphone carried by the farmer in the greenhouse as a Bluetooth receiver. By using a smartphone application, we can obtain Received Signal Strength Indicators (RSSIs) from each beacon. Based on the positions and RSSIs of the beacons, we can estimate the farmer's position.

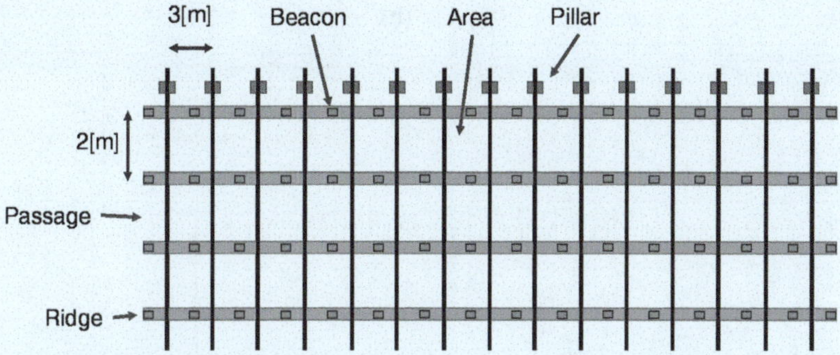

Fig. 16 Beacon placement in a tomato greenhouse

Our position estimation method is area-based estimation. As shown in Fig. 16, we divide each passage into small areas based on the pillars that support the greenhouse, define an X-axis and a Y-axis, and place multiple beacons on the ridges. In this environment, we measure the RSSIs from beacons and estimate the farmer's position, i.e., which area the farmer is in.

Our position estimation method consists of three steps:

1. select two beacons from different ridges whose RSSIs are higher than the others, and select an area that includes the middle point between the two beacons as the farmer's primary position,
2. replace the X-axis position of the farmer with the mode of X-axis positions around the current time to reduce X-coordinate spike noise, and
3. replace the Y-axis position of the farmer with the mode of Y-axis positions from when the farmer enters one end of the passage until the farmer arrives at the other end, i.e., while the farmer is in a single passage, according to the restriction that a farmer in the middle of one passage cannot move to another passage, as the tomato plants are too high to jump over.

For further details of this method, please refer to [6].

Figure 17 shows an example of the farmer position estimation results. The solid lines denote the estimated position and the dashed lines show the manually obtained ground truth. It is clear that these two lines almost overlap. The accuracy of estimation in this 25-min experiment was 86% and 100% along the X-axis and Y-axis, respectively.

4.2.2 Farmer Action Information

To obtain the farmer's action information, smartwatches are placed on the right and left wrists to measure the motion sequences of the farmer's right and left arms from the accelerometers and gyroscopes in the smartwatches. First, we extracted the motion sequences involved in harvesting a tomato from the measured acceleration

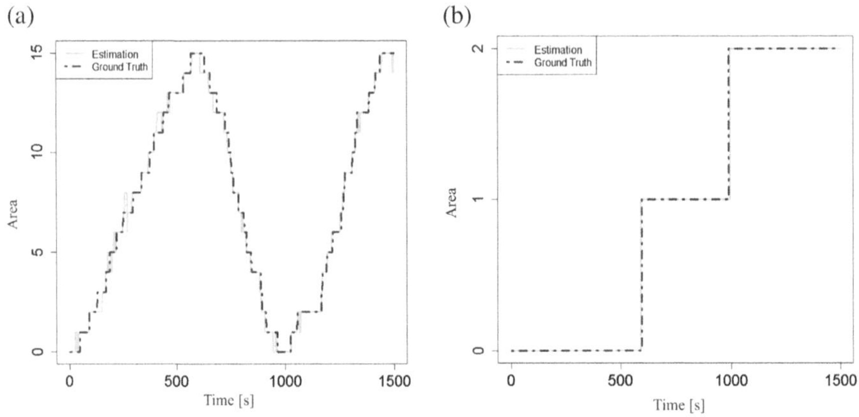

Fig. 17 Results of farmer position estimation (**a**) X-axis estimation result. (**b**) Y-axis estimation result

sequence of the farmer's right arm (used to operate the scissors), as it was observed that the farmer made a unique motion with his right arm when cutting the stem off a tomato fruit.

To extract harvesting motion sequences from the measured acceleration sequence, we used the dynamic time warping (DTW) algorithm [16]. DTW is a well-known algorithm for computing the distance between two sequences with nonlinear time deformations.

Using training motion sequences smoothed by the simple moving average (SMA) algorithm, we manually extracted the motion sequences associated with harvesting a tomato fruit as template sequences. SMA is a widely used smoothing method that replaces a value x_t with the average value $\bar{x}_t$:

$$\bar{x}_t = \frac{1}{W_t} \sum_{i=t-W_t+1}^{t} x_i \tag{3}$$

(in our experiment, $W_t = 10$). From the smoothed test motion sequences, we then automatically extracted motion sequences similar to the template sequences as harvesting motion sequences using DTW. Precision and recall ratios of 75% and 94%, respectively, were achieved using this approach. For further details of this method, please refer to [7].

4.2.3 Farm Work Information and Its Application

The position information and action information were combined into farm work information through the following three steps:

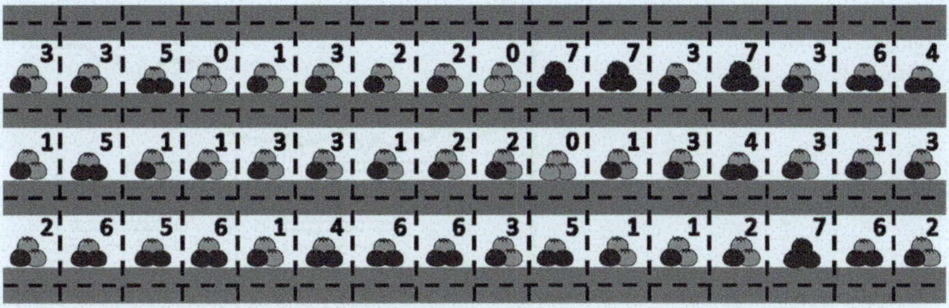

Fig. 18 Farm work information visualization

1. extract all items of information about one farmer from both the position information and action information using the "who" attribute, which is registered before starting harvesting,
2. sort the extracted information in order of time using the "when" attribute, which is recorded by the smartphone and the smartwatches simultaneously with the RSSIs and motion sequences, and
3. add the "where" attribute to the action information according to the closest position information at that time.

Figure 18 shows a sample application of farm work information visualizing the number of tomatoes harvested in each area in a day. The farmer stated that the unevenness of the yield shown in this illustration agreed with his intuition.

5 Analysis of Agricultural Information

As more agricultural information is recorded by ICT systems and stored in databases, it will become a huge repository of information. It is difficult to manually extract the characteristic values from such vast amounts of information. Moskvina and Zhigljavsky [15] and Ide and Inoue [9] proposed spectrum analysis methods to extract the change points from time series data. Their approaches have been adopted for network access analysis, machinery trouble diagnosis, and climate change evaluation [10, 25]. The authors [21] have employed change point analysis based on the singular spectrum transformation (SST) [9] to extract feature values from field environmental information measured by the monitoring system described in the Sect. 2. Some examples of this analysis method are introduced here.

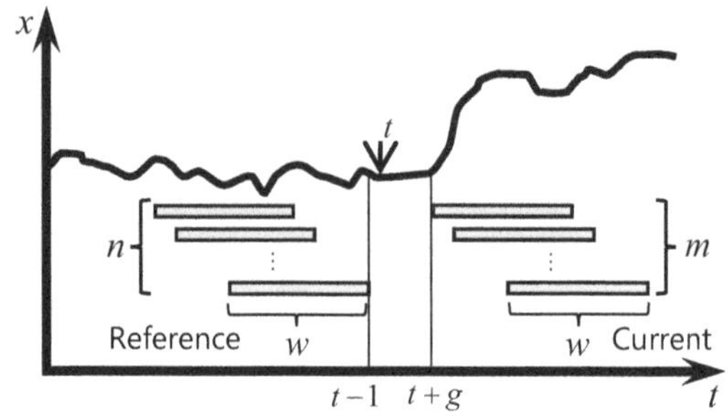

Fig. 19 Basic concept of singular spectrum transformation

5.1 Singular Spectrum Transformation

The SST proposed by Ide and Inoue [9] identifies change points or phase transformation points from the time series data measured by the monitoring systems described above. Consider the set of time series data shown in Fig. 19. The change point score, which denotes the difference in the patterns of the reference and current time series data, is defined as

$$z(t) = 1 - \frac{\boldsymbol{U}^{\top}\bar{\boldsymbol{u}}_1}{\left\|\boldsymbol{U}^{\top}\bar{\boldsymbol{u}}_1\right\|}\bar{\boldsymbol{u}}_1, \tag{4}$$

where $\boldsymbol{U}$ is l pieces of the eigenvector in descending order of the eigenvalue for the segmented reference time series data in Fig. 19. Similarly, $\bar{\boldsymbol{u}}_1$ is the eigenvector for the maximum eigenvalue for the segmented current time series data. $z(t) = 0$ indicates that there is no difference between the reference and current pattern, whereas $z(t) = 1$ represents a significant difference. Using the value of $z(t)$, we can extract change points from time series data. To evaluate the validity of this method, the change point analyses for a simple data-set and environmental monitoring datasets were examined. Figure 20 shows the results for artificial time series data having two different waves. The input parameters were selected as $w = g = 12$, $n = m = 5$, and $l = 1$ for this calculation. However, the result is strictly influenced by the input parameters. These parameters must be determined according to the pattern of the time series data. Here, fast Fourier transform is adopted to determine the input parameters w and g, because the data such as the field environmental data shown in Fig. 5 has clear periodic pattern due to the daily weather and plant growth behavior and farm work. Other parameters can be selected so as to fit the given data.

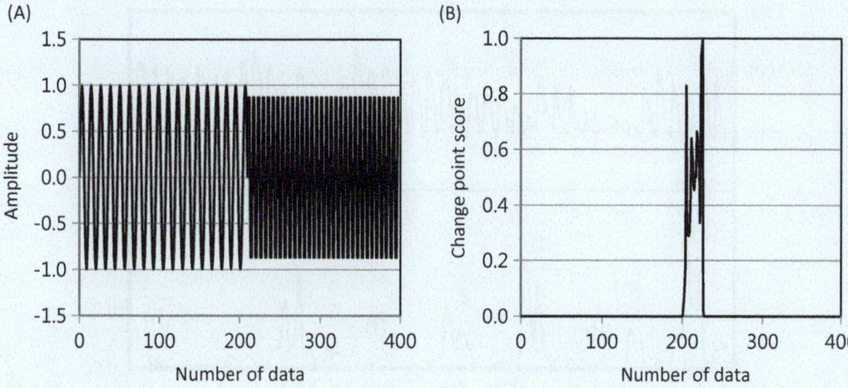

Fig. 20 Evaluation of change point analysis for a simple time series. (**a**) Input data. (**b**) Change point analysis result

5.2 *Change Point Analyses for Field Environmental Data*

We applied the change point analysis method to identify environmental impacts caused by farm work activities. Figure 21 shows the change point analysis results with respect to CO_2 concentrations in the tomato greenhouse. The parameters in the calculation were selected as $w = g = 24$, $n = m = 5$, and $l = 1$. The CO_2 concentration increased overnight with the respiration of the tomato plants, whereas it decreased with the action of photosynthesis during the day and changed to approximately 400 ppm (the same as in ambient air) when the greenhouse windows were opened. Applying the change point analysis, eight clear change points (labelled (a)–(h)) were obtained. To evaluate the validity of the change points, we examined the recorded farm work information. Table 5 lists the farm work information around these change points. Comparing the change points with farm work information, there is good agreement except for change points (g) and (h), which lack any farm work information. For example, the change points appear after the spraying of fertilizer. As a further example, consider change point (b). The variation in CO_2 concentration was unusually small during the day and night in this period. The reason was found to be that all the windows were left open because the automatic window controller had been accidentally switched off. Using this change point analysis, farmers can identify not only human errors or problems with their facilities but also environmental changes brought about by weather conditions or plant activities. This analysis could be used to develop a warning system.

Next, we introduced a method to evaluate the air temperature management in the greenhouse. Figure 22 shows the change point analysis results for air temperature changes in the greenhouse. The parameters in the calculation were selected as $w = g = 24$, $n = m = 5$, and $l = 1$, as for the calculation of CO_2 concentrations. The air temperature varied irregularly compared with the CO_2 concentration, because it was influenced by the ambient environmental situation. We found that the minimum

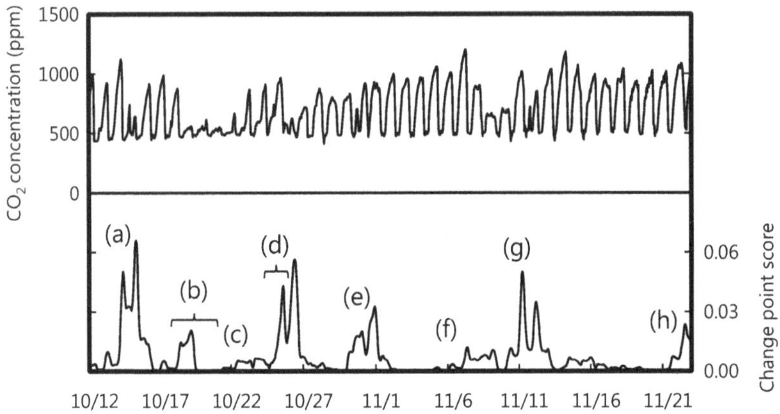

Fig. 21 Change point analysis results for CO_2 concentration in the tomato greenhouse

Table 5 List of farm work records around the change points

Date	Farm operations	Label
2010/10/13	Foliar fertilizer application	(a)
2010/10/19 2010/10/21	Window was opened for 3 days, The automatic window controller was turned off. (Human error)	(b)
2010/10/22	Foliar fertilizer application	(c)
2010/10/25 2010/10/26	Disbudding, Thinning out, Fertilizer application	(d)
2010/10/31	Foliar fertilizer application	(e)
2010/11/06	Foliar fertilizer application	(f)
2010/11/11	Start heating	(g)
2010/11/22	No data	(h)

air temperature in each day changed considerably until 11 Nov., 2010, on which date the heating started. The points with large change point scores coincided with the points where the minimum air temperature changed significantly. This means that the maximum air temperature should be controlled by opening and closing the windows. However, no clear change points could be found under the controlled environmental condition.

These results can be used to evaluate environmental changes caused by farm work and plant activities. We will continue to study the effectiveness of change point analysis for other field environmental information measured by ICT monitoring systems.

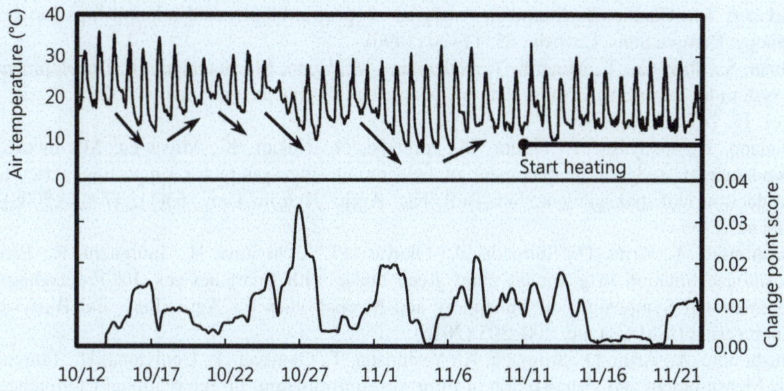

Fig. 22 Change point analysis results for air temperature change in the greenhouse

6 Conclusion

In this chapter, we introduced several ICT systems and applications related to sensing, visualizing, and analyzing data to support the advancement of agriculture. The aims and effects of these ICT systems are to provide valuable information and notifications to assist decision-making processes in farming. These developments and improvements will be extended according to the farmers' demands, environments, and situations. However, such systems cannot be expanded without a clear illustration of their benefits, even if the installation and management costs will be reduced in the near future. To establish next-generation agriculture based on high sustainability and security, we have to focus on the reduction of production costs and environmental impacts during agricultural production and also have to provide and share agricultural information for stakeholders including consumers. These efforts will be contributed to shift sustainable society.

Acknowledgements This work was supported by the 29th CASIO research grant (2011), the AgriSNS research project commissioned from the Ministry of Economy, Trade and Industry in Japan, and the Japan Society for the Promotion of Science of KAKENHI Grant Numbers 25292517, 15H01695, and 15K07677. Further, valuable comments and materials for the development of the field monitoring system were provided by Professor Dr. Takehiko Hoshi at Kinki University. We would like to express our thanks for this valuable support.

References

1. Bours, R., Muthuraman, M., Bouwmeester, H., van der Krol, A.: Oscillator: a system for analysis of diurnal leaf growth using infrared photography combined with wavelet transformation. Plant Methods **8**, 29 (2012)
2. Fukatsu, T., Hirafuji, M.: Field monitoring using sensor-nodes with a web server. J. Rob. Mechatronics **17**(2), 164–172 (2005)

3. Gilabert, M., Gandia, S., Melia, J.: Analyses of spectral-biophysical relationships for a corn canopy. Remote Sens. Environ. **55**, 11–20 (1996)
4. Guan, S., Shikanai, T., Minami, T., Nakamura, M., Ueno, M., Setouchi, H.: Development of a system for recording farming data by using a cellular phone equipped with GPS. Agric. Inf. Res. **15**, 241–254 (2006)
5. Hadano, R., Okayasu, T., Hirata, M., Yamabe, N., Nakaji, K., Mitsuoka, M., Inoue, E.: Fundamental study on development of field monitoring system for supporting agricultural production and management. Sci. Bull. Fac. Agric. Kyushu Univ. **63**(1), 57–63 (2008). In Japanese
6. Hashimoto, Y., Arita, D., Shimada, A., Okayasu, T., Uchiyama, H., Taniguchi, R.: Farmer position estimation in a tomato plant green house with smart devices. In: Proceedings of International Symposium on Machinery and Mechatronics for Agriculture and Biosystems Engineering (ISMAB), pp. 200–205 (2016)
7. Hashimoto, Y., Arita, D., Shimada, A., Yoshinaga, T., Okayasu, T., Uchiyama, H., Taniguchi, R.: Measurement and visualization of farm work information. In: International Conference on Agriculture Engineering (CIGR AGEng) (2016)
8. Hirafuji, M.: Creating comfortable, amazing, exciting and diverse lives with CYFARS (CYber FARmerS) and agricultural virtual corporation. In: Proceedings of the Second Asian Conference for Information Technology in Agriculture, pp. 424–431 (2000)
9. Ide, T., Inoue, K.: Knowledge discovery from heterogeneous dynamic systems using change-point correlations. In: SIAM International Conference on Data Mining, pp. 571–575 (2005)
10. Itoh, N., Kurths, J.: Change-point detection of climate time series by nonparametric method. In: Proceedings of the World Congress on Engineering and Computer Science 2010, pp. 445–448 (2010)
11. Jiang, J.A., Tseng, C.L., Lu, F.M., Yang, E.C., Wu, Z.S., Chen, C.P., Lin, S.H., Lin, K.C., Liao, C.S.: A GSM-based remote wireless automatic monitoring system for field information: a case study for ecological monitoring of the oriental fruit fly, bactrocera dorsalis (hendel). Comput. Electron. Agric. **62**, 243–259 (2008)
12. Lucas, B.D., Kanade, T.: An iterative image registration technique with an application to stereo vision. In: Proceedings of Imaging Understanding Workshop, pp. 121–130 (1981)
13. de Mairan, J.J.: Observation botanique. In: Histoire de l'Académie Royale des Sciences, pp. 35–36. Imprimerie royale, Paris (1729)
14. Minervini, M., Scharr, H., Tsaftaris, S.A.: Image analysis: the new bottleneck in plant phenotyping. IEEE Signal Process. Mag. **32**(4), 126–131 (2015)
15. Moskvina, V., Zhigljavsky, A.: An algorithm based on singular spectrum analysis for change-point detection. Commun. Stat. Simul. Comput. **32**, 319–352 (2003)
16. Müller, M.: Dynamic Time Warping, pp. 69–84. Springer, Berlin/Heidelberg (2007)
17. Murakami, N.: Work recording system for supporting safety and security agricultural produce. J. Jpn. Soc. Agric. Mach. **68**(2), 17–19 (2006). In Japanese
18. Nanseki, T., Sugahara, K., Fukatsu, T.: Farming operation automatic recognition system with RFID. Agric. Inf. Res. **16**, 132–140 (2007). In Japanese
19. Nugroho, A., Okayasu, T., Hoshi, T., Inoue, E., Hirai, Y., Mitsuoka, M., Sutiarso, L.: Development of a remote environmental monitoring and control framework for tropical horticulture and verification of its validity under unstable network connection in rural area. Comput. Electron. Agric. **124**, 325–339 (2016)
20. Okayasu, T., Miyazaki, T., Marui, A., Yamabe, N., Mitsuoka, M., Inoue, E.: Development of field monitoring and work recording system in agriculture. In: Proceedings of International Symposium on Machinery and Mechatronics for Agriculture and Biosystems Engineering (ISMAB) (2010)
21. Okayasu, T., Mitsuoka, M., Prima, N.A., Yoshida, H., Nanseki, T., Inoue, E.: Change point analysis for environmental information in agriculture. In: Proceedings of Title World Congress on Computers in Agriculture, Asia Federation for Information Technology in Agriculture (2012)

22. Roberto, F.N., Aluízio, B. (eds.): Phenomics: How Next-Generation Phenotyping is Revolutionizing Plant Breeding. Springer International Publishing, Cham (2015)
23. Shi, J., Tomasi, C.: Good features to track. Technical report, Cornell University (1993)
24. Strachana, I., Pattey, E., Boisvert, J.: Impact of nitrogen and environmental conditions on corn as detected by hyperspectral reflectance. Remote Sens. Environ. **80**, 213–224 (2002)
25. Tokunaga, T., Ikeda, D., Nakamura, K., Higuchi, T., Yoshikawa, A., Uozumi, T., Fujimoto, A., Morioka, A., Yumoto, K., Group, C.: Onset time determination of precursory events in time series data by an extension of singular spectrum transformation. Int. J. Circuits Syst. Signal Process. **5**, 46–60 (2011)
26. Wang, N., Zhang, N., Wang, M.: Wireless sensors in agriculture and food industry: recent development and future perspective. Comput. Electron. Agric. **50**, 1–14 (2006)

Learning Analytics for E-Book-Based Educational Big Data in Higher Education

Hiroaki Ogata, Misato Oi, Kousuke Mohri, Fumiya Okubo, Atsushi Shimada, Masanori Yamada, Jingyun Wang, and Sachio Hirokawa

1 Introduction

Recently, digital textbooks or electronic textbooks (i.e., e-books) were introduced to schools [1] in many countries (e.g., Japan, Korea, Taiwan, and Singapore); this is especially true of K12 schools. For example, the Japanese Ministry of Education, Culture, Sports, Science, and Technology compiled "The Vision for ICT in Education," a comprehensive policy that promotes the utilization of information and communication technology (ICT) in education [2]. As part of this policy, the Japanese government planned the introduction of e-books in all K12 schools by 2020 as well. In Korea, the research on e-books started in 1997; further, in 2007, the Korean Education and Research Information Service [3] announced the e-book usage plan.

Most of the pilot studies focus on the introduction of e-books in schools. However, very little attention was paid to the analysis of e-book activity logs, although it is imperative to investigate how these logs can be used to improve e-book contents and the quality of learning and education. Once the logs of K12 e-book learning activities accumulate on a server, educational big data can be collected. The analysis of the big data, which include information from e-books and learning management systems (LMSs), is necessary for supporting and enhancing several learning activities [4].

The e-book policies of many countries focus only on the introducing the technology of e-books in K12 schools [3, 5–7]. However, this study discusses the

H. Ogata (✉) • M. Oi • K. Mohri • F. Okubo • A. Shimada • M. Yamada • J. Wang • S. Hirokawa
Learning Analytics Center, Kyushu University, 744 Motooka, Nishi-ku,
Fukuoka 819-0395, Japan
e-mail: hiroaki.ogata@gmail.com

H. Yasuura et al. (eds.), *Smart Sensors at the IoT Frontier*,
DOI 10.1007/978-3-319-55345-0_13

introduction of e-books at the university level. We believe that several advantages make the introduction of e-books in universities easy, as listed here:

1. ICT skills: University students need to use ICT in their campus life (e.g., to submit a report through an LMS, register courses in a web-based system, and view the scores of various courses in a web-based system). Therefore, university students should have better ICT skills than K12 students.
2. Internet accessibility: Some universities provide faster broadband Internet access than K12 schools. Therefore, it may be easier for university students than school students to download e-books on campus.
3. Learning materials: Recently, some professors have started creating their own learning materials (e.g., by using PowerPoint and Keynote), revising them by themselves, and using them in their courses. Therefore, it is easier to put these materials into the e-book system, as opposed to the uploading of entire books in K12 schools.
4. The flexibility of course design and contents: In Japan, it is not easy for K12 teachers to change the course design and contents. However, professors in universities can change the design and contents of their courses whenever they deem it necessary. Therefore, it is easier to integrate e-book activities in university courses, which is a very important factor encouraging the introduction of e-books.
5. The management of teaching and learning skills: University professors and students should have the scientific skills needed to analyze their own teaching and learning log data, respectively. Further, university students should be more self-directed than school students in their learning. Due to these points, it is easier to introduce e-books and utilize log data for university students than for K12 students.

Due to the above reasons, we think that the university context provides an ideal test bed for the introduction of e-books. Therefore, we argue that the effectiveness of e-books and utilization of e-book logs must be first tested at the university level. This study describes the ongoing research on the analysis of e-book-based educational big data in Kyushu University.

2 The M2B System

In order to improve teaching and learning, Kyushu University introduced a single-platform learning system (Mitsuba, M2B) constructed from a learning management system (Moodle), an e-portfolio system (Mahara), and an e-book system (BookLooper). This project is supported as part of "Research and Development on Fundamental and Utilization Technologies for Social Big Data" by the National Institute of Information and Communications Technology. The project started on July 1, 2014, and it will end by the end of March 2018.

(A) Bookstore (B) Bookshelf (C) Viewer

Fig. 1 E-book interface. (**a**) bookstore, (**b**) bookshelf, (**c**) viewer

Date	Devices	User ID	e-book ID	Title (e-book)	Operation	Page	Status	M Start	M End
2014-11-12 08:54:42.0	AIR	xxxxxxxxxx	00000000N	A-11-暗号	OPEN	0	0	0	0
2014-11-12 08:54:46.0	AIR	xxxxxxxxxx	00000000N	A-11-暗号	PORTRAIT	1	1	0	0
2014-11-12 08:54:48.0	AIR	xxxxxxxxxx	00000000N	A-11-暗号	NEXT	2	1	0	0
2014-11-12 08:54:49.0	AIR	xxxxxxxxxx	00000000N	A-11-暗号	NEXT	3	1	0	0

Fig. 2 Samples of e-book logs

Since 2013, Kyushu University has been adopting an approach called *Bring Your Own Personal Devices* (BYOD) for all the students, and the entire campus has high-speed broadband wireless Internet access. This infrastructure enables students to browse e-book materials before, during, and after lectures. In addition, in order to educate "active learners" by using this infrastructure, Kyushu University started the Faculty of Arts and Science in 2014. "Active learning" is learning behavior and think about what they have done or are doing spontaneously [8, 9]. M2B is used, for example, to support the following:

- Teachers use Moodle to manage student attendance, provide quizzes, and receive reports.
- Both teachers and students keep notes on e-portfolios after lectures by using Mahara.
- Students use e-books via the BookLooper (Fig. 1) to study the learning material provided by the teachers using their preferred device (Windows or Macintosh computer, iPhone or iPad, and Android devices).

Figure 2 presents sample e-book logs. In logs, there are many types of operations, for example, OPEN means that the student opened the e-book file and NEXT means that he or she clicked the next button to move to the subsequent page. Further, PORTRAIT signifies that the student turned the computing device to the portrait position.

As shown in Table 1, as of December 31, 2015, approximately 6,710,000 log data from BookLooper and 4,730,000 log data from Moodle were collected from various academic courses (e.g., information science, programming, Earth and planetary science, and history) with the cooperation of approximately 100 teachers.

Table 1 Data by using BookLooper and Moodle as of December 31, 2015

	BookLooper			Moodle
Term period	Oct. 2014–Feb. 2015	Apr. 2015–Aug. 2015	Oct. 2015–Dec. 2015	Oct. 2015–Dec. 2015
No. of students	300	2687	2687	19,293 (No. of teachers 10,490)
No. of courses	5	38	22	112
No. of e-books	148	183	131	–
No. of log data	580,000	5,490,000	6,710,000	4,730,000
Data size	3.2 GB		1.9 GB	

The educational data logs from Moodle and BookLooper are quantitative educational data, and they are used to meet the following objectives:

- Learning:
  - Analyzing the details of behavior of "active learners" to make the students more active.
  - Based on the relationships between log patterns and academic achievements, detecting the students who may drop out and those who will perform excellently.
- Teaching:
  - Based on the logs made during a class session, improving course designs, which include collaborative learning and flipped classroom approaches.
  - Based on the students' patterns of viewing e-books (e.g., understanding which page was frequently viewed), improving teaching materials and the structure of the e-books.
  - The educational data log from Mahara contains qualitative data, and it is used to support quantitative analyses by supplying subjective data from students and teachers: what they think and have questions about a class session by the students and the answers from the teacher of the class session.

3 The Integration and Visualization of Learning Logs for Learning Analytics

Learning management (Moodle), e-portfolio (Mahara), and e-book (BookLooper) systems collect various types of learning logs from students. Understanding these logs is a crucial task of learning analytics. For this purpose, one of the possible methodologies is visualizing the learning logs; however, there are some difficulties associated with visualization, such as how the original logs can be interpreted, how multiple logs from different systems can be combined, and so on. An essential operation is to identify a unique student among the systems by aggregating user ID information. Another important task is to integrate the learning logs in different spatiotemporal data sizes, for example, monthly, daily, and hourly sizes. This section introduces several examples of visualization results.

First, Fig. 3 shows an hourly report of the number of students who opened the e-book system and read some material. Besides, the number of students was aggregated according to two aspects, in-class and out-class. This visualization was realized by the combination of e-book and LMS logs. The number of students and their reading material information can be collected by the e-book system; however, the system does not know whether a student is reading the material in a class. On the other hand, class information, such as the starting and ending times of each class, is

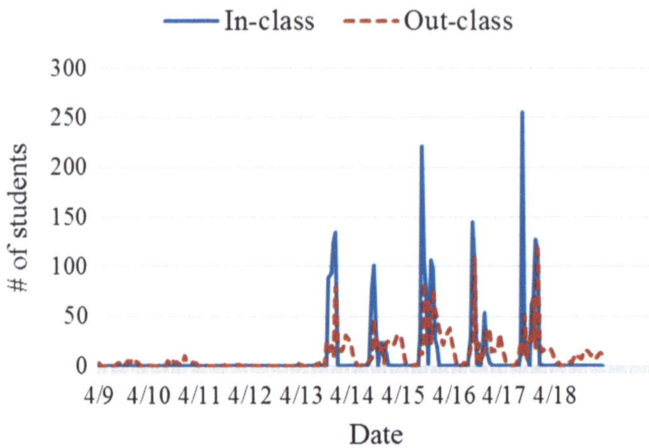

Fig. 3 A daily report of the number of students who studied with e-book system

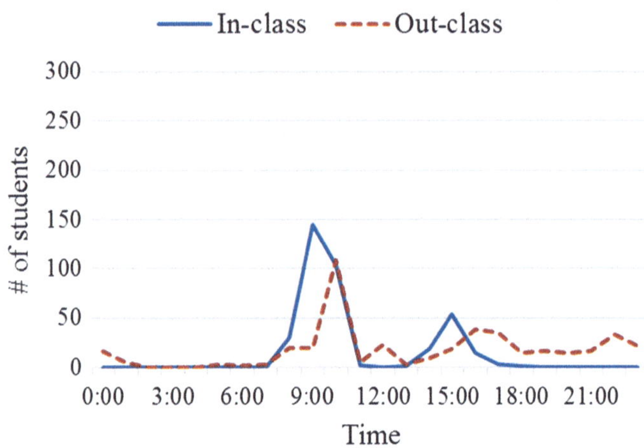

Fig. 4 An hourly report of students' activities

stored in the LMS. In this manner, the system knows when a student attends a class. Therefore, additional information can be added to the e-book logs, irrespective of whether the material is opened inside or outside the class.

On the beginning days of Fig. 3, the school term was not started in Japan. Therefore, on these days, only few learning logs were collected. Once the classes began, a large number of learning logs were generated by students' daytime activities. An interesting result was that the number of students who used the e-book system outside the class was more than the expected number. Furthermore, from Fig. 4, we can see that some students studied even during the nighttime. As described, log integration provides a new vista for understanding the students' activities.

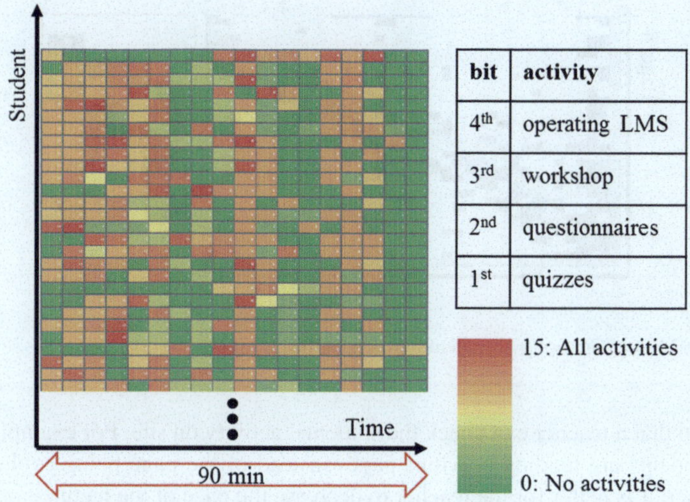

Fig. 5 Visualization of students' activities in a class

Second, students' activities in a 90-min class can be visualized as shown in Fig. 5. In this example, the learning logs from LMS were integrated. In fact, four types of learning logs, operating LMS, attending workshop, answering questionnaires, and taking quizzes, were analyzed according to the activity performed by each student every 5 min. The color chart of the blocks in Fig. 5 was defined by a 4-bit (16-level) combination. Each activity was assigned to one of the four bits, the fourth bit for "operating LMS," third bit for "attending workshop," second bit for "answering questionnaires," and first bit for "taking quizzes." If a student performs all the activities during a 5-min period, the activity value becomes 15 because all the bits are equal to 1.

In Fig. 5, the horizontal axis represents the time sequence, separated every 5 min. For example, a student's activity corresponds to a single line consisting of 18 blocks (5 min × 18 blocks = 90 min). During one class, students were asked to follow the teacher's instruction. Therefore, the blocks in one column should have the same color. We can see that most of the students followed the teacher's instruction. However, some students seem to have performed other things. The student in the top row did not do anything on the LMS from the second to eighth block (approximately 35 min). This visualization strategy is helpful for checking students' activities as to whether they could follow the teacher's instruction.

Third, the real-time monitoring of students' reading activities can be realized, as shown in Fig. 6. The e-book logs were collected on the cloud server in real time; subsequently, the number of students reading the material page by page was aggregated every minute. Finally, the aggregation results were displayed in the form of a heat map in 16 levels as shown in Fig. 5. The heat map is updated every

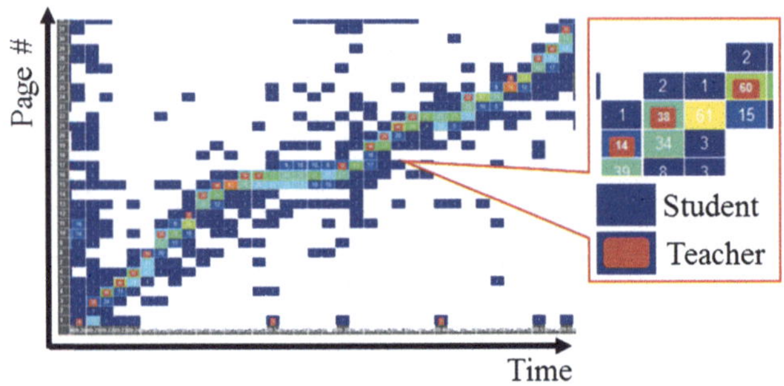

Fig. 6 Real-time monitoring of students' reading activities

minute so that a teacher can check the students' activity on site. For example, when many students are reading previous pages instead of the page being explained by the teacher, it is better for the teacher to decrease the pace of the lecture.

4 Visualizing Preview and Review Patterns by Analyzing e-Book Logs

4.1 Visualization of Preview and Review Patterns

To ensure effective learning, it is important to cover the same content before and after learning it in class [10]. Hereafter, we refer to learning before class as "preview" and that after as "review." In order to investigate learning behaviors and achievements, most of the previous studies used subjective measures such as the answers of questionnaires [6, 11, 12]. In contrast, we used e-book logs as an objective measure to address this issue. First, we visualized preview and review behaviors from 1 week before a class session to 1 week after it (Fig. 7), with special emphasis on when and how frequently the students performed previews and reviews [13].

For this analysis, 400,000 e-book logs were collected using BookLooper from the first-year students ($n = 100$) in an information science course of Kyushu University, which consisted of nine class sessions over a 3-month period. The students could access learning materials (i.e., e-books), which could be used any time at any place. If the students accessed an e-book that was to be used as a textbook in a class session (0) before the session, the logs of the e-book were defined as preparation (preview) logs (−). If the students accessed the e-book after the class session, its logs were defined as review logs (+).

For this visualization, we used three types of measurements. First, change indicates the number of times a student changed e-books over the course of 1 h. Change was calculated for each e-book for each hour. Duration indicates the number of seconds for which a student accessed a given e-book over 1 h. Finally, page flip

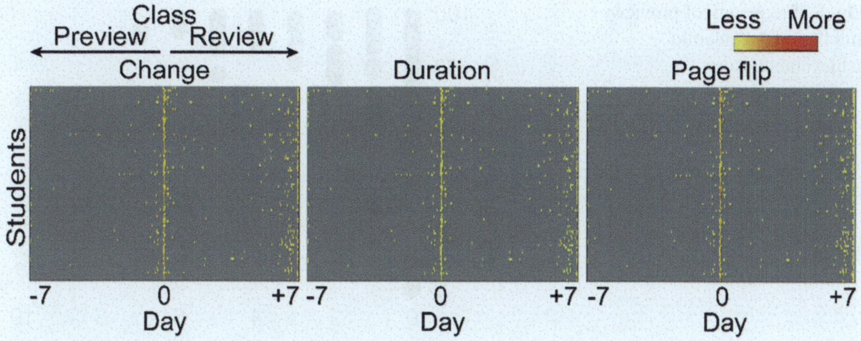

Fig. 7 Preview and review patterns of each student from −7 to +7 days

indicates the number of pages of the e-book that a student flipped through over 1 h. Figure 7 indicates that although most of the students performed reviews, only a few students performed previews.

4.2 The Relationship Between e-Book Logs and Academic Achievement

4.2.1 Correlation Between the Frequency of Previews and Academic Achievement

In order to examine whether performing previews and reviews ensures the better academic achievement of students, we first calculated Spearman's rank correlation between the frequency of previews for nine class sessions and the final academic achievement scores of the courses [14]. Figure 8 shows the frequencies of previews and the final scores. The analysis indicates the presence of significant positive correlation, one-tail: $rs = .52$, $p < .001$. This result supports the side of previews of our hypothesis.

4.2.2 Preview and Review Patterns and Academic Achievement

In order to examine the relationship between preview/review performances and academic achievement in more detail, we analyzed three types of measurements (i.e., change, duration, and page flip) during previews and reviews [13]. In these analyses, we used the following procedure. First, we coded the quartiles of students' midterm and term-end examination scores (first quartile: A; second quartile, B; etc.). The data from 17 students who did not take the midterm and final (term-end) examinations for a course or use e-books were discarded from further analysis. Subsequently, the students were categorized into six groups according to a

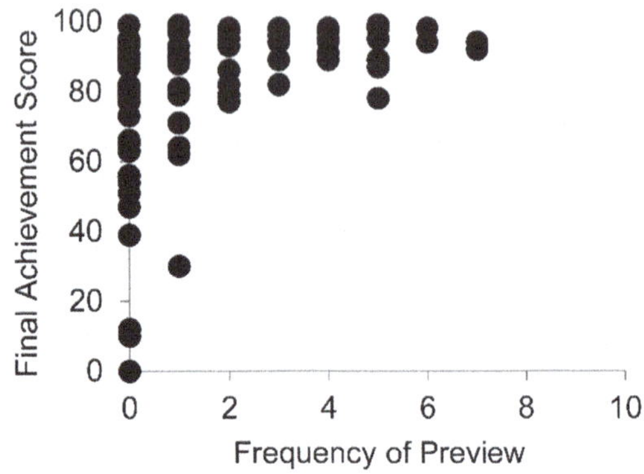

Fig. 8 Frequency of preview and the final academic achievement score

Table 2 The six groups and the numbers of students of each group

Midterm	Term-end			
	A	B	C	D
A	10	6	4	6
B	9	11	6	4
C	3	10	1	7
D	-	-	2	4

A red, *B* green, *CD* blue, *U1* pink, *U2* yellow, *L* gray

combination of midterm and term-end coded scores. Table 2 shows the six groups and number of students in each group. The students who received the same scores for their midterm and term-end examinations were subcategorized into A (A-A) and B (B-B). Since C-C and D-D students were too few to be considered as separate groups, they were combined into a single group, CD. Further, the students who improved their scores were categorized into two groups: Students in group U1 got a B, C, or D for the midterm examination and an A for the term-end, while students in group U2 got a better score, but not an A (hence, they got a B or C), for the term-end than the midterm examination. The last group, L, got worse scores for the term-end than the midterm examination.

We calculated the sum of previews and that of reviews for each student and each measurement from all the e-book logs; subsequently, we averaged the values for each group and each measurement (Fig. 9).

In order to examine whether the students who achieved higher academic achievement showed higher values in any measurement of preview/review (i.e., change, duration, and page flip), we conducted one-way analyses of variances (ANOVAs) with the group (U1, U2, A, B, CD, and L) as a between-participant factor on the sums of previews and reviews for all the three measurements. In previews alone, change and page flip revealed significant effects of groups, change, $F(5, 77) = 3.43, p = 0.007$; page flip, $F(5, 77) = 3.76, p = 0.004$. Post hoc analyses with Bonferroni adjustment (with significance level at 5%) revealed that group A showed significantly more frequent change and more page flips than groups U2, CD, and L.

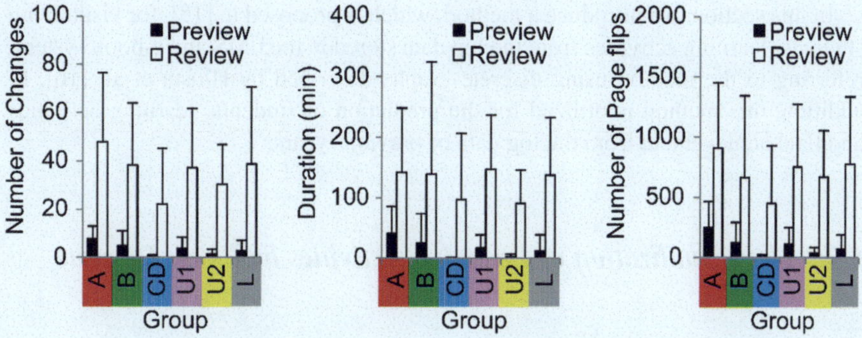

Fig. 9 The averages of preview and review for each group for each measurement

These results reveal that regardless of academic achievement, all students performed reviews in a similar manner, at least when using e-books. In contrast, for previews, the students who showed the highest academic achievement (group A) revealed significantly higher values than those who showed lower academic achievement (groups U2 and L), across change and page flip.

These results suggest that previews may be more relevant to academic achievement than reviews. However, we also note that in this course, the students took quizzes in every class and knew that their scores on the quizzes would be part of their final grade in the course. These characteristics of the course may have caused the students with higher motivation to perform previews.

5 The Visualization and Prediction of Learning Activities

5.1 Background

In Kyushu University, students must bring their own PCs and use the well-known learning management system (LMS) Moodle and the e-book system BookLooper provided by KYOCERA MARUZEN System Integration Co., Ltd. These ICT-based education systems enabled us to collect automatically many types of log data corresponding to the learning activities of students, both inside and outside the class.

These collected data can be utilized for identifying the typical learning patterns of particular students, for example, those who are likely to fail or drop out of class, that is, students referred to as "at-risk" students. It is an important task to detect "at-risk" students early. For this purpose, it is useful to supply information to teachers so that a teacher can grasp the learning activities of students visually and advise them to avoid failing the class.

In this section, we introduce a method, which is proposed in [15], for visualizing students' learning activities from the log data stored in the LMS and e-book system, referring to the method using discrete graphs described by Hlosta et al. [16]. In addition, this method is utilized for the prediction of students' learning activities and final achievement from the log data of previous years.

5.2 *The Visualization of Learning Activities by Using Discrete Graphs*

We consider the designated class held over 14 weeks, during which each lecture is presented by using several slides in the e-book system, with each slide being associated with a single lecture alone. Students use the slides for their preparation and/or review sessions of each lecture. They are required to submit a report and answer a quiz related to a week's lecture through the LMS. The students in the class are graded in terms of categories A, B, C, D, and F in the usual manner, with A being the best grade and F indicating failure.

For such a class, the following four types of data are stored in the LMS and e-book system for each student each week:

1. Attendance or absence.
2. The submission of a report or failure to do so.
3. The sum of the time spent browsing slides for preparation and/or review is longer or shorter than 10 min.
4. A quiz score is higher or lower than 70%.

Based on the combination of achievement (+) or failure (−) for the four items, the learning logs of a student can be represented in $2^4 = 16$ types osf states a week, as shown in Table 3.

An edge of the graph between state p of the nth week and state q of the $n - 1$th week is constructed if there exists a student who performed such learning activities. The edge is colored light yellow if only one student meets the condition; as the number of students increases, the color of the edge approaches deep orange.

We collected the learning logs of 100 students attending the "information science" class that started in October 2014 and applied the proposed method. In Fig. 10, the graph in the left visualizes the learning logs of 100 students attending

Table 3 A correspondence of the state numbers and the four kinds of learning logs

State number	0	1	2	3	4	5	6	7	8	9	10	11	12	13	14	15
1. Attendance	−	+	−	+	−	+	−	+	−	+	−	+	−	+	−	+
2. Browsing time	−	−	+	+	−	−	+	+	−	−	+	+	−	−	+	+
3. Report	−	−	−	−	+	+	+	+	−	−	−	−	+	+	+	+
4. Quiz score	−	−	−	−	−	−	−	−	+	+	+	+	+	+	+	+

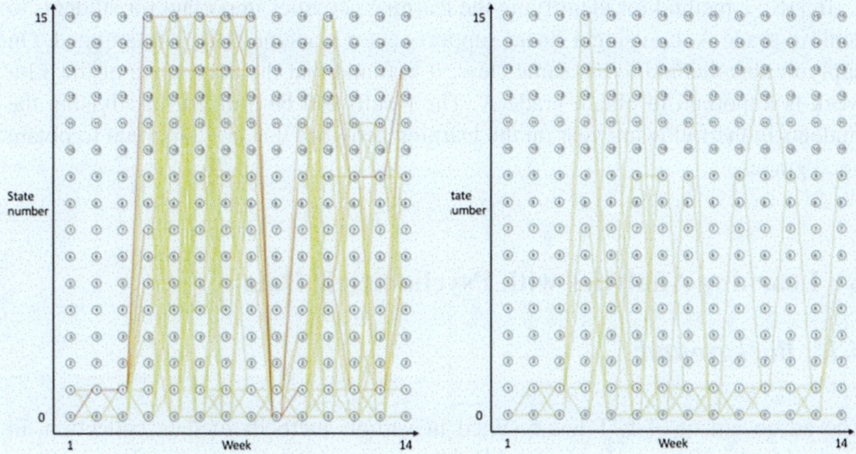

Fig. 10 The graphs constructed from the learning logs of all students (*left*) and of the students who obtained grade F (failure) (*right*)

the class. This graph indicates that a student who achieved items 1, 3, and 4 (i.e., state 13 or 15) was likely to continue achieving these items from the third to the eighth week. The graph in the right of the figure visualizes the learning logs of the ten students who obtained grade F. Comparing the two graphs, we can find the feature of learning activities of the students who fail the class. In fact, most edges in the graph in the right appear in the lower half, since students who obtain grade F were unlikely to achieve the four items.

5.3 The Prediction of Learning Activities

It is an important task to predict a student's learning activities and final achievement in a class. The predication can be utilized to detect "at-risk" students or lead students to a better final achievement.

We propose two ways to predict a student's learning activities by using the discrete graph constructed from the same class held in the previous year.

- From the state of the ith week, we can predict the state of the i+1th week that the student is likely to reach by tracing the edge colored deep orange of the same states in previous year's graph. It is noted that using this method, we can predict the learning activities for the coming week alone, i.e., the graph has the Markov property [17].
- Using the graph of the learning logs of students who obtained grade F, we can find the learning activities that should be avoided by students (or using the logs of grade A students, the activities that should be performed can be found).

In [18], a method for identifying the learning activities important for students to achieve grade A, by using a linear support vector machine [19], is mentioned. On applying this method to the same class, it is found that the attendance of the 12th week is important to obtain grade A. The results can be utilized for advising the students in the following year on the learning activities that are important to obtain good grades.

6 Learning Analytics with Psychometric Data

6.1 Background

The advancement of ICT has resulted in various methods of data collection, in particular ubiquitous technologies [4]. Ubiquitous technologies allow us to collect not only access logs but also location data. However, psychometric data such as learning style and motivation, as well as learning logs, should be collected in order to analyze learners' behaviors to provide effective learning support. The awareness of self-regulated learning (SRL), which is one of the most important perspectives in educational research, is helpful. Yamada et al. indicate that self-efficacy, which is one of the factors of SRL, has significant correlation with learning behaviors such as highlighting and annotation [20]. Goda et al. conducted a research study on the relationship between SRL and learning performance [21, 22]. They suggest that psychometric data on SRL are useful to predict the degree of help seeking and learning performance [21, 22]. If a relationship between SRL and learning behaviors is found, its results may be used to support learners effectively. Goda et al. suggest that one of the SRL skills is adaptive help seeking, which leads to high academic performance [21]. They conclude that a sense of help seeking can be useful in predicting learners' academic performance. This section aims to investigate learning behaviors' influence on learning performance, examining the relationship between learning behaviors, SRL factors, and learning performance.

6.2 Participants and Class

This study was conducted in two information technology courses. One is a 15-week course (course one) and the other an 8-week course (course two). The participants were 127 first-year university students in an information technology class (93 and 34 students for courses one and two, respectively). The teachers distributed digital learning materials to the students with the use of a digital learning material reader (DLMR) and encouraged the students to read the materials in advance for every class. The DLMR allowed the students to access the learning materials on devices such as laptops and smartphones and use marking and annotation functions

Table 4 Descriptive data of MSLQ

Variables	Pre-post	Mean	SD	Min.	Max.
SE	Pre	32.54	8.95	9	60
IV	Post	36.42	9.21	12	63
	Pre	45.21	6.29	29	63
CS	Post	44.31	8.15	16	63
	Pre	60.12	7.79	21	81
SR	Post	60.63	8.69	38	90
	Pre	36.26	5.63	15	48
TA	Post	36.49	6.22	17	63
	Pre	15.17	4.45	4	27
	Post	16.07	4.34	4	24

whenever and wherever Internet was available. In every class, learners were engaged in programming practice, following a comprehension test in every class. They were required to answer questionnaires before the first class (pre questionnaire) and at the end of the last class (post questionnaire).

6.3 Data Collection and Analysis

For data collection, two methods were used: a questionnaire and log. The Motivated Strategies and Learning Questionnaire (MSLQ) [23], which consists of five factors (self-efficacy, SE; internal value, IV; cognitive strategies, CS; self-regulation, SR; and test anxiety, TA), has 44 items in total, and is rated on a seven-point Likert scale from 1 (negative) to 7 (positive), was used for the subjective evaluation of learners' SRL skills. The students were asked to complete the MSLQ both before and after classes. The differences between their responses on the pre- and post-questionnaires were analyzed. The second method of data collection comprised a log that recorded the number of pages, as well as the students' behavior of marking and annotation. Learning performance is the final score.

6.4 Results

The number of collected data is 121, which answered pre- and post-MSLQ. Tables 4 and 5 show the descriptive data of MSLQ (mean of sum-up score in each factor), learning behaviors (frequency over 15 weeks), and the final score. In order to investigate the relationship between each SRL factor, learning behaviors, and the final score, stepwise multiple regression analysis was conducted, setting the final score as the dependent valuable and each MSLQ factor and the learning behaviors as independent valuables. The results are displayed in Table 6.

Table 5 Descriptive data of learning behaviors and the final score

Variables	Mean	SD	Min.	Max.
Slide	1,462.93	1,259.09	0	5365
Marker	7.84	12.25	0	83
Annotation	4.02	8.94	0	62
The final score	85.84	11.72	39.93	101.25

Table 6 The results of multiple regression analysis with stepwise

Variables	Coef.	SE	β	p
Self-efficacy	0.369	0.122	0.298	$p < 0.01$
Internal value	−0.329	0.141	−0.231	$p < 0.05$
Marker	0.160	0.078	0.167	$p < 0.05$
Slide	0.003	0.001	0.354	$p < 0.001$

$F(4, 116) = 9.31$, $p < 0.001$, $R^2 = 0.243$, adjusted $R^2 = 0.217$

The results of multiple regression analysis revealed that SE, IV, the frequent use of markers, and the frequent reading of slides significantly affected the final score. Although SE, marker, and slides had positive effects on the enhancement of learning performance, IV's effect was negative. Considering R^2 and significance, model fitness seems to be acceptable to some extent; however, three variables, IV, marker, and annotation, should be considered, from the view of model application, due to the large standard deviation.

6.5 Implementation in This Section

This section explains our research findings regarding the relationship between psychometric data and learning logs, in particular, the relationship with SRL. These findings suggest that classes should be designed according to the factors and learning behaviors mentioned in the results. Further, the design should consider the role of learning analytics, which helps education and learning improvement and makes the learners aware of self-efficacy, the use of marker, and annotation but not internal value. For the improvement of this model, more concise analytics should be developed, for example, comparative analytics with high and low groups of IV, marker, and annotation. In addition, in order to understand the key points to support learners, the overall relationship among all the variables should be investigated. There is high possibility that learning analytics mixed with psychometric data can find effective variables to support and improve education and learning.

7 An Ontology-Based Visualization Support System for E-Book Users

7.1 Background

In Kyushu University, three learning support systems, the learning content management system Moodle, e-profile system Mahara, and e-book system BookLooper, are used to support daily classroom teaching. The log data collected from these three systems are analyzed to further study the learning performance of students. However, the development of a knowledge framework is usually not supported in these systems; in addition, in these systems, it is difficult to identify the relevant knowledge items possessed by a learner before and after a learning activity.

When a learner requires several knowledge items, these items should be compared, and, at the same time, the existing relations between them should be realized and understood; the acquired knowledge items and their relations form the knowledge framework of the learner. The effective assimilation of new knowledge into an existing knowledge framework is defined as the achievement of "meaningful learning" in Ausubel's learning psychology theories [8, 24, 25]. The theories suggest that knowledge is finally incorporated into the human brain when it is organized in hierarchical frameworks, and learning approaches that facilitate this type of organization significantly enhance the learning capability of all learners. Otherwise, in the case of rote learning, knowledge tends to be forgotten quickly unless rehearsed repeatedly. Moreover, retained knowledge cannot contribute to the enhancement of a learner's knowledge framework and has a low possibility of being used in future problem solving [26]. Therefore, e-learning systems try to perform the complicated task of moving beyond rote learning and helping learners construct their knowledge framework effectively. In this study, an ontology-based Visualization Support System for e-book users (VSSE), which uses a hierarchical map structure to manage the knowledge items of a curriculum, was developed to encourage the development of comparison skills and foster meaningful learning in students.

7.2 A Semiautomatically Built Course-Centered Ontology

In an e-book system, learners normally read several pages of a file as part of one activity. For example, after studying pages 10–13 of a given file in BookLooper, which cover seven new knowledge items, the learner can log on to VSSE, the Visualization Support System for e-book users, to check the new knowledge points just studied. Our system will try to encourage the learner to understand the relations between these seven new knowledge items visually. Furthermore, the system will utilize the quiz results of the learner to identify the acquired knowledge points (KPs) and subsequently encourage him or her to compare the new KPs with the related acquired ones visually.

To facilitate this visualization function, the system requires a description of the information about all the knowledge items and their relations. In this study, this information is recorded using a map, which has nodes as key concepts and links as the relationships between the concepts [27]. Ontology is one of the main techniques adopted in maps for knowledge representation. Therefore, we present a method to develop semiautomatically a course-centered ontology to describe all the required information from the knowledge contained in the courses.

First, the information from the "Syllabus system" of Kyushu University is extracted automatically to create the basic framework of the course-centered ontology, for example, for Kikan Education in the Arts and Science Department, 3730 courses are registered in the syllabus system. However, for most of the courses in the system, less than ten keywords are described; the basic ontology framework built on these keywords is sufficient to provide the visualization support described in the previous section. Therefore, in the second step, we encourage professors/instructors to provide information manually to maintain the ontology. A tool that can automatically transfer information between an Excel file and an ontology web language (owl) file is developed to support the modification of the ontology. By applying and adjusting the ontology design method described by Wang et al. [28], a course-centered ontology of an existing computer science course (called COCS) is developed as a demo. In our previous research study [28, 29], the experimental results suggest that with the support of a system, which provides a visual environment to encourage the comparison of related "knowledge points" so as to foster meaningful learning, participants achieved significantly better learning achievement than without the system support. In our current study, a KP is defined as "a minimum learning item which can independently describe the information of one certain piece of knowledge in a specific course"; a learner can understand a KP by its own expression or can acquire it by practice. By analyzing the teaching materials of this computer science course, approximately 200 KPs and 20 types of relations are extracted and defined in COCS. In the last step, a tool that can automatically identify the location (including the file ID and page number) of the knowledge items in the BookLooper system and place these location details into the ontology is developed. Although time-consuming maintenance is required, a course-centered ontology can be developed semiautomatically by following these three steps.

7.3 A Visualization Learning Support System Providing a Knowledge Comparison Environment

We design and develop a system that automatically manipulates the course-centered ontology in order to provide visualization learning support for the construction of learner knowledge frameworks. A learner' view of the computer science course in the VSSE is shown in Fig. 11.

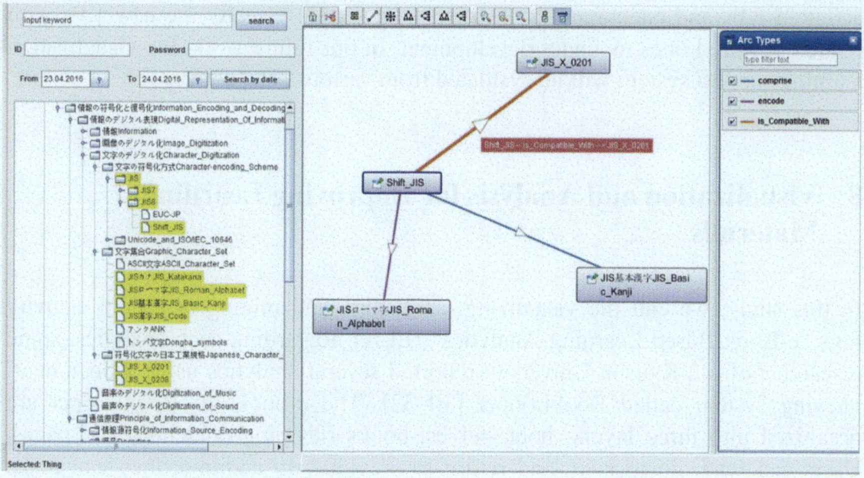

Fig. 11 The main interface of VSSE for learners

On the left side of this view, all the concepts of COCS are displayed using a tree structure. Users can find the KP they are searching by opening all the concepts level by level. Moreover, a search function is provided at the top-left side of the screen. Using this search function, the learner can set a period (e.g., from April 23, 2016, to April 24, 2016) and push the search button; as a result, the KPs involved in the pages that the learner had read during that period will be highlighted. Besides, when the learner searches for items by keyword, the items containing the given keywords in the tree structure will be highlighted to enable further checking.

Regarding the map displayed in the center of the panel, when the user double clicks a leaf representing a KP, the right-hand-side relation panel will display the selected KP and all its related KPs lined by the relations defined in COCS. For instance, in Fig. 1, the individual representing the KP "shift_JIS" is selected. As a result, users can obtain a visual representation of important information, as shown in the relation panel.

Moreover, the users can see a list of essential properties of each KP (represented by the data properties of one individual in COCS) by moving the mouse on a node shown in the relation panel. Similarly, for every arc shown in the relations panel, the relation statement will be displayed (e.g., the displayed relation axiom between "shift_JIS" and "JIS_X_0201" in Fig. 1). Therefore, users can conveniently obtain the essential properties of every KP and all its related KPs from the relations panel. This information is extracted automatically from the OWL file of COCS. Furthermore, in case too many relations are shown in the relation panel, a user can filter them using the arc-type panel.

The functions explained in the above paragraphs are expected to provide visualization support for the construction of learner knowledge frameworks. Currently, another function, which intends to utilize the quiz results of learners to identify the acquired KPs and encourage the learners to compare visually the new KPs with related acquired ones, is under development. In our future work, the visualization learning support system will be evaluated from various perspectives.

8 Visualization and Analysis for Improving Learning Materials

In this study, we call the visualizing, analyzing, and mining of e-book activity logs "e-Book-Based Learning Analytics" (ELA). Regarding such analytics, some researchers from Kyushu University reported several analytics using a document-viewing system called BookLooper [30–32]. The e-books of BookLooper are organized into three layers: bookshelves, books (learning contents), and pages. Users can read, go to next, and return to previous. In addition, they can make bookmarks and take a memo. Table 7 presents the actions and their explanations, as listed by Yin et al. [32].

For ELA, two methods were followed to improve learning materials and find the learning styles of students. The first is the visualization method based on learning behaviors, such as "NEXT" and "PREV." Figure 12 shows the visualization graph. From the visualization results, we found two learning styles: Digital Sequential Learning (DSL) and Digital Backtrack Learning (DBL). While the DSL style refers to students who proceed to the next page and rarely go back to previous pages once they finish reading one page, DBL is followed by those who frequently backtrack in their reading. For example, if current knowledge refers to previously discussed knowledge, then the students following DBL go back to previous pages to review or reflect. According to [32], the DBL learning style is better than DSL

Table 7 Action explanation

Action name	Explanation
NEXT	While a user goes to next page, he will click "NEXT" button, and the action name will be saved as "Next"
PREV	While a user goes to previous page, he will click "PREV" button, and the action name will be saved as "Prev"
MARKER	While a user wants to highlight some row in the learning content, he will click "Marker" button, and the action name will be saved as "Marker"
MEMO	While a user wants to write some memo in the learning content, he will click "Memo" button, and a textbox will be shown. After he finished writing memo, the action name will be saved as "Memo"
ZOOM-In ZOOM-OUT	While a user wants to zoom-in or zoom-out the page in the learning content, he will click "ZOOM" button, and the action name will be saved as "ZOOM-IN" or "ZOOM-OUT"

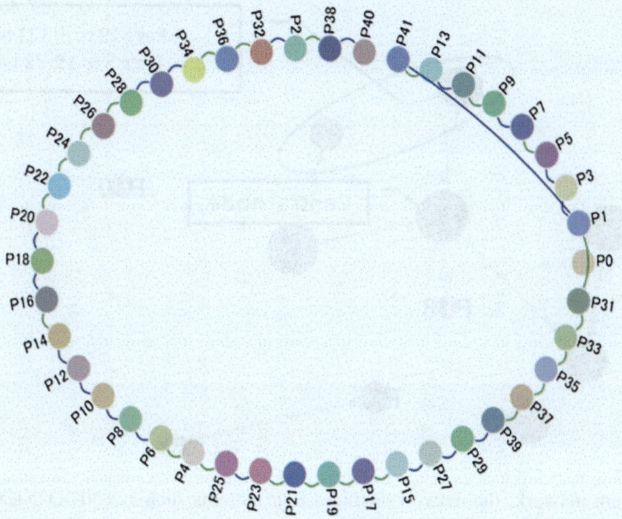

Fig. 12 Visualized learning behavior

because students who follow DBL have been found to obtain high scores in final examinations. Based on these results, the second method is considered.

The second method is social network analysis with n-gram. Researchers report that the analysis method is very useful in finding central concepts among learning logs [33, 34]. Figure 13 shows some example sequences and the corresponding 2-g sequences. It is noted that 2-g refers to four patterns: "NEXT-NEXT," "NEXT-PREV," "PREV-NEXT," and "PREV-PREV." The network graph is created based on two conditions: (1) the difference between the pages in one learning material of information science is more than 10, and (2) the frequency in 2-g is more than 10. The node size is calculated based on degree centrality, and the edge size is calculated based on the difference between pages.

From the results of the 2-g network, we found meaningful relationships between pages. As shown in Fig. 13, the central node shows page 6, which is connected to four other nodes: pages 18, 19, 21, and 23. This means that many students "went to pages 18, 19, 21, and 23 after they read page 6" or "went back to page 6 after they read pages 18, 19, 21, and 23." By finding these relationships, instructional designers and teachers can understand whether the learning material should be improved through their own judgment because the page order might not be appropriate. In addition, there is the possibility that analyzing other actions such as "ZOON-IN" and "ZOOM-OUT" may lead to the finding that learning material should be improved.

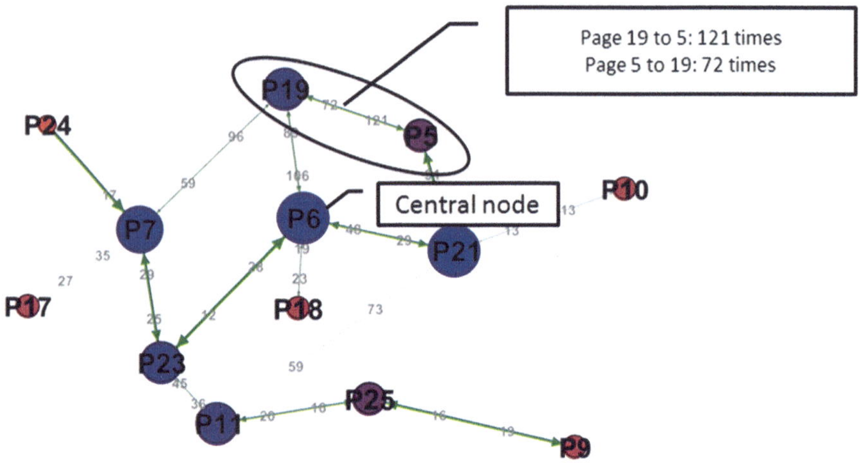

Fig. 13 Bi-gram network: the network includes four patterns such as "NEXT-NEXT," "NEXT-PREV," "PREV-NEXT," and "PREV-PREV"

9 Conclusion

This study describes a research project that accumulated and analyzed educational big data by using an M2B system (i.e., Moodle, Mahara, and BookLooper). From the initial experiment, this system may predict the final score if the course in the first four lectures by using e-book logs. In future works, we will allow teachers and students to download their own data; the system will provide them with data analysis tools to manage their learning and teaching skills. From the technological point of view, we will tackle research issues such as data integration, real-time data mining, visualization, recommendation, and predictions. In addition, we will integrate e-book and SCROLL [35, 36] in order to enhance learning experiences.

Acknowledgments The research is supported by "Research and Development on Fundamental and Utilization Technologies for Social Big Data" (178A03), the Commissioned Research of the National Institute of Information and Communications Technology (NICT), Japan; Grant-in-Aid for Scientific Research (S) No. 16H06304; Grant-in-Aid for Scientific Research (B) No. 25282059; Grant-in-Aid for Challenging Exploratory Research No. 26560122; Japan Science and Technology Agency (JST) PRESTO; and the Education Enhancement Program of Kyushu University.

References

1. Nakajima, T., Shinohara, S., Tamura, Y.: Typical functions of e-textbook, implementation, and compatibility verification with use of ePub3 materials. Procedia Comput. Sci. **22**, 1344–1353 (2013)
2. MEXT, Japanese Ministry of Education, Culture, Sports, Science and Technology.: The vision for ICT in education. http://www.mext.go.jp/b_menu/houdou/23/04/_icsFiles/afieldfile/2012/08/03/1305484_14_1.pdf (2011)

3. Shin, J.H.: Analysis on the digital textbook's different effectiveness by characteristics of learner. Int. J. Educ. Learn. **1**(2), 23–38 (2012)
4. Yin, C., Okubo, F., Shimada, A., Kojima, K., Yamada, M., Fujimura, N., Ogata, H.: Smart phone based data collecting system for analyzing learning behaviors. Proceedings of International Conference of Computers on Education 2014, Nara, Japan, pp. 575–577 (2014)
5. Fang, H., Liu, P., Huang, R.: The research on e-book-oriented mobile learning system environment application and its tendency. International Conference on Computer Science and Education, Singapore, pp. 1333–1338 (2011)
6. Ihmeideh, F.M.: The effect of electronic books on enhancing emergent literacy skills of pre-school children. Comput. Educ. **79**, 40–48 (2014)
7. Song, H.D., Jun, J.S., Ryu, J.H.: The Effects of Digital Textbooks in Student Learning. Seoul Metropolitan Board of Education, Seoul (2007)
8. Ausubel, D.P.: The Psychology of Meaningful Verbal Learning. Grune and Stratton, New York (1963)
9. Bonwell, C.C., Eison, J.A.: Active Learning: Creating Excitement in the Classroom, ASHE-ERIC Higher Education Report No1. The George Washington University, School of Education and Human Development, Washington, DC (1991)
10. Shinogaya, K.: Learning strategies: a review from the perspective of the relation between learning phases. Jpn. J. Educ. Psychol. **60**, 92–105 (2012)
11. Shinogaya, K.: Students' strategies in preparation and lectures: direct and moderating effects of teachers' teaching strategies. Jpn. J. Educ. Psychol. **62**, 197–208 (2014)
12. Woody, W.D., Daniel, D.B., Baker, C.A.: E-books or textbooks: students prefer textbooks. Comput. Educ. **55**, 945–948 (2010)
13. Oi, M., Okubo, F., Shimada, A., Yin, C., Ogata, H.: Analysis of preview and review patterns in Undergraduates' e-book logs. Proceedings of ICCE 2015, Hangzhou, China, pp. 665–669 (2015)
14. Ogata, H., Yin, C., Oi, M., Okubo, F., Shimada, T., Kojima, K., Yamada, M.: Analyses of learning behavior of active learners using logs of digital teaching materials. Bull. KIKAN Educ. **2**, 48–60 (2016)
15. Okubo, F., Shimada, A., Yin, C., Ogata, H.: Visualization and prediction of learning activities by using discrete graphs. Proceedings of ICCE2015, Hangzhou, China, pp. 739–744 (2015)
16. Hlosta, M., Herrmannová, D., Váchová, L., Kužílek, J., Zdrahal, Z., Wolff, A.: Modelling student online behaviour in a virtual learning environment. Workshop Proc. LAK 2014, Indianapolis, USA (2014)
17. Norris J.R.: Markov Chains. Cambridge Series in Statistical and Probabilistic Mathematics. Cambridge University Press, Cambridge, UK (1998).
18. Okubo, F., Hirokawa, S., Oi, M., Shimada, A., Kojima, K., Yamada, M., Ogata, H.: Learning activity features of high performance students. Proceedings of Cross-LAK2016, Edingburgh, UK, pp. 24–29 (2016)
19. Cortes, C., Vapnik, V.: Support-vector networks. Mach. Learn. **20**, 273–297 (1995)
20. Yamada, M., Yin, C., Shimada, A., Kojima, K., Okubo, F., Ogata, H.: Preliminary research on self-regulated learning and learning logs in a ubiquitous learning environment. Proceedings of the 15th IEEE International Conference on Advanced Learning Technologies (ICALT 2015), Hualien, Taiwan, pp. 93–95 (2015)
21. Goda, Y., Yamada, M., Matsuda, T., Kato, H., Saito, Y., Miyagawa, H.: Effects of help seeking target types on completion rate and satisfaction in e-learning. Proceedings of INTED 2013, Valencia, Spain, pp. 1399–1403 (2013)
22. Goda, Y., Yamada, M., Matsuda, T., Saito, Y., Kato, H., Miyagawa, H.: Procrastination and other learning behavioral types in e-learning and their relationship with learning outcomes. Learn. Individ. Differ. **37**, 72–80 (2015). doi:10.1016/j.lindif.201411.001
23. Pintrich, R.R., DeGroot, E.V.: Motivational and self-regulated learning componentsof classroom academic performance. J. Educ. Psychol. **82**, 33–40 (1990) http://dx.doi.org/10.1037/0022-0663.82.1.33
24. Ausubel, D.P.: Educational Psychology: A Cognitive View. Holt, New York (1968)

25. Ausubel, D.P., Novak, J.D., Hanesian, H.: Educational Psychology: A Cognitive View, 2nd edn. Holt, Rinehart and Winston, New York (1978)
26. Novak, J.D.: Meaningful learning: the essential factor for conceptual change in limited or appropriate propositional hierarchies (liphs) leading to empowerment of learners. Sci. Educ. **86**(4), 548–571 (2002)
27. Lee, J.H., Segev, A.: Knowledge maps for e-learning. Comput. Educ. **59**(2), 353–364 (2012)
28. Wang, J., Mendori, T., Juan, X.A.: Language learning support system using course-centered ontology and its evaluation. Comput. Educ. **78**, 278–293 (2014)
29. Wang, J., Mendori, T., Xiong, J.A.: Customizable language learning support system using ontology-driven engine. Int. J. Dist. Educ. Technol. **11**(4), 81–96 (2013)
30. Mouri, K., Okubo, F., Shimada, A., Ogata, H.: Profiling high-achieving students using e-book-based logs. Proc. of the first international workshop on Learning Analytics and Knowledge (LAK 16), Edingburgh, UK, pp. 1–6 (2016)
31. Shimada, A., Okubo, F., Yin, C., Kojima, K., Yamada, M., Ogata, H.: Informal learning behavior analysis using action logs and slide features in e-textbooks. Proceedings of IEEE International Conference on Advanced Learning Technologies, Hualien, Taiwan, pp. 116–117 (2015)
32. Yin, C., Okubo, F., Shimada, A., Oi, M., Hirokawa, S., Yamada, M., Kojima, K., Ogata, H.: Analyzing the features of learning behaviors of students using e-books. Workshop proceedings of International Conference on Computers in Education 2015, Hangzhou, China, pp. 617–626 (2015)
33. Mouri, K., Ogata, H., Uosaki, N., Liu, S.: Visualization for analyzing ubiquitous learning logs. Proceedings of International Conference on Computers in Education (ICCE 2014), Nara, Japan, pp. 461–470 (2014)
34. Mouri, K., Ogata, H.: Ubiquitous learning analytics in the real-world language learning. Smart Learn. Environ. **2**(15), 1–18 (2015)
35. Ogata, H., Li, M., Bin, H., Uosaki, N., El-Bishoutly, M., Yano, Y.: SCROLL: supporting to share and reuse ubiquitous learning logs in the context of language learning. Res. Pract. Technol. Enhanc. Learn. **6**(3), 69–82 (2011)
36. Ogata, H., Bin, H., Li, M., Uosaki, N., Mouri, K., Liu, S.: Ubiquitous learning project using life-logging technology in Japan. Educ. Technol. Soc. J. **17**(2), 85–100 (2014)

Security and Privacy in IoT Era

Orlando Arias, Kelvin Ly, and Yier Jin

1 Introduction

Totaling an estimated 15 billion devices, there are roughly two connected devices per living human [1]. This is thanks to trends in this past decade, which show a drastic increase in the number of Internet of Things (IoT) and wearable devices in the market. This trend is expected to continue, with an estimate of 26 billion connected devices by the year 2020, the majority of which being IoT and wearable devices [2].

IoT and wearable devices mainly consist of sensor nodes with the ability of transmitting data. Very little processing often takes place within this type of devices, relying on remote services or nodes to perform the computational workload. The information collected by these devices can range from a simple heartbeat, to temperature and humidity data, to energy consumption patterns, all while providing functionality such as health monitoring and home automation. Because of the type of information these devices gather and store, they become prime targets for attackers. Further, given their always-on network connectivity some of these devices exhibit, these devices can be targets for malware, increasing their potential for harmful usage.

Although some manufacturers are aware of the privacy and security implications in IoT and wearable devices, in most cases, security is either neglected, treated as an afterthought, or implemented incorrectly. The few devices that implement security mechanisms usually employ software-level solutions, such as firmware signing and signed binaries. These are methods reminiscent of those used in regular computing [3–12]. These solutions, however, do not consider the difference in usage patterns between IoT, wearable, and industrial devices when compared to traditional

---

O. Arias • K. Ly • Y. Jin (✉)
University of Central Florida, Orlando, Florida
e-mail: oarias@knights.ucf.edu; rangertime@knights.ucf.edu; yier.jin@eecs.ucf.edu

H. Yasuura et al. (eds.), *Smart Sensors at the IoT Frontier*,
DOI 10.1007/978-3-319-55345-0_14

computing systems. This has proven to be insufficient at times. Furthermore, concentrating on software-based security mechanisms often leaves the underlying hardware platform unintendedly vulnerable, allowing for new attack vectors.

In order to understand the security and privacy issues associated with current IoT, wearable and industrial devices, their design flow, and their implication, we categorize types of vulnerabilities we have encountered during our research. We also examine trends in manufacturing while providing a discussion of their effects in the final device. We then present four case studies: the Google Nest Learning Thermostat, the Nike+ Fuelband SE Fitness Tracker, the Haier SmartCare home automation system, and the Itron Centron CL200 electric meter. These devices were chosen because of their popularity and importance to the industry. Furthermore, we believe that these devices are good representatives of their respective categories. We will provide a security evaluation of these devices and present their vulnerabilities, demonstrating how their software-based solutions were insufficient to fully protect the device.

2 Design Practices and Taxonomy of Vulnerabilities

Throughout our study of Internet of Things (IoT) devices, wearable devices and IoT devices, we have found common patterns in their design flow. Although these patterns simplify the design process for manufacturers, it also leaves room for security oversights. In this section, we discuss common design patterns we have encountered while also presenting their consequences. We then categorize these consequences into common security vulnerabilities that are found in these embedded devices.

2.1 Common Design Patterns

Time to market is an important metric for companies looking to introduce their products while remaining competitive. This usually results in a shortened research and development phase which brings about patterns on the design. We now discuss some of these patterns.

Reliance on Vendor Designs. There are cases where the lack of familiarity with the hardware being used has led to over reliance on vendor designs. That is, products are directly based on a design or application solution a vendor has provided. Whereas for targeted applications, this may be sufficient, when the only available designs are for general-purpose computing devices or development boards, it may lead to the unintentional exposure of interfaces that are meant for debugging or reprogramming purposes.

For example, Texas Instruments provides the EVM430-F6779 kit [13]. This kit is a demonstration platform and development board for smart meter and related applications. It is based around an MSP430F6779 microcontroller and a peripheral set necessary to build a three-phase electric meter. Texas Instruments provides documentation [13] on how to design a smart meter around this platform; however it provides no details on security. As a development board, this platform comes equipped with the necessary debug facilities meant for testing. If left in a production run, an attacker can easily leverage these interfaces to leak internal sensitive information or even install malicious firmware to control device operation.

Software Source Models. At firmware level, some of the higher-end devices commonly utilize Linux-based software stacks. However, other open-source projects such as FreeRTOS [14] are also popular choices. Other manufacturers opt for proprietary solutions, such as Wind River's vxWorks [15] or Blackberry's QNX [16]. Smaller devices are often designed using a hardware vendor's toolkit, such as Texas Instruments' DriverLib [17]. The general idea is to utilize a pre-existing framework, saving time and development costs on the device.

Whether the software development model directly affects security is a hard question to answer. Open-source software provides the attacker with the means to easily find vulnerabilities to utilize as an attack vector. However, under an open-source model, a manufacturer does not have to rely on a vendor for security fixes. Closed-source software requires extra effort for an attacker to reverse engineer, providing a layer of resistance against finding vulnerabilities. However, manufacturers need to rely on vendors once a vulnerability is found.

Weak or Bad Cryptographic Implementations. If a device is designed to be remotely updated, it must be able to verify the downloaded image for both integrity and authenticity. This usually involves a cryptographic algorithm, sometimes many. Cryptographically securing a product is a complicated task, as proven by the countless vulnerabilities found in software, not only because of the mathematics involved but because of implementation errors [18–23]. Two of these vulnerabilities are of critical importance to our research as it shows how weakly implemented cryptographic systems can be bypassed, providing for a way to remotely attack the device. These exploits describe how an attacker can remotely compromise a Belkin WeMo Home Automation device by exploiting the faulty usage of SSL, allowing remote firmware installation by spoofing a distribution server or by spoofing SSL servers via arbitrary certificates.

Debug Interfaces on Production Runs. It is often cheaper to write images to flash chips when assembling the device, rather than purchasing preprogrammed parts. Furthermore, the device must be functionally tested before it leaves production. This implies that the circuit board must expose programming interfaces and test points for the different components present within. Although at times unlabeled, these often unpopulated interfaces are not removed after testing. An attacker can utilize them to inject his own code on the unit or alter their functional behavior. The software component may also fall prey to this issue, as compilers can generate binaries that include debugging symbols, expressing the constructs that generated a certain block of machine code. Leaving these debugging symbols in production runs aids an attacker in reconstructing the original sources, allowing for easier vulnerability detection.

Supply Chain Threats. Hardware Trojans also pose a serious threat to IoT security. These malicious modifications to integrated circuits can leak key data to an attacker, cause a device to operate outside specified parameters, or otherwise render the device inoperable. Hardware Trojans further pose the threat of not being detected by normal testing methodologies, requiring expensive specialized tests to detect them. For example, a malicious adversary could insert a hardware Trojan in a cryptographic IP core utilized in a system on chip (SoC) used in an IoT device [24]. When triggered, this Trojan weakens the entropy of the random number generator used to generate keys. If these keys are used to encrypt sensitive data that is being transmitted by the device, the amount of computational effort required by the attacker to decrypt the data is severely reduced.

2.2 Security Threat Taxonomy

We now group common security vulnerabilities found in embedded devices. These categories range from software-based issues to hardware-based errors. We enumerate the types of vulnerabilities below and discuss their implications.

Board Level Exploitation. During manufacturing, test points and debug ports are added to devices in order to ensure their functionality before being shipped. This is necessary as it is part of quality assurance during production. Furthermore, it is often cheaper to perform in-board programming of the device rather than purchase preprogrammed chips. Unfortunately, leaving open test points and programming or debug ports on the circuit board provides an avenue for an attacker with physical access to probe the device and test its functionality. For example, exploits on the Xbox 360 allow an attacker to downgrade the system to a vulnerable kernel version through a timing attack [25] utilizing the onboard debug facilities.

Chip Level Exploitation. Commercial off-the-shelf (COTS) components are designed with general-purpose usage in mind. This is specially the case with microprocessors and microcontrollers. These devices offer commonly used functions and peripherals with the aim to make them as flexible as possible. As such, documentation on the operation of these devices is often public knowledge. As such, COTS components are not designed to contain a per-device root of trust internally embedded.

However, vendors such as Texas Instruments are capable of designing and fabricating customized parts for application-specific scenarios. These parts, such as some of their OMAP-based parts, sport an on-die root of trust. Unfortunately, chip-level exploitation of integrated circuits defeats this kind of protection. Semi-invasive and invasive probing can reveal the secrets contained within the root of trust of the device. Modern technology facilitates the reverse engineering and leakage of sensitive information stored on-chip. For example, by "bumping" the internal memory on an Actel ProASIC3 FPGA, researchers were able to extract the stored AES key [26]. Furthermore, vendors such as Chipworks are capable of performing most reverse engineering tasks on a device [27].

Boot Process Vulnerabilities. Devices that, due to processor and system limitations, chainload an operating system may present security vulnerabilities. Chainloading refers to running sequentially larger pieces of software until the target software has been reached. This is done since devices do not usually have all of their hardware or software mechanisms initialized during boot. However, an attacker may leverage issues in the boot process of a device to inject a malicious payload. Any protection mechanism that is not active from the time of boot can be leveraged by an attacker to insert a malicious payload.

The boot sequence is one of the main targets of attack, as many of the high-level protection mechanisms are unable to be executed during the boot process. Since these mechanisms are not present, it leaves the system open for attack, which makes this a critical area to protect. For example, the attack on the iPhone's bootloader leads to a chain-of-trust exploit [28].

Implementation Errors. Encryption and hash functions are used in smart devices to secure passwords and other sensitive information, in addition to playing a key role in device communication and authentication. These functions are mathematically proven to be secure and robust; however, side-channel attacks and information-based cryptanalysis methods are threatening their integrity. In addition, improper implementations of these functions and the utilization of cryptographically weak encryption algorithms threaten the security of these devices. For example, the Sony PlayStation 3 firmware was downgraded due to a series of vulnerabilities in weak cryptographic applications [29, 30]. Interestingly, while the problems have been repeated in modern smart devices, the mitigation methods have already been proposed decades ago [31].

Software-level vulnerabilities in smart devices are similar to those in traditional embedded systems and general computing systems. Because smart device software stacks are often derived from the general computing domain, any software vulnerabilities found in the general computing area will also affect these devices. Therefore, software patches are required to update smart devices against known software-level attacks. Recent examples include a stack-based buffer overflow attack in glibc [32]. Methods to mitigate software exploitation attacks often follow those developed in general computing areas [33, 34]. However, as discussed in [35], these solutions may not fit in smart devices due to the resource constraints.

Remote Access Channels. Smart devices are often equipped with channels that allow for remote communication and debugging after manufacturing. These channels are also used for over-the-air (OTA) firmware upgrades. Though these channels are extremely useful, their implementations are not always secure.

During development, manufacturers may leave in APIs which allow arbitrary command execution, or developers may not properly secure the communications channel. Through this attack vector, attacks may be able to remotely obtain the status of the device, or even control the device. A modern example of a backdoor in a remote channel is the Summer Baby Zoom WiFi camera, which has hardcoded credentials for administrator access [36].

3 Case Study 1: Smart Thermostat

Boot process hijacking invalidates software-level protection schemes before they are properly installed and loaded. In this case, attackers try to break the normal boot process through the vulnerabilities within the chain of trust and install customized userland images or kernel modules. Malicious payloads can be inserted into the kernel modules and/or userland filesystems. One example of this type of attack is the compromise of the Google Nest Thermostat [37–39].

The Nest Thermostat is a smart device designed to control a standard heating, ventilation, and air conditioning (HVAC) unit based on heuristics and learned behavior. Coupled with a WiFi module, the unit is able to connect to the user's home or office network and interface with the Nest Cloud, thereby allowing for remote control of the unit. The thermostat is divided into two main components, a backplate which interfaces with the HVAC unit and a front plate which presents the main user interface. The largest part count is found in the front plate of the thermostat, which is driven by a Texas Instruments Sitara AM3703 system on chip (SoC) [40], interfacing directly with a Micron ECC NAND flash memory module, a Samsung SDRAM memory module, and an LCD screen. Figure 1 shows the device's internal components and the overall device configuration.

Upon normal powering on process, the Sitara AM3703 starts to execute the code in its internal ROM. This code initializes the most basic peripherals, including the general-purpose memory controller (GPMC). It then looks for the first stage bootloader, `x-loader`, and places it into SRAM. Once this operation finishes, the ROM code jumps into `x-loader`, which proceeds to initialize other peripherals and SDRAM. Afterward, it copies the second stage bootloader, `u-boot`, into SDRAM and proceeds to execute it. At this point, `u-boot` initializes the remaining subsystems and executes the `uImage` in NAND flash with the configured environment. The system finishes booting from NAND flash as initialization scripts are executed and services are run, culminating with the loading of the Nest Thermostat proprietary software stack. Figure 2 shows the normal boot sequence of the device. The device boot configuration is set by six external pins, `sys_boot[5:0]`. After power-on reset, the value of these pins is latched into the `CONTROL.CONTROL_STATUS` register. Table 1 describes the boot selection process for a selected set of configurations.

After performing basic initialization tasks, the on-chip ROM may jump into a connected execute in place (XIP) memory, if the `sys_boot` pins are configured

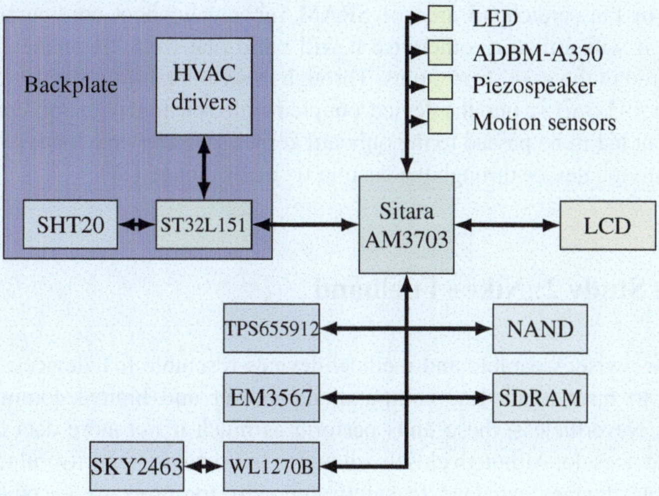

Fig. 1 Device map of the Nest Thermostat [39]

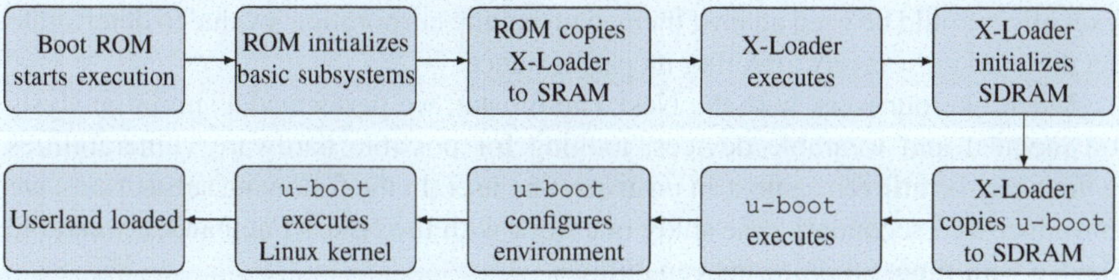

Fig. 2 Standard Nest Thermostat boot process

Table 1 Selected boot configurations

`sys_boot[5:0]`	First	Second	Third	Fourth	Fifth
001101	XIP	USB	UART3	MMC1	
001110	XIPwait	DOC	USB	UART3	MMC1
001111	NAND	USB	UART3	MMC1	
101101	USB	UART3	MMC1	XIP	
101110	USB	UART3	MMC1	XIPwait	DOC
101111	USB	UART3	MMC1	NAND	

as such. This boot mode is executed as a blind jump to the external addressable memory as soon as it is available. Otherwise, the ROM constructs a boot device list to be searched for boot images and stores it in the first location of available scratchpad memory. The construction of this list depends on whether or not the device is booting from a power-on reset state. If the device is booting from a power-on reset, the boot configuration is read directly from the `sys_boot` pins and latched into the `CONTROL.CONTROL_STATUS` register. Otherwise, the ROM will look in the scratchpad area of SRAM for a valid boot configuration. If it finds one, it will utilize it; otherwise it will build one from "permanent devices" as configured in the `sys_boot` pins. Through this vulnerability, attackers can send a modified `x-loader` into the device, coupled with a custom `u-boot` crafted with an argument list to be passed to the onboard kernel. Arbitrary payloads can then be inserted into the device through the custom `u-boot` image [39].

4 Case Study 2: Nike+ Fuelband

Architecture-wise, wearable and medical devices resemble IoT devices; however, they tend to have much less computational power and limited communication interfaces. Nevertheless, these units perform as much if not more data collection than IoT devices do. Although closely related to IoT devices, security vulnerabilities on wearable devices can lead to safety concerns for users. A pacemaker with wireless capabilities was proven to be vulnerable and could be used to affect the health of the patient [41]. Information leaks from fitness devices owned by corporate executives could be used against them, causing the corporation's value to deteriorate on the market, severely affecting its performance.

Much like our work with the Nest Thermostat, we performed a similar analysis on medical and wearable devices, looking for possible hardware vulnerabilities which may be utilized against an unsuspecting user. In the following subsections, we introduce as a secondary case study our work with the Nike+ Fuelband, a wearable device with fitness monitoring capabilities.

4.1 High-Level Overview

The Nike+ Fuelband is a low-power Bluetooth 4.0-enabled fitness wristband designed to measure daily physical activity, such as the amount of steps taken and sleep patterns, and estimate the amount of calories burned (see Fig. 3). This is done by means of reading data from the onboard three-axis accelerometer, which is subsequently stored within the unit. By means of software provided by the manufacturer, the unit can communicate with a Windows- or OS X-based computer, as well as Android and iOS devices. The collected data can then be analyzed, tracked, and shared with the Nike+ online community. Periodic synchronization with the device can be achieved with the mobile applications, and real-time feedback is performed with the onboard LED matrix display. The device is powered by two lithium-polymer batteries, advertised to provide up to 4 days of continuous usage.

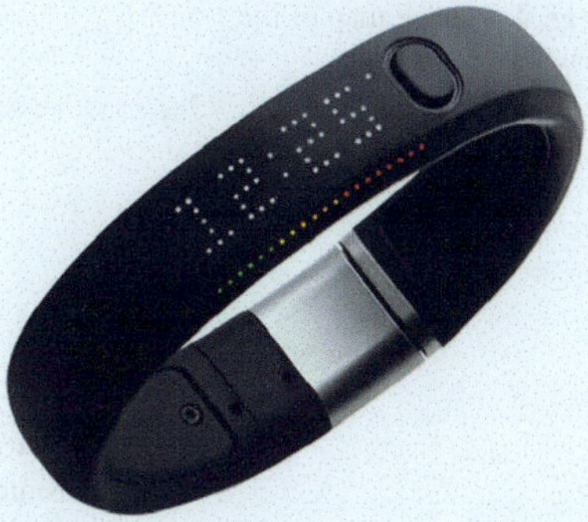

Fig. 3 Nike+ Fuelband SE fitness tracker (credit: Nike)

4.2 *Device Security*

The Nike+ Fuelband contains a Bluetooth interface which it uses to communicate with a smartphone. Some settings of the Fuelband can be configured through these means and information from the band can be sent back to the smartphone using this channel. Firmware updates, however, are performed by means of the Nike+ application on a Windows- or OS X-based personal computer. Most of the communications from the Smartband are done through the smartphone or personal computer application. Upon boot, the firmware is checked against a checksum before it is run ensuring a valid image.

4.3 *Device Descriptive Overview*

The main processing unit in this device is the ST Microelectronics STM32L151QCH6 microcontroller. Built upon an ARM Cortex-M3 core, this microcontroller is described in greater detail in Sect. 4.4. An LIS3DH 3-axis MEMS accelerometer from the same manufacturer interfaces with the STM32 by means of a Silego SLG46300 programmable mixed signal array. The 120-LED matrix is driven by an AMS AS1130 driver, which simplifies some LED matrix-related operations. Power management is provided by the ST Microelectronics RS12, which also facilitates communications over USB 2.0. Bluetooth communication is achieved by means of a Cambridge Silicon Radio CSR1010 Bluetooth Low Energy module. Figure 4 shows the device map of the unit.

4.4 *The STM32L151QCH6: A Closer Look*

The ST Microelectronics STM32L151QCH6 system on chip (SoC), hereafter referred to as STM32, is an ultralow-power platform offering a 12-channel DMA

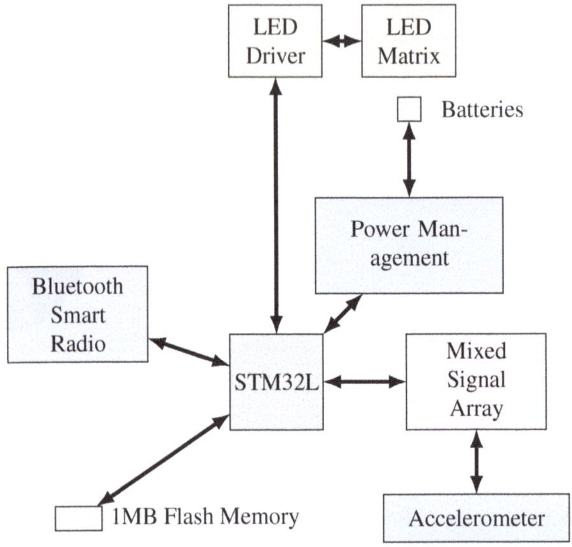

Fig. 4 Device map of the Fuelband

controller, 23 capacitive sensing channels, and a CRC calculation unit. The SoC further includes a 96bit unique ID, a preprogrammed bootloader supporting both USB and USART programming, and116 fast input/output pins which are mappable to a 16-interrupt-vector table. Storage-wise, the STM32 in question offers 256Ki B of flash storage with ECC support, 32Ki B of SRAM, 8Ki B of ECC supporting EEPROM, and a 128B backup register. Included peripherals range from an LCD driver to communication interfaces supporting USB 2.0, USART, SPI and I^2C [42].

The included ARM Cortex-M3 core supports both the Thumb and Thumb-2 instruction set architectures. Advanced low-power optimizations are achieved by means of multiple power and clock domains, architecture-defined sleep modes, and support for advanced low-power technologies such as state retention power gating. A JTAG mechanism is provided by means of serial wire debug, which provides real-time access to system memory without halting the processor.

A simplified memory map of the STM32 is illustrated in Fig. 5. The highlighted block of addresses in the figure are multiplexed between flash and system memory, depending on the status of the external `BOOT0` pin (see Sect. 4.5).

4.5 *Boot Process and Device Initialization*

Upon device power on, the STM32 executes the code stored in its internal ROM, initializing the device's basic peripherals. Execution then continues from internal flash memory, which proceeds to finish device setup into a working model. Specific to the Nike+ Fuelband, this entails activation of the Bluetooth radio, mixed signal

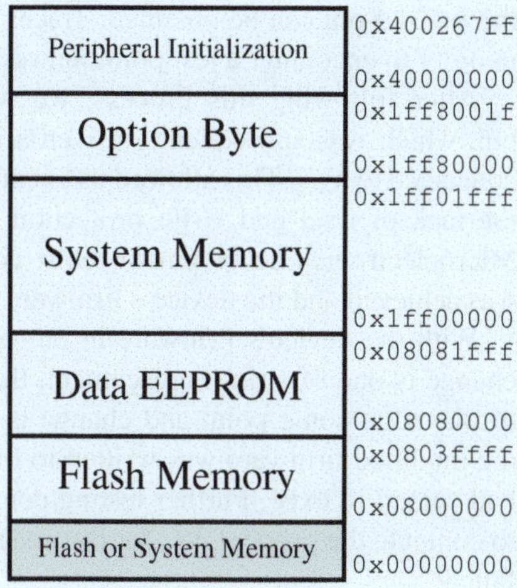

Fig. 5 Simplified memory map of the STM32L151QCH6

array, and LED driver, along with the calibration of the accelerometer. At this point, the device is ready for regular usage.

The STM32, however, implements a secondary boot mode, which is triggered by holding the `BOOT0` pin to a logic 1 as the device starts. If started this way, the device initializes a basic set of peripherals and configures the USB subsystem. Then, if a USB cable is detected while being driven by the proper clock signal, the internal PLL reconfigures the system clock to 32 MHz and the USB subsystem clock to 48 MHz. The system proceeds to execute the DFU bootloader with USB interrupts enabled, as to allow for communication. Using this mechanism, the STM32 can be sent commands which allow for read and write operations to memory, changing memory protection modes and status retrieval.

4.6 Attack Vector on the Nike+ Fuelband

Although the STM32 documentation states that the microprocessor contains the necessary capabilities to lock external reads and writes against the internal flash, thus isolating the device's firmware from the external world, this protection was not employed on the Nike+ Fuelband. As such, the contents of flash can be freely modified by an attacker with access to the device.

The Nike+ Fuelband contains a standard USB connector which is used for both device charging and synchronization. This connector can also be used to write new firmware onto the device; however, the necessary access to the `BOOT0` pin is not externally provided. As such, the device must be opened in order to trigger the alternate boot sequence. Further complicating the issue is the fact that the microcontroller is packaged as a ball grid array (BGA) and thus no direct access to the BOOT0 pin can be obtained. Traces on the circuit board must then be followed in order to encounter a test point indirectly exposing the pin in question.

After following this process, we were able to indirectly locate the BOOT0 pin, which was subsequently driven a logic 1 state by means of a 100Ω resistor connected to V_{DD}. This allowed us to enter the alternate boot mechanism and exploit the lack of read and write protection on the device. By means of standard ST Microelectronics development tools, communication over USB with the STM32 was achieved and the device's firmware was obtained.

With the device's firmware in our hand, we set on to modify it. The simplest change is one of string replacement, that is, find a string in the program that gets displayed at some point and change it to something else. With the change made, the modified firmware was written to the device, only to find normal functionality had ceased to exist. Further testing demonstrated that this was caused by a failure to compute the proper CRC for the image. Since the image was modified, the check failed.

Closer examination of the disassembled firmware image demonstrated that it utilized the CRC engine within the STM32 microcontroller in order to verify itself as genuine by checking the result of the CRC computation against a stored value. This value was found within the image itself and thus easily modifiable. With the proper checksum added, the modified firmware was sent to the device and proven to work.

5 Case Study 3: Haier SmartCare

Commercial IoT devices which directly target end users are often designed with emphasis on device functionality. Security features are often added in an ad hoc manner where remote attacks are treated as the main threats. Therefore, commercial IoT devices often suffer from hardware-level vulnerabilities [37] which may be remotely exploited. In order to demonstrate these security vulnerabilities and help designers/consumers better understand the design backdoors, the Haier SmartCare home automation system is selected as a case study in this paper.

5.1 High-Level Overview

The Haier SmartCare is a smart device designed to control and read information from various sensors placed throughout a user's home which include a smoke detector, a water leakage sensor, a sensor to check whether doors are open or closed, and a remote power switch. These sensors are connected through the ZigBee protocol. The primary function of this device is to allow the user to better monitor their homes when they are away and to get alerts based on sensor information (Fig. 6).

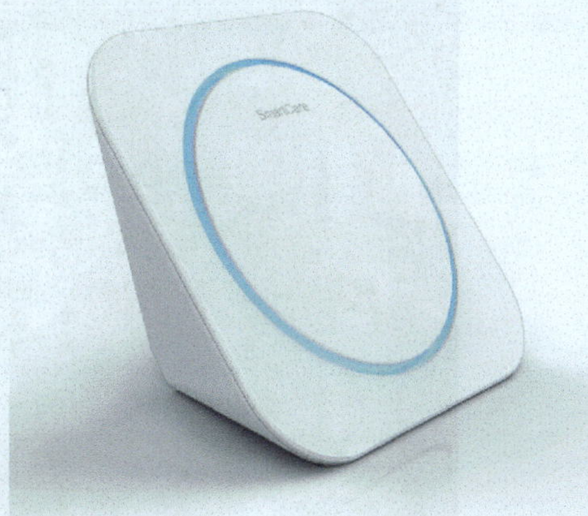

Fig. 6 Haier SmartCare device (Credit: Haier)

In order for users to connect to the device, they must first download a mobile application from the manufacturer's website. Next, they must connect the SmartCare to their network using an Ethernet connection. Following, they must connect their mobile device to the same local network as their SmartCare. Once it is connected, they must open the mobile application and create an account through the manufacturer's cloud service, which allows users to view their sensor data outside of their local network. Once this has been established, the users will be able to interact with the sensors from their SmartCare through the mobile application.

5.2 *Hardware Analysis*

The first step in our vulnerability analysis was to analyze the components on the SmartCare's hardware platform. The main processing unit is a TI AM3352BZCZ60, which is a part of TI's Sitara line of processors. The processor contains an ARM Cortex A8 with NEON extensions. The processor also supports the use of operating systems such as Linux and Android. Upon analyzing the data sheet for the processor, we were able to locate traces for UART on the device. The SmartCare PCB is shown in Fig. 7.

By leveraging the UART connection, we are able to read serial data from the device. By setting the correct parameters in the terminal emulator and connecting a serial-to-USB device to the SmartCare, we were able to view its start-up sequence. In the beginning of the boot process, the device prompted us as to whether we would like to stop the automatic boot sequence. Upon stopping the process, we were dropped into a U-Boot shell. It is here where we were able to modify specific boot parameters for the device, such as where to start reading from memory and what the initial shell will be. By modifying the initial shell among other variables,

Fig. 7 SmartCare hardware platform

Fig. 8 SmartCare hashed root password

```
root@am335x-evm:/etc# cat /etc/shadow
root:RE0tDm4UmFgUo:16674:0:99999:7:::
```

attackers will be able to gain low-level access to the device. After modifying the parameters, we initiated the boot process. Once the device had finished booting up, we were dropped into a rudimentary shell.

5.3 Into the Shell

After reading the boot output of the device, it was apparent that this device was running Linux. Being on a Linux device, it is necessary to know what kind of permissions we have; running `id` showed us that we were on the root account of the device. Looking through the BusyBox utility showed us that the device is capable of running a `telnet` server, allows for TFTP file transfer, and is able to fetch files from the web through `wget`.

Being on the root shell of the device also gave us the opportunity to look at the password hashes on the device, shown in Fig. 8.

By referencing documentation on Linux shadow file structures, we were able to deduce that this device was using DES encryption on the password while also not using a salt. This means that the password is truncated to a maximum of eight characters, then hashed. In order to obtain the root password for the device, the root password hash had to be cracked. The first attempt at cracking utilized a dictionary attack. In a dictionary attack, each password in the dictionary is hashed and subsequently checked against the hash in question. If the hashes match, then the password has been found; otherwise it will continue to check and hash each password in the list until it has reached the end. In this attack, a large word list containing approximately 32 million passwords was checked against.

Though 32 million passwords were checked against, none of them matched the root password of this device. The next option was a brute force attack, where every possible combination of characters is checked and hashed in order to find the root password. The total keyspace for a DES password using printable ASCII characters is $\sum_{i=0}^{8} 95^i$. This is a somewhat large keyspace and may take hours or even days to go through every iteration on high-performance hardware. Given that this method of attack is much more computationally intensive, we tried to optimize the cracking procedure leveraging high-performance hardware with parallel processing capabilities. In our case study, we used two AMD R9 290 graphics cards to speed up the process.

In our run, it took around five hours to get the root password. Since the root password for the device was known, the next course of action was to move onto another layer of attack. That is, we wanted to find out how we could attack other SmartCare devices using the secret learned from the device.

5.4 *SmartCare Network Analysis*

The new attack we tried to perform was a network-based remote attack. The first step in performing the network analysis was to scan the ports on the SmartCare to see if it is listening or transmitting on any of them. By performing a network scan, we were able to identify that the device may have had a `telnet` server running. Connecting to the device over telnet, we encountered a login prompt. Using the root credentials that were found earlier, we were able to get a root shell, which is shown in Fig. 9.

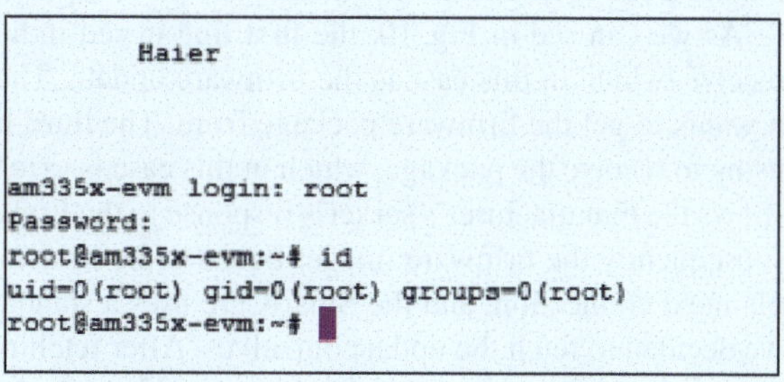

Fig. 9 SmartCare `Telnet` login prompt

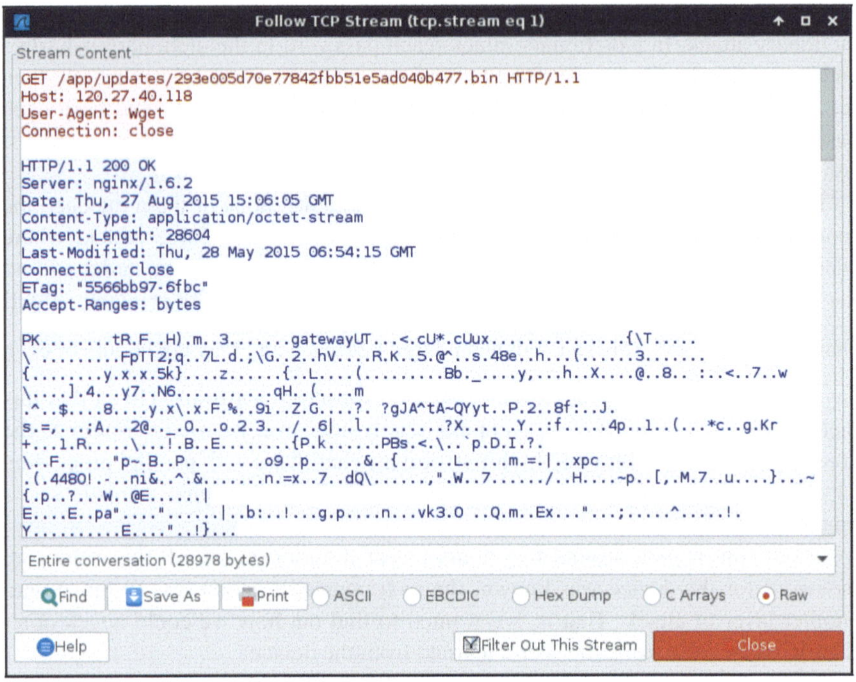

Fig. 10 SmartCare fetching update from manufacturer's server

Since we were able to get a root shell over a local network, the next step was to see what kind of traffic this device generates. In order to analyze its network traffic, we had to perform a man-in-the-middle attack. This involved us using our computer as the gateway for the network the SmartCare was on. Through the gateway we were able to provide Internet access. Using a packet sniffing program, we were able to see what kind of traffic the device generates.

Once the network was up and running, we started the packet sniffer and looked at the network traffic. While most of the traffic going to and coming from the server was encrypted at the beginning, the device later fetched a firmware update over a plaintext HTTP connection, which is shown in Fig. 10.

As we can see in Fig. 10, the first line in red indicates the package it wants to receive, which in this case is the firmware update. The second line indicates where it wants to get the firmware package from. The third line indicates the method it is using to receive the package, which in this case is `wget`. The blue section following shows the manufacturer's server's response to the firmware update fetch request and subsequently the firmware image. Because the firmware update was fetched over a plaintext connection, and the SmartCare uses a standard utility to fetch the update, we decided to fetch the update ourselves. After fetching the update using `wget` and performing a file analysis on it, we were able to find that the firmware update was simply a ZIP archive.

Unzipping the archive allowed us to see the SmartCare's main binary along with `bash` scripts for updating the device and one of the SmartCare's main initialization scripts. Based on the initialization script, the device will set itself up, and then run the device's main binary. Knowing this information, the next step in our analysis was to see how the device handles firmware updates, which involves reverse engineering the SmartCare's binary.

5.5 SmartCare Binary Analysis

Using binary analysis software, we were able to search through the binary and see how it handles updates. The device utilizes the MQTT protocol in order to communicate securely with the manufacturer's server through an encrypted channel. MQTT is a publisher/subscriber protocol, where there is a broker which takes in information from publishers and pushes the information to subscribers. The subscribers subscribe to topics, which are posted by the publishers. In our case, the SmartCare is a subscriber which communicates to the manufacturer's server to fetch the names of firmware updates, the correct hashes for the updates, commands from the user, and the current time. It also acts as a publisher, sending sensor information back to the manufacturer's server.

In terms of actually performing the firmware update, the device will fetch the package using the information gathered over MQTT. Once received, the device will run an MD5 checksum on the package and compare this hash to the hash provided by the manufacturer over MQTT. If both hashes match, the device will go through with the update. If the hashes do not match, the device will reboot, and start the entire process again. The whole verification mechanism is still under investigation for possible security vulnerabilities.

6 Case Study 4: Itron Centron CL200 Meter

Similar to commercial IoT devices, smart devices are also widely used in industrial applications. These devices, if compromised, may have a more serious impact than compromised commercial IoT devices. To better understand the security protections in place for industrial IoT devices, we selected the Itron Centron smart meter as the other case study. Figure 11 shows the smart meter.

Fig. 11 Itron Centron CL200 smart meter (credit: Itron)

6.1 High-Level Overview

The primary functionality of this device is to measure a customer's energy usage and report the collected information through an RF channel to a nearby meter reader or to a local substation. This information is then used to charge the customer for their energy usage and may also be used to get statistics on community energy usage.

6.2 Hardware Analysis

Similar to our work on the home automation device, the first step in our analysis was to analyze the hardware platform of the smart meter. Inside of the device, we were able to see a heavy-duty plastic cover, which guarded the main hardware platform. When looking at the hardware platform, we identified that it measures line voltage, measures reference voltages, checks the energy flow direction and energy pulse data, and checks the line frequency. Attached to the main hardware platform is a daughterboard, which is used when a company wants to implement functionality on the meter without having to replace the entire device.

In this case, the daughterboard is used to collect energy usage information along with tamper data and the ID of the board itself (see Fig. 12). Located on the daughterboard is an ATMega microcontroller, a tamper sensor, and a 1 KB EEPROM. Through the microcontroller, we were able to re-enable JTAG and re-enable write access for on-chip memories.

Fig. 12 Smart Meter CL200 daughterboard

6.3 Device ID Modification

For our analysis, our objective was to modify the smart meter ID in order for a meter reader to read the incorrect ID for the device. Upon further analysis, the ID was being stored in the external EEPROM. In order to figure out the ID of the meter, we had to read the ID on the meter itself, which is found on the front of the device underneath the gray cover. By analyzing the EEPROM dump, we were able to find where the ID was stored and change the ID to any arbitrary value.

6.4 Demonstration

Now that we had modified the ID of the meter, we needed to read the ID of the meter remotely to demonstrate that a smart meter reader will pick up the wrong ID from a modified device. Utilizing a software-defined radio (SDR), we were able to run a TCP server on the SDR and connect it to another program which parses wireless information and displays the ID, the tamper bit status, and the energy usage for the meter. Through the experimental platform, we were able to demonstrate that due to the lack of proper protection, one compromised smart meter can "represent" itself as any other smart meter. Figure 13 shows the SDR output in which two smart meters

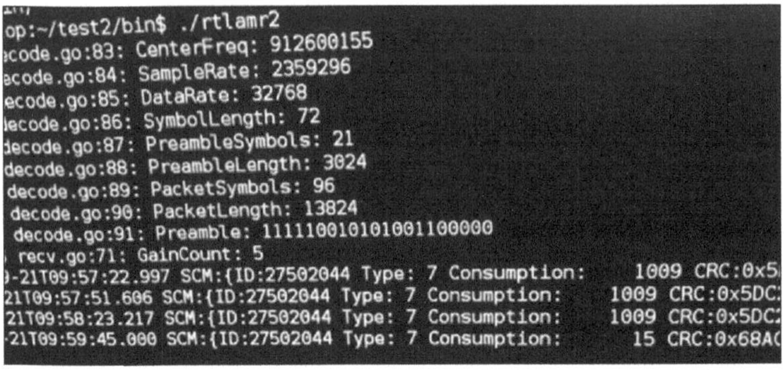

Fig. 13 Demonstration of the security vulnerability on the meter

share the same ID but different power consumption values. At the bottom of the figure, there is a meter which identifies as the other; however its power consumption is different than those above it. Through this vector, energy theft becomes possible.

7 Discussions

7.1 Security Impact to Network

A compromised IoT device can be utilized to further attack other units in an unsuspecting victim's network. Effects could range from simple backdoor injection to leaking user information and credentials to even causing physical harm to the user. As shown with the case of the Nest Thermostat, it can be used as a beachhead to other nodes within the network, allowing for discovery and attack of those nodes.

Furthermore, rogue services may be installed on the device, aiming to disrupt regular network operations. For instance, a rogue DHCP server may be utilized to inject DNS requests to a poisoned server which would return false information, allowing for traffic shaping. Address resolution protocol (ARP)-based attacks are also possible, with the compromised device masquerading as the router, allowing for the capture and redirection of a target computer's network traffic.

Security issues with backdoored IoT devices are exacerbated by the fact that local network credentials need to be stored within the unit, thus becoming accessible to an attacker. Leveraging the extraction of network credentials allows for the introduction of extraneous devices into the local network, granting for new methods of exploitation against other nodes. In the case of the Nest Thermostat, the network credentials are stored in regular text files, and even if these were encrypted, the algorithms necessary to obtain the clear text would necessarily be present on the device, granting the attacker the means to collect them.

7.2 *Safety Concerns*

Safety concerns arise when compromised IoT and wearable devices see on-field deployment. Due to the services these units provide, from communications to medical applications, a compromised device could then be used to cause physical harm to its user[41]. The Nest Thermostat could be employed to overstress the HVAC unit it is connected to, causing it to malfunction. Furthermore, all the information stored within the device can be utilized by the attacker to build a profile of the victim, aiding on the determination of a daily routine, the usage of which can result in facilitating the burglarizing of the victim's property.

7.3 *Privacy Concerns*

Almost all IoT and wearable devices, upon setup, will start collecting user information. For example, the Nest Thermostat will collect information such as the location of the thermostat, whether it is being used in a home or business, the postal code of the area, and device information from the HVAC system to determine its capabilities. The onboard sensors on the thermostat will also collect temperature data and humidity and ambient light data, by means of the onboard passive infrared sensor, whether somebody is moving in the room. Any direct temperature adjustments to the device are also recorded and utilized in algorithms to learn and compute comfort levels under different situations. Whenever the HVAC unit is activated, the thermostat will record the time and duration for which this happened. Using this information, the thermostat builds a profile for the users in order to help them feel comfortable while also providing energy savings. The Nike+ Fuelband will store the user's heartbeat and sleeping patterns, which can then be learned by the attacker. The information could potentially be used against the user, or against any entity the user is part of.

Although there are laws and standards defining data collection policies, some of these have proven to be ineffective and often antiquated, as demonstrated by information leaks from companies [43–45]. User information collected by the Nest Thermostat is stored within the unit and uploaded to the Nest Cloud. Local log files are sent to Nest as well and removed from the unit as to save space. System and software logs contain information such as the user's Zip code, device settings, HVAC settings, and wiring configuration. Forensic analysis of the unit yields that the Nest Thermostat has code to prompt the user for information about their place of residence or office. Reports indicate that Nest plans to share this information with energy providers in order to aid with efficient power generation [46]. As for the Nike+ Fuelband, the information collected and stored by the unit is then sent to a personal computer or mobile device, from where it can be publically shared with other users. Even if the information is not shared, an unauthorized third party still has access to the data from a compromised device and can use it for their own purposes. Although IoT manufacturers have gone through considerable efforts to ensure the secure transmission of this data, it is all for naught if it can be leaked at the source.

8 Related Work

Current IoT and wearable device literature often treats IoT from a network perspective or provides solutions that are inherently incompatible with the needs of a manufacturer. Few works have been published discussing the security of IoT devices themselves [47, 48]. In the ensuing sections, we summarize some of the previous work that has been presented in this area.

8.1 IoT Secure Protocols and Network Protection

An early survey about the IoT has shown that security and privacy are the main concerns that need to be addressed before IoT devices are widely adopted [49]. Proposed solutions for security rely on network protocols to ensure IoT security. Meanwhile, encrypted communication is treated as the effective solution for privacy protection. However, these proposed approaches do not consider the unique properties of IoT devices. The authors in [50] summarized all current security threats to the IoT network, but these threat models are mostly derived from network security. They claim that hardware-level attacks, such as differential power analysis (DPA) [51], are of high cost and therefore less harmful. Similarly, the authors in [4] treat IoT as an extremely interconnected network and list possible solutions to secure the IoT network including protocol and network security, data and privacy, identity management, trust and governance, fault tolerance, cryptography and protocols, identity and ownership, and privacy protection. All these methods try to regulate the communication between IoT devices under the assumption that all IoT devices are operating properly. The authors in [5] tried to solve IoT security through different IoT topologies: centralized architectures [6] and distributed architectures [7, 8]. Again, the network-based solutions only emphasize high-level structures without considering whether the available resources in IoT devices can afford these topologies.

Another research focuses on the secure communication between IoT nodes. For example, the authors in [9] focus on secure communication between IoT devices and present an Identity Authentication and Capability-based Access Control (IACAC) model to protect IoT from man-in-the-middle, replay, and denial-of-service (DoS) attacks. The authors in [10, 11] expand the definition of IoT to include four nodes in a typical IoT network: person, intelligent object, technological ecosystem, and process. The authors claim that IoT security cannot be solved at a single-layer, but should require the analysis of the interactions between these nodes. A 2D version of the systemic approach was developed, which was expanded to a 3D version highlighting new functional plans of security [12].

Following this route, communication protocols were then developed to secure the interactions between IoT nodes such as 6LoWPAN [52] and Constrained Application Protocol (CoAP) [53]. The CoAP was constructed based on Datagram Transport Layer Security (DTLS) [54] and IPsec [55]. To counter the attacks at the transport layer, protocols were enhanced to use either HTTP/TLS or CoAP/DTLS by proposing a mapping between TLS and DTLS [56] or using secure tunneling on the transport layer [57]. However, these communication layer security analyses and protection methods ignore device-level vulnerabilities and often impose unrealistic constraints on device deployment.

8.2 *Hardware-Based Protection*

Besides network-level protection, researchers from the industry have also tried to develop highly secure processor/SoC architectures for IoT protection. ARM TrustZone is an industry landmark in providing a basis of trust for various applications such as secure payment, digital rights management (DRM), enterprise, and web-based services. TrustZone technology provides infrastructure foundations that allow a SoC designer to choose from a range of components that can perform specific functions within the security environment [58]. Intel proposed the concept of enclaves recently [59, 60]. An enclave contains software code, data, and a stack that are protected by hardware-enforced access control policies. Samsung KNOX has also been developed with protection in mind [61]. KNOX provides a safe execution environment in a KNOX-enabled device where the userland is verified and a KNOX container holds sensitive data, such as corporate contacts and e-mails in a cellphone. If the device is deemed to be compromised by altering the bootloader, an e-fuse is blown inside the SoC driving the unit, thus branding it as untrusted. However, these hardware-based secure architectures are developed with passive protection in mind, whereas they do not detect and mitigate hardware- and software-level attacks. Samsung KNOX is possibly an exception to this; however, it remains to be proven whether or not it is possible to bypass any checks to the e-fuse protection in the bootloader. TrustZone environments have been proven to be compromised as shown in [62–64] by exploiting bugs in the software stack. Furthermore, these solutions do not transfer well to low-power embedded units. For example, at the time of writing, Samsung KNOX is only available in select Android-based cellular phones and tablets.

9 Device Security Enhancement

9.1 Security Solutions Common to IoT and Wearable Devices

Verifying the firmware at update time is a step toward securing IoT devices; however, this is often done by the onboard software. As with the Nest Thermostat and the Nike+ Fuelband, the onboard software is trusted to be authentic. The implementation of this check, however, must be sound. For example, schemes that utilize random numbers must ensure the usage of a cryptographically secure random number generator; any used cryptographic certificates must be validated by a trusted certificate authority [22]. A weakly implemented cryptographic algorithm is no better than a lack of a cryptographic algorithm.

However, as we have demonstrated with our case studies, it is insufficient to authenticate an update image. The software stack must also be authenticated before it can reliably determine if an update is valid or not. With the devices compromised, we are free to bypass any checks on the update image, thus rendering the protection mechanism ineffective. A proper chain of trust in the hardware infrastructure of the device can aid the process of determining an authentic software stack [65].

The attack in both the Nest Thermostat and the Nike+ Fuelband could have been avoided had a proper chain of trust been implemented. Inherently, this needs the type of hardware support which is not available in either the Sitara AM 3703 used in the Nest Thermostat or the STM32 microcontroller used in the Nike+ Fuelband.

The exposure of debug interfaces in these devices further presents a risk. These are often left as residues from development prototypes or as test points used during manufacturing. These debug interfaces can also serve as the means to service IoT or wearable devices on the field, as to ease repairs. As such, we can see why they may be needed. However, these interfaces must be protected against attackers. For example, FRAM devices in the MSP430 lines provide means to both secure JTAG access and protect certain memory segments from access using a built-in IP Encapsulation Module [66]. Other microcontrollers and microprocessors offer the same kind of functionality, implementing means to restrict access to its debug units. As such, manufacturers are able to still expose these interfaces for testing purposes and lock them before they are deployed. Ideally, however, any debug interfaces should be removed from production runs or have proper protections.

9.2 Specific Solutions for IoT and Wearable Devices

Often, IoT devices provide a full operating system in which binaries are loaded into a userland. This simplifies the interface to the hardware and provides high-level application programming interfaces (APIs). The Nest Thermostat, for example, employs an embedded Linux stack which is used to launch the proprietary Nest application which relays commands to the backplate of the unit and controls the communications channels. As we demonstrated in our case study, binaries can be injected into the filesystem of the unit and executed in devices that utilize this model. As such, extra protection must be added to devices that load binaries into a userland. A possible approach is to only load and execute cryptographically signed binaries. This requires the kernel to have a custom loader that verifies these binaries as they are prepared for execution. If the signature verification fails, then the binary is not run and the device is set into a failsafe mode, notifying the user of possible tampering.

In devices whose architecture is self-contained, that is, microcontroller-based systems, it becomes necessary to secure all update channels. External reprogrammability of the microcontroller and any debug interfaces it may feature must be disabled. The microcontroller must also be programmed before being placed in the circuit board, as to avoid adding unnecessary interfaces which could expose functionality.

9.3 Overhead of Security Solutions

There is usually a certain degree of overhead associated with any protection mechanism. Cryptography necessarily adds computational overhead to any protection scheme that utilizes it. Any device which utilizes encryption or any other cryptographic function will require binaries with functions for necessary checks and have higher memory and CPU requirements. However, current industry solutions include parts which are capable of accelerating these processes, much like the microcontroller utilized in the Nike+ Fuelband which can accelerate CRC32 computations [42]. This reduces the software overhead needed to perform these checks, but slightly increases the area and power consumption of these parts. It should be noted, however, that for most parts, power can be gated to the SoC subsystems that are not being utilized, thus reducing power consumption in the device.

10 Conclusion

As our case studies demonstrated, a nonsecure hardware platform will inevitably lead to a nonsecure software stack. A vulnerability in the design of the unit can result in its compromise. Furthermore, without being able to authenticate the running software, it cannot be trusted to make decisions about its own validity. Due to the short time market engineers are given to finish a product, we believe that most of the current IoT and wearable devices suffer from similar issues. Software protection becomes ineffective if the hardware is vulnerable to attack. This raises safety and privacy issues with users, is their information safe?

Moving forward, we will continue to probe other IoT devices for security, with the goal of finding vulnerabilities in their hardware. Ultimately, this will lead us to a better understanding of design issues and how to correct them. We will attempt to build prototypes of smart devices that utilize our proposed chain of trust to test for their viability and ability to prevent malicious attacks.

References

1. Evans, D.: The internet of things – how the next evolution of the internet is changing everything. White Paper. Cisco Internet Business Solutions Group (IBSG) (2011)
2. Middleton, P., Kjeldsen, P., Tully, J.: Forecast: the internet of things, worldwide, 2013. Gartner (2013)
3. Welch, D., Lathrop, S.: Wireless security threat taxonomy. In: IEEE Systems, Man and Cybernetics Society Information Assurance Workshop, 2003, pp. 76–83 (2003)
4. Roman, R., Najera, P., Lopez, J.: Securing the internet of things. Computer **44**(9), 51–58 (2011)
5. Roman, R., Zhou, J., Lopez, J.: On the features and challenges of security and privacy in distributed internet of things. Comput. Netw. **57**(10), 2266–2279 (2013)
6. Williams, A.: How the internet of things helps us understand radiation levels (2011). [Online]. http://readwrite.com/2011/04/01/ow-the-internet-of-things-help
7. Viehland, D., Zhao, F.: The future of personal area networks in a ubiquitous computing world. Int. J. Adv. Pervasive Ubiquit. Comput. **2**(2), 30–44 (2010)
8. Schaffers, H., Komninos, N., Pallot, M., Trousse, B., Nilsson, M., Oliveira, A.: Smart cities and the future internet: towards cooperation frameworks for open innovation. In: The Future Internet. Lecture Notes in Computer Science, vol. 6656, pp. 431–446. Springer, Berlin/Heidelberg (2011)
9. Mahalle, P.N., Anggorojati, B., Prasad, N.R., Prasad, R.: Identify authentication and capability based access control (IACAC) for the internet of things. J. Cyber Secur. Mobil. **1**, 309–348 (2013)
10. Challal, Y.: Internet of things security: towards a cognitive and systemic approach. PhD thesis (2012)
11. Riahi, A., Challal, Y., Natalizio, E., Chtourou, Z., Bouabdallah, A.: A systemic approach for IoT security. In: 2013 IEEE International Conference on Distributed Computing in Sensor Systems (DCOSS), pp. 351–355 (2013)
12. Riahi, A., Natalizio, E., Challal, Y., Mitton, N., Iera, A.: A systemic and cognitive approach for IoT security. In: 2014 International Conference on Computing, Networking and Communications (ICNC), pp. 183–188 (2014)
13. EVM430-F6779 – 3 phase electronic Watt-Hour EVM for metering, [Online]. http://www.ti.com/tool/EVM430-F6779
14. Freertos reference manual: api functions and configuration options, Technical Report., Real Time Engineers Limited (2009)
15. Barbalace, A., Luchetta, A., Manduchi, G., Moro, M., Soppelsa, A., Taliercio, C.: Performance comparison of VxWorks, Linux, RTAI and Xenomai in a hard real-time application. In: Real-Time Conference, 2007 15th IEEE-NPSS, pp. 1–5 (2007)
16. Qnx operating systems. http://www.qnx.com/products/neutrino-rtos/index.html, (1982–2014)
17. MSP Driver Library, [Online]. http://www.ti.com/tool/mspdriverlib
18. CVE-2014-0160. Common Vulnerabilities and Exposures [Online]. https://cve.mitre.org/cgi-bin/cvename.cgi?name=CVE-2014-0160
19. CVE-2014-2783. Common Vulnerabilities and Exposures [Online]. http://www.cve.mitre.org/cgi-bin/cvename.cgi?name=CVE-2014-2783

20. CVE-2014-2001. Common Vulnerabilities and Exposures [Online]. http://web.nvd.nist.gov/view/vuln/detail?vulnId=CVE-2014-2001
21. CVE-2013-7373. Common Vulnerabilities and Exposures [Online]. http://web.nvd.nist.gov/view/vuln/detail?vulnId=CVE-2013-7373
22. CVE-2013-6951. Common Vulnerabilities and Exposures [Online]. http://web.nvd.nist.gov/view/vuln/detail?vulnId=CVE-2013-6951
23. CVE-2013-6950. Common Vulnerabilities and Exposures [Online]. http://web.nvd.nist.gov/view/vuln/detail?vulnId=CVE-2013-6950
24. Becker, G., Regazzoni, F., Paar, C., Burleson, W.P.: Stealthy dopant-level hardware trojans. In: Cryptographic Hardware and Embedded Systems – CHES 2013. Lecture Notes in Computer Science, vol. 8086, pp. 197–214 (2013)
25. Xbox 360 timing attack (2007) [Online]. http://beta.ivc.no/wiki/index.php/Xbox_360_Timing_Attack
26. Skorobogatov, S.: Fault attacks on secure chips: from glitch to flash. In: Design and Security of Cryptographic Algorithms and Devices (ECRYPT II) (2011)
27. http://www.chipworks.com/
28. Apple iphone bootloader attack, (2008) [Online]. http://rdist.root.org/2008/03/17/apple-iphone-bootloader-attack/
29. Bushing, marcan, segher, and sven: Console hacking 2010: Ps3 epic fail. In: 27th Chaos Communication Congress, (2010) [Online]. https://events.ccc.de/congress/2010/Fahrplan/attachments/1780_27c3_console_hacking_2010.pdf
30. Lemos, R.: Sony left passwords, code-signing keys virtually unprotected. eWeek (2014) [Online]. http://www.eweek.com/security/sony-left-passwords-code-signing-keys-virtually-unprotected.html
31. Schneier, B.: Cryptographic design vulnerabilities. Computer **31**(9), 29–33 (1998)
32. Critical security flaw: glibc stack-based buffer overflow in getaddrinfo() (cve-2015-7547) (2015) [Online]. https://access.redhat.com/articles/2161461
33. Cowan, C., Pu, C., Maier, D., Walpole, J., Bakke, P., Beattie, S., Grier, A., Wagle, P., Zhang, Q., Hinton, H.: Stackguard: automatic adaptive detection and prevention of buffer-overflow attacks. In: Usenix Security, vol. 98, pp. 63–78 (1998)
34. Cowan, C., Beattie, S., Johansen, J., Wagle, P.: Pointguard tm: protecting pointers from buffer overflow vulnerabilities: In: Proceedings of the 12th Conference on USENIX Security Symposium, vol. 12, pp. 91–104 (2003)
35. Arias, O., Wurm, J., Hoang, K., Jin, Y.: Privacy and security in internet of things and wearable devices. IEEE Trans. Multi-Scale Comput. Syst. **1**(2), 99–109 (2015)
36. Fowler, B.: Some top baby monitors lack basic security features, report finds (2015) [Online]. http://www.nbcnewyork.com/news/local/Baby-Monitor-Security-Research-324169831.html
37. Hernandez, G., Arias, O., Buentello, D., Jin, Y.: Smart nest thermostat: a smart spy in your home. In: Black Hat USA (2014)
38. Potter, R., Jin, Y.: Don't touch that dial: how smart thermostats have made us vulnerable. In: RSA Conference (2015)
39. Arias, O., Wurm, J., Hoang, K., Jin, Y.: Privacy and security in internet of things and wearable devices. IEEE Trans. Multi-Scale Comput. Syst. **1**(2), 99–109 (2015)
40. Texas Instruments: AM3715, AM3703 Sitara ARM Microprocessor (2011)
41. Halperin, D., Heydt-Benjamin, T., Ransford, B., Clark, S., Defend, B., Morgan, W., Fu, K., Kohno, T., Maisel, W.: Pacemakers and implantable cardiac defibrillators: software radio attacks and zero-power defenses. In: IEEE Symposium on Security and Privacy (SP), pp. 129–142 (2008)
42. ST Microelectronics: STM32L15xQC, STM32L15xRC-A, STM32L15xVC-A, STM32L15xZC Ultra-low-power 32 b MCU ARM-based Cortex-M3, 256 KB Flash 32 KB SRAM, 8 KB EEPROM, LCD, USB, ADC, DAC, no. 026119 Rev 5 (2015)
43. BARBARO, M., T. Z. Jr., A face is exposed for aol searcher no. 4417749. The New York Times (2006) [Online]. http://query.nytimes.com/gst/abstract.html?res=9E0CE3DD1F3FF93AA3575BC0A9609C8B63

44. Reynolds, I., Fujioka, C.: Update 2-sony removes data posted by hackers, delays playstation restart. Reuters (2011) [Online]. http://www.reuters.com/article/2011/05/07/sony-idUSL3E7G701T20110507
45. Whittaker, Z.: Amazon's zappos in massive data breach 24 million affected. ZDNet (2012) [Online]. http://www.zdnet.com/article/amazons-zappos-in-massive-data-breach-24-million-affected/
46. Mombrea, M.: Google's real plan behind the purchase of the nest thermostat (2014) [Online]. http://www.itworld.com/consumerization-it/416110/googles-plan-rake-cash-nest-thermostat
47. Ziegeldorf, J.H., Morchon, O.G., Wehrle, K.: Privacy in the internet of things: threats and challenges. Secur. Commun. Netw. **7**(12), 2728–2742 (2014)
48. Thierer, A.D.: The internet of things and wearable technology: addressing privacy and security concerns without derailing innovation. Rich. JL & Tech. **21**, 6–15 (2015)
49. Atzori, L., Iera, A., Morabito, G.: The internet of things: a survey. Comput. Netw. **54**(15), 2787–2805 (2010)
50. Babar, S., Mahalle, P., Stango, A., Prasad, N., Prasad, R.: Proposed security model and threat taxonomy for the internet of things (IoT). In: Recent Trends in Network Security and Applications. Communications in Computer and Information Science, vol. 89, pp. 420–429. Springer, Berlin/Heidelberg (2010)
51. Kocher, P., Jaffe, J., Jun, B.: Differential power analysis. In: Advances in Cryptology – CRYPTO'99, pp. 789–789 (1999)
52. Mulligan, G.: The 6lowpan architecture. In: Proceedings of the 4th Workshop on Embedded Networked Sensors, EmNets'07, pp. 78–82 (2007)
53. Shelby, Z., Hartke, K., Bormann, C., Frank, B.: Constrained application protocol (coap), draft-ietf-core-coap-13. In: The Internet Engineering Task Force (IETF) (2012)
54. Rescorla, E., Modadugu, N.: Datagram transport layer security. RFC 4347 (2006)
55. Kent, S., Seo, K.: Security architecture for the internet protocol. RFC 4301 (2005)
56. Brachmann, M., Keoh, S.L., Morchon, O., Kumar, S.: End-to-end transport security in the ip-based internet of things. In: 21st International Conference on Computer Communications and Networks (ICCCN), pp. 1–5 (2012)
57. Seggelmann, R.: SCTP: strategies to secure end-to-end communication. PhD thesis, University of Duisburg-Essen (2012)
58. ARM: Building a secure system using trustzone technology. ARM Limited (2009)
59. McKeen, F., Alexandrovich, I., Berenzon, A., Rozas, C., Shafi, H., Shanbhogue, V., Savagaonkar, U.: Innovative instruction ans software model for isolated execution. In: Hardware and Architectural Support for Security and Privacy (2013)
60. Anati, I., Gueron, S., Johnson, S.P., Scarlata, V.R.: Innovative technology for CPU based attestation and sealing. In: The 2nd International Workshop on Hardware and Architectural Support for Security and Privacy (HASP) (2013)
61. Samsung: Samsung KNOX: mobile enterprise security (2015)
62. Keltner, N., Holmes, C.: Here be dragons: a bedtime tale for sleepless nights. In: RedCon (2014)
63. Rosenberg, D.: Reflections on trusting trustzone. In: BlackHat USA (2014)
64. Wei, T., Zhang, Y.: To swipe or not to swipe: a challenge for your fingers. In: RSA Conference (2015)
65. Arbaugh, W., Farber, D., Smith, J.: A secure and reliable bootstrap architecture. In: Proceedings of the IEEE Symposium on Security and Privacy, 1997, pp. 65–71 (1997)
66. Texas Instruments: MSP430 programming via the JTAG interface (2015)

The manufacturer's authorised representative in the EU is Springer Nature Customer Service Centre GmbH, Europaplatz 3, 69115 Heidelberg, Germany. If you have any concerns regarding our products, please contact ProductSafety@springernature.com

Printed and bound by CPI Group (UK) Ltd, Croydon, CR0 4YY

12/08/2026

02194532-0005